THE QUAD ANTENNA

BY BOB HAVILAND, W4MB

Library of Congress Card Number 93-071168

ISBN 0-943016-05-3

Editors: Terry Littlefield, Alfred Wilson
Illustrations: Lois Racy
Layout and Design: K&S Graphics and Hal Keith
Phototypographer: Linda Romanello
Cover Photo: Larry Mulvehill, WB2ZPI

Published by CQ Communications, Inc.
76 North Broadway
Hicksville, New York 11801 USA

Printed in the United States of America

Table of Contents

Introduction

This book was first envisioned as a series of articles on the quad antenna family by R.P. Haviland, W4MB. The first three chapters appeared as articles in *Ham Radio* magazine in May, June, and August of 1988. The fourth chapter was published in the Winter 1991 issue of *Communications Quarterly*. The articles generated so much mail requesting additional information, that we felt a book containing all of Haviland's work on the quad antenna would be a valuable addition to every amateur's library.

Haviland was first licensed as W9CAK back in 1931. Although he says he doesn't remember what specifically sparked his interest in becoming a ham, Bob worked his way up from building and listening to crystal radio sets, to studying about amateur radio, to obtraining his ticket.

For many years, Haviland worked on satellite communications at General Electric and participated in a number of ITU and CCIR conferences relating to space communications. He even had the opportunity to operate a number of ITU stations such as 4U1ITU, LA1ITU, EA6ITU, and VU2ITU.

Haviland's first paper relating to amateurs and satellites, written with a colleague, was published in *The Scientific Uses of Earth Satellites* well before the beginning of the space age. Since that time, Haviland has written many articles on antennas of all types for all the major amateur publications. He has been a featured speaker at the Dayton Hamvention® , discussing such topics as quad antennas and supergain.

Haviland first began his quad work about 20 years ago when he moved to a location that wouldn't support a full-size Yagi and decided the quad would provide a smaller-sized alternative. Along the way, he began a full-blown study of the quad antenna. Haviland has devoted many years of investigation, experimentation, and computer time to this study. His labors have produced a comprehensive body of work covering all aspects of this popular antenna. But, as he comments, there is still much more digging to be done, and there are many more secrets to be unlocked. It is his wish that this book will provide a vehicle for further analysis and study.

In addition to the time spent researching the quad antenna, Haviland has put many hours into studying various antennas and antenna phenomenon using the MININEC program. Haviland is currently continuing his computer studies with the NEC program and another program specifically designed to handle insulated antennas. When he's not busy writing about, studying, or designing antennas, W4MB can be found on whatever band offers the best DX ragchews—a pastime he finds very satisfying!

Alf Wilson, W6NIF, and Terry Littlefield, KA1STC, editors

Preface

Virtually all antenna types were developed to provide some particular performance feature. The search for the solution to an unusual problem led to the development of the original quad. The problem arose as follows:

In 1942 missionary station HCJB in Quito, Ecuador placed a new 10-KW 12-MHz transmitter in operation, feeding a four-element Yagi array. The combination of the 10,000 foot elevation, the low pressure associated with storms, and rain drops accumulating on the ends of the Yagi radiator produced spectacular fireworks. Corona discharge was common, and occasionally 4-foot long standing arcs were produced. The resulting heat would melt the end of the radiator, or even cause the aluminum to burn.

The engineer of the station, Clarence J. Moore, first worked out a temporary, but partial, solution by installing the hollow copper balls used in flush toilets at the end of the elements. The extra diameter and smooth contour reduced the potential gradient, decreasing the number of damaging events.

Moore continued to work on the problem. He finally developed the concept of bending the ends of two dipoles toward each other until they touched, creating the quad loop. Once there were no ends and no high gradient, the arc problem disappeared. Further tests and experiments led to the development of radiator/reflector and director/radiator/reflector combinations of "classic quads."

At the end of World War II, hams all over the world rushed to get back on the air. Materials were often scarce. Quads appeared, first based on data supplied by over the air reports then by magazine articles. Variations were developed, such as the open loop used in the bi-square beam. The delta loop came later; this time to solve a perceived mechanical problem.

The origins of the complete quad family go back to Hertz, the first radio experimenter and inventor of the antenna. Most of his work used dipoles, but loops were also investigated. In the commercial field they were used only in direction finding. Apparently, this occurred because commercial radio developed from the low frequencies, where loops are inefficient in realistic sizes.

The theory of the general quad is quite recent. Only the special case of small loops with equal currents in all parts has an appreciable history.

The material in this volume came from a study started when a change from a two-element to a three-element quad gave the expected forward gain increase, but at the expense of very poor front-to-back ratio. My search for the reasons led to a study of available literature, experiments, and, finally, a computer analysis. As the saying goes, "one thing led to another."

I have attempted to include adequate references to theoretical studies. However, there are no references for the extensive literature of practical quad construction. For these, see, for example, the list in *The ARRL Antenna Handbook*, or *From Beverages Thru OSCAR*—a comprehensive bibliography of amateur radio literature available from Didah Publishing of Nashua, New Hampshire.

R.P. Haviland, W4MB

Chapter 1

General Concepts

A comprehensive study of this popular antenna

Quad Versus Yagi

The Yagi is the most popular type of amateur directive antenna. Surveys on the HF bands show that over half of all stations use some form of Yagi. Three-element triband designs with traps for band isolation are the most common.

The quad, second in popularity, is usually a two-element design using the square configuration. Many are single band; others are two or three single-band antennas mounted on a common boom. True multi-band quad elements have been designed, but are very rarely used.

The lower portions of the HF band tend to favor delta or triangular loop quads that need only a single high support. The circular loop is popular on VHF. However, total usage is about one-fourth that of the Yagi—some 12 to 15 percent of all installations. (This includes all types of antennas, not just Yagis.) Commercial use of quads is limited and they are relatively low sales items on the amateur market.

The pros and cons of quads are often debated. The most common reason given for using one is that the quad is a good directive antenna which can be built from locally available materials. The quad is fun to experiment with and has a reputation for good performance at low height. Long-time quad users seem to stay with the design because they like the performance.

Three reasons usually given for not using a quad are: susceptibility to damage during high winds, lack of space, and lack of design data for the high-performance types. In actuality, these are weak reasons for rejecting this antenna. First, the wind damage factor can be reduced by solving aerodynamic and mechanical design problems. Second, for a given gain, the quad can have a smaller turning radius than a Yagi. This allows it to fit into smaller places, and makes it easier to obtain gain at lower frequencies. Finally, there is an enormous amount of literature covering all elements of quad theory, design, performance, and construction.

Unfortunately, quad literature tends to be fragmented. The theory and most precise design and performance data are written by scientists and engineers, while the practical construction data appears in amateur radio publications. The scientific data are difficult to implement in a practical sense, because computers are needed to derive useful design values.

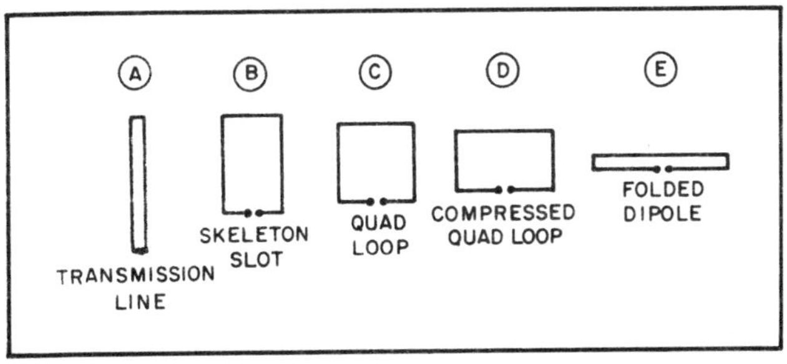

Figure 1. The square (4-sided) quad family, conceptually formed by separating the sides of a two-wire open transmission line shorted at the far end. As a first approximation, all members behave as a transmission line.

Shapes of the Quad Families

One of the easiest ways to approach the quad concept is to start with a length of open-wire transmission line, shorted at one end and with a generator at the other, as shown at A in Figure 1. If the spacing between the sides of the line is increased progressively, the shapes of B to E develop successively. These shapes are representative of the rectangular quad family. From A to E, the shapes are:

A: shorted line
B: skeleton slot
C: quad loop
D: compressed quad
E: folded dipole

Instead of keeping the sides of the figures parallel, we could choose the midpoint of the sides as the point of inflection and create a second family with diamond-shaped members (Figure 2, A to E). The center shape, the diamond quad, is the only one commonly named.

When three points of inflection are used instead of four, the result is two families of triangles as shown in Figure 3, A to E, and Figure 4, A to E. The center triangles in each figure are generally called delta loops, and are distinguished by whether they are fed at an apex or the midpoint of a side.

You can use more than four points of inflection. At the upper limit, the sides become smooth curves—the ellipses and circle of Figure 5, A to E. At the extremes, A is still a shorted transmission line section and E is a folded dipole.

It is not necessary that the shapes remain symmetrical, that the conductor always lie along the perimeter of the figure, or that no part of the figure be reentrant. Three possible shapes are shown in Figure

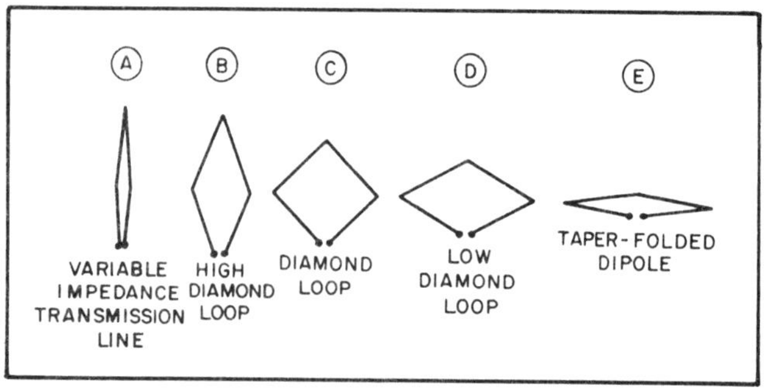

Figure 2. The diamond quad family, formed by changing the separation of transmission line sides at the midpoint. Again, all members behave as a transmission line. Properties are very close to those of the square quad family.

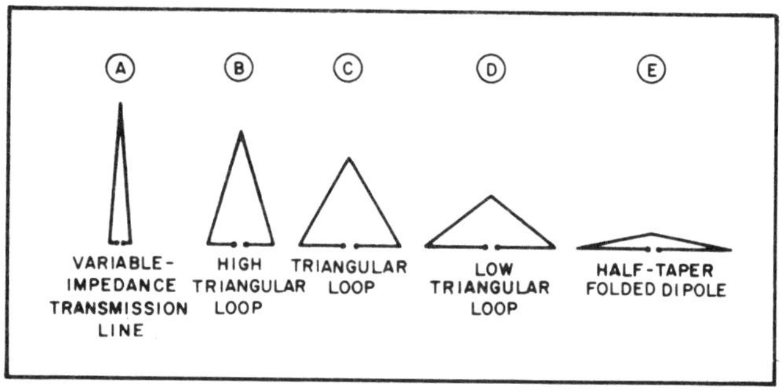

Figure 3. Triangular loop family, formed by changing the transmission line spacing at the feed point. Transmission line behavior indicates that the performance will be similar to the square and diamond loops.

6. A is an acute triangle, B is a bent-side or bat-wing quad, and C is a "line shortened" delta loop.

The element need not be confined to a plane. The G4ZU Birdcage and the Swiss quad, examples of nonplanar elements, are basically variations of the bat-wing element of Figure 6 with the apex of the bent element pulled out at right angles to the paper.

Finally, simple and symmetrical shapes aren't the only ones that may be used. Figure 7 shows some other designs. The design at A uses a transmission-line section to reduce size, while B and C use a form of capacitive hat for the same purpose. D uses a form of open-wire feed to give choice of polarization.

Conceptual Approach to Quad Performance

The end points of the major quad families have well-known characteristics, providing a basis for a conceptual approach to quad performance. Let's start by taking a look at the shorted transmission line.

First, suppose the generator frequency for the line of Figure 1A is varied. At very low frequencies, any practical line is basically a short circuit and has little

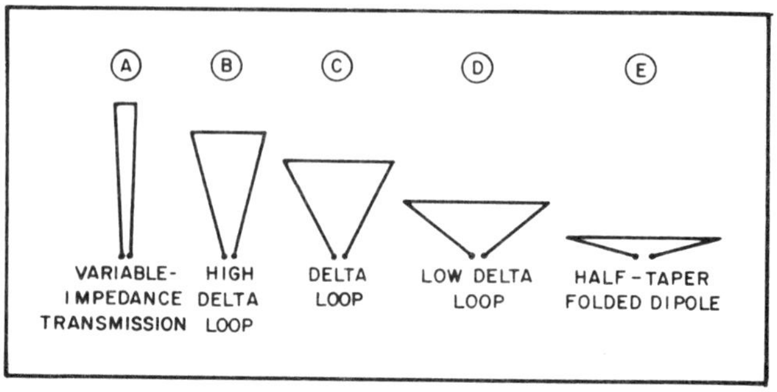

Figure 4. Triangular or the delta-loop family. Transmission line spacing change is at the far end of the line.

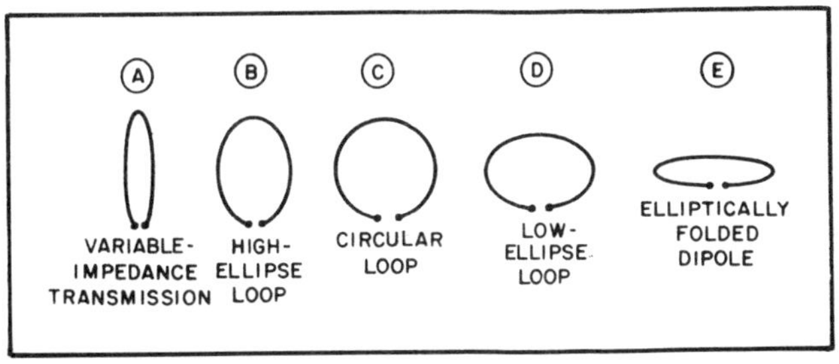

Figure 5. Circular-loop family, formed by smoothly changing the transmission line spacing. For a given length of conductor, the circle has the greatest area, and will show higher gain than the square. (For the same given perimeter, the area of a circle is 27 percent greater than that of a square. Ed.) However, the figures with greater separation of the current maxima can show still higher gain; this is true for all loops.

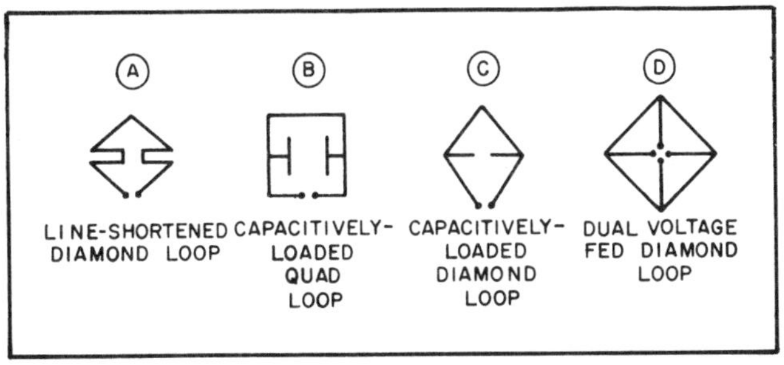

Figure 7. More examples of special shapes, all using some form of loading to reduce antenna size. Line sections can be replaced by inductors, and the capacitive sections by actual capacitors. The fourth shape uses the capacitive lines as the feed point, to allow horizontal or vertical polarization.

effect. The resistive component of impedance is small, essentially zero. The reactance is also small and inductive.

Both resistance and inductive reactance increase with frequency. This increase continues until the line becomes one-quarter wave long. The input impedance is very high and purely resistive at the frequency where this occurs. The line acts as a parallel resonant circuit. This is also known as the first resonance point.

For still higher frequencies, the resistance decreases, and the reactance decreases from its infinite value, but is now capacitive. When the line is one-half wave long, the reactance again becomes zero. The input resistance would be zero for a lossless line. This is the first series-resonance point. At slightly higher frequencies, the resistance and reactance increase, and the reactance is (again) inductive.

The pattern of alternating low and high impedance, and inductive and capacitive reactance, repeats at increasing frequencies. Figure 8 shows the pattern with the peaks and valleys marking additional resonances.

At the other (shape) extreme is the folded dipole of Figure 1E. At low frequencies, its input resistance is very low. Its reactance is high and capacitive. The antenna looks like a small capacitor in series with a small resistor.

Reactance decreases and resistance increases with increasing frequency, until the antenna is somewhat less than a half wavelength overall. At this point, the reactance is zero and the resistance is very nearly 300 ohms—just four times the resistance of a resonant single-wire antenna. This is the first resonance point.

At still higher frequencies, the resistance continues to increase. The reactance also increases, but is inductive in sign. With the antenna just less than a full wave overall, reactance again becomes zero, and impedance is very high.

This pattern also repeats at close to each half wavelength. Figure 8 shows a typical pattern of variation. The points of low impedance and zero reactance mark the second, third, etc. current-feed resonances, and the high impedance points to the voltage-feed resonances.

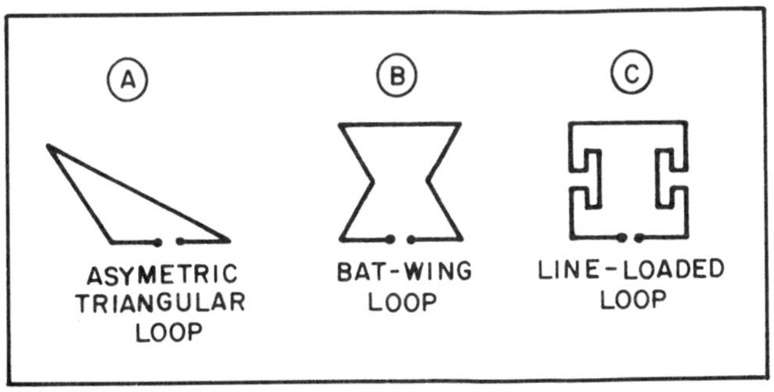

Figure 6. Examples of other possible shapes. In each case, the area in the figure is less than that for a corresponding simple symmetric member. This reduces gain below maximum (possible), but allows attainment of some other objective, such as small size or use of a given support.

Frequency/Size/Pattern Relationship

Before we continue, let's consider the effect that changing frequency (or size) has on the radiation pattern.

At frequencies small compared to first resonance, the current on all parts of the wire will be nearly equal in magnitude and phase, as indicated for the loop in Figure 9. For a point on the loop axis at right angles to the plane of the paper, the field contributions of the eight points of current shown appear as in the small

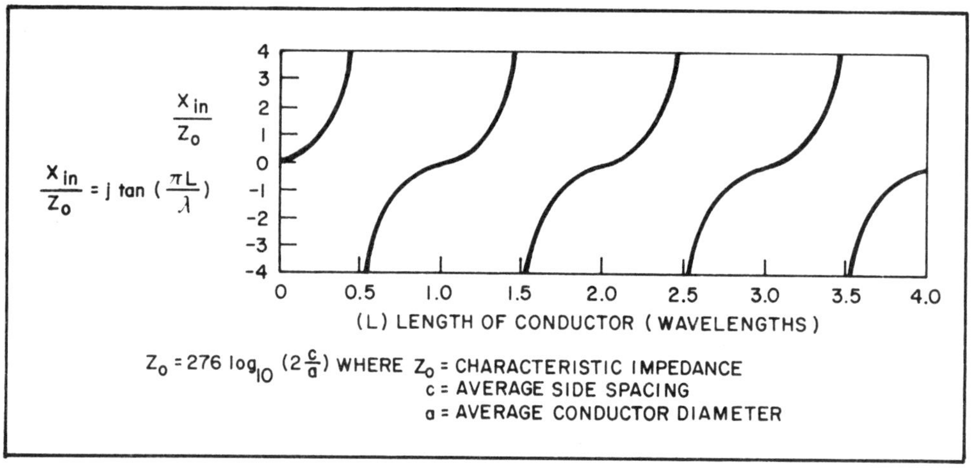

Figure 8. Generalized terminal reactance of a two-wire transmission line. The characteristic impedance is determined by wire size and spacing. From the standard formula, $Z_0 = 276 \log 2(c/a)$, *c* being the separation and *a* the wire diameter. Use the average separation for nonparallel-sided lines. Lacking other data, use this curve to give (approximately) the feed-point reactance for any member of the quad family of loops.

vector diagram, and add up to zero. In the plane of the paper, the components from the near and far side of the loop have an out-of-phase component, because of the time required for a radio wave to travel across the loop. A doughnut-shaped pattern results. The hole axis is at right angles to the plane of the paper.

These very small loops will radiate well, but will be difficult to feed because they have very low resistance—not much more than the conductor itself. Loss in the matching system will be high.

At frequencies close to first resonance, maximum current will occur at the feed point and half way around the loop from the feed point, as shown in Figure 10. The major field components from the upper and lower parts of the loop, on axis and at right angles to plane of the paper, are now in phase. In the plane of the paper, the components on the line of maximum current are out of phase. Maximum radiation is at right angles to the loop, and is horizontally polarized.

There are small components of current on the vertical parts of the loop. These are equal and opposite in phase and form a vertically polarized pattern. Most quad analyses neglect this component because it is much smaller than the main lobe. Its practical importance isn't clear, but it may be responsible for the quad's reputation as a good performer at low heights and under marginal conditions.

For frequencies close to second resonance, maximum currents occur at each one-quarter point around the loop (Figure 11). Because currents on opposite sides are equal but opposite in phase, there's no net radiation on the loop axis. Maximum radiation occurs in the plane of the paper, with lobes of 90 degrees each. This is the general pattern for all resonances higher than the first, with the number of lobes equal to twice the order of the resonance. The pattern is the same for all simple shapes.

These considerations indicate that the primary area of interest for the quad family of Figures 1 through 6 will be at or near first resonance, where the lobe structure is simple and on the axis of symmetry. For all of these shapes, the total conductor length is very close to one wavelength.

Although quads with conductor lengths of two or

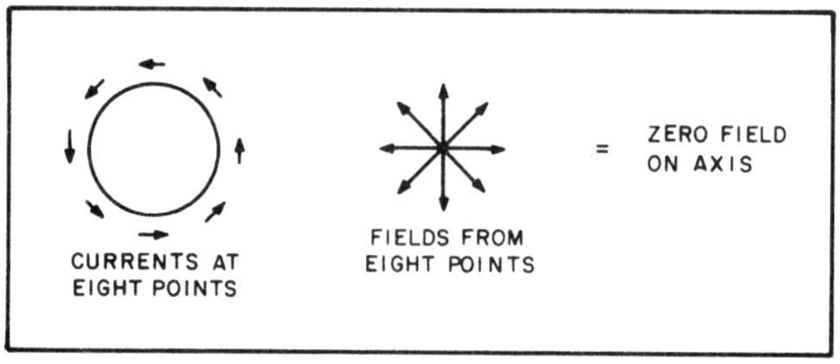

Figure 9. Vector field pattern of small loop, showing that the fields from individual wire segments cancel for on-axis radiation. The resultant field is doughnut shaped.

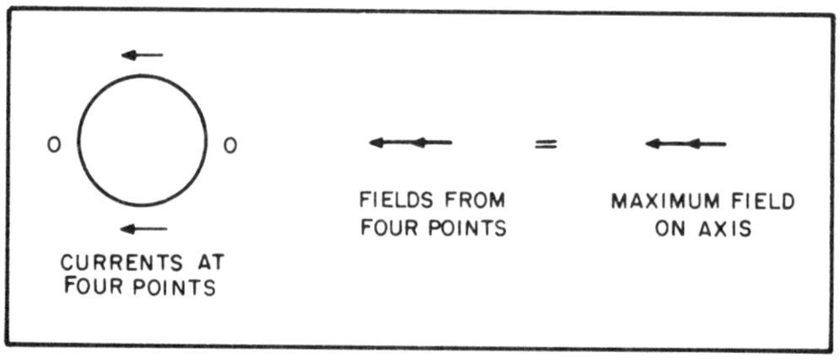

Figure 10. Vector field pattern for a one-wavelength loop. Components from the loop top and bottom add to form a major lobe on axis. The components at 45 degrees do not cancel completely, so there is a small doughnut pattern of vertically polarized radiation.

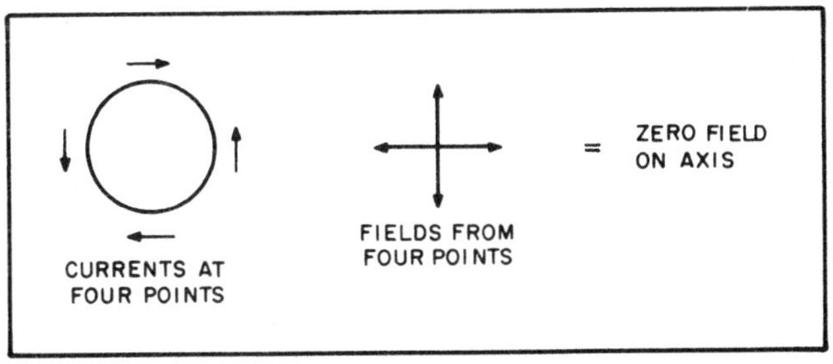

Figure 11. Vector field pattern for a two-wavelength loop. Components on axis cancel, but add to form a double cone-shaped pattern. Cancellation of on-axis radiation occurs for wire lengths of three or more integral wavelengths also.

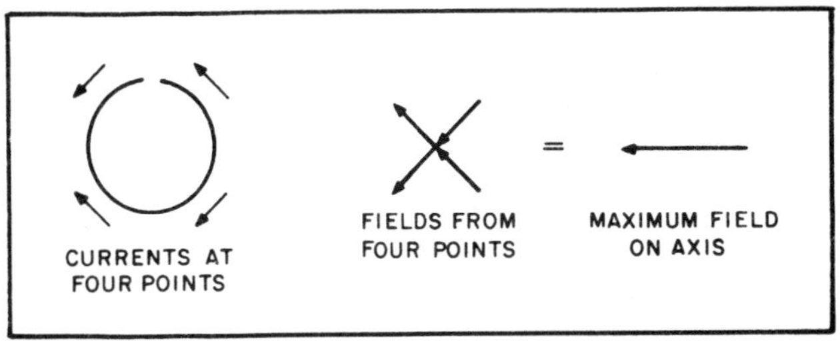

Figure 12. Vector field pattern for two-wavelength loop formed from an open transmission line. The components add to produce an on-axis lobe. This lobe exists for longer wires also, but there are also added lobes.

more wavelengths have useful radiation patterns, I will consider them later. However, one exception I'd like to point out occurs if the basic transmission line is open circuited (see Figure 12). The first, or serial, resonance point isn't too useful. The next (or parallel) one gives a simple lobe structure on the axis of the loop, which is useful in itself or as a switched open/shorted two-frequency loop. This was the basis of the now rarely used bi-square beam.

Drive Resistance Relationship

The input resistance of a shorted one-half wave transmission line is almost zero (see discussion referring to Figure 1). For the folded dipole, the drive resistance is about 300 ohms. Considering that the currents in two wire segments interact to a lesser degree the further they are apart, we expect the input resistance of the intermediate shapes to fall between these limits. For the shapes designated as C, the input resistance should be near 150 ohms. The resistance will be lower for the skeleton slot, or B types, and higher for the squashed D types. The exact pattern of variation must be worked out.

Drive Reactance and Resonance Relationship

Antenna resonance is defined by zero input reactance. All of the shape variations will have a specific point or points of resonance, and we can't expect them to be independent of conductor shape. For a shorted, air-insulated line, resonance will be very close to the physical half-wavelength point. The resonant frequency of a folded dipole will be some 5 percent lower. Because we don't know the current change due to variations in the separation of current points, we must determine the effect of changing shapes.

The magnitude of reactance change in moving away from resonance involves cosine functions. For small deviations from resonance, it's reasonable to expect that the reactance can be approximated by a simple linear function. The reactance, then, can be specified by two values—the frequency of resonance and the slope of the reactance curve. We must see if an additional simple function for shape can be developed.

Gain Relations Near Resonance

We can look at the gain from two viewpoints. The first is based on element length and separation. The gain of the shorted line section is close to zero, because the radiating section is approximately zero length. As we move toward other shapes, the radiating length increases, but the spacing between the two high-current sections decreases. As the shape approaches that of the skeleton slot, there will be two well-separated high-current sections of reasonable length. The gain should approach twice that of a simple dipole.

At the other extreme, the gain of the folded dipole is the same as a simple dipole. Thus, we expect a gain decrease when going from the very tall, narrow slot toward the dipole. Overall, the gain should increase slowly as the sides of the dipole are pulled apart, reaching a maximum at some separation, and then falling to zero gain. However, without some other guidance, we can't state the gain of intermediate shapes.

A second way of looking at this gain variation is from the view of the effective area. Considering just the high-current portion, the area measured in average amperes times electrical degrees of length will be nearly zero for the shorted line, unity on a per unit scale for the folded dipole, and basically two for the shapes (C). Thus we expect the gain of an optimum member of the quad family to approach 3 dB above isotropic. The circular loop should have the best gain, the rectangular one slightly lower, and the delta loop lower still. However, all nearly symmetrical shapes should show some gain with respect to a dipole.

Further consideration of the area indicates that the gain of the resonant open-circuited (two wavelength) square should approach 6 dB above an isotropic at parallel resonance. We would expect many of the odd and re-entrant shapes to have gain, but the ones with small area will show a loss compared to a dipole.

Quad Elements in Beams

Any of the above shapes can be used in either a driven or parasitic beam. One way of looking at such beams is to assume that the radiation effect of each element is the same as if it comes from a point source located at the center of electrical symmetry of the element. This leads to the concept of pattern multiplication, where the overall pattern is the product of the element pattern and an array pattern, as determined by the relative spacing, current, and phase.

Because array patterns are well known (and quite simple in many cases), beam patterns will be easy to develop when the element patterns are worked out. There is, however, a corollary to this. Any unusual benefit from the quad family will have to come from the elements themselves and not from the fact that they are assembled into an array.

In this simplified consideration, some gain above a dipole has appeared. It's expected that a quad array can be shorter than an array of dipoles (Yagi) and still achieve the same gain. Although the vertical height is much greater, the horizontal span and the turning radius are less. For a given gain, quads require less space.

Neglected Factors

I made a number of implicit assumptions in the preceding paragraphs. One is that the conductor has no loss and is very thin. Currents were assumed to be sinusoidal. Small pattern factors, like the lobes in the plane of the elements, were ignored. These are important in a detailed analysis, but should represent only variations from simply derived values and not major changes.

Chapter 2

Circular and Octagonal Loops

Analysis shows good performance with interchangeable data

Although the terminology is contradictory, it's convenient to consider the circular loop, and arrays built of loops, as the first members of the quad family. One reason for this is that all other versions can be regarded as departures from the "ideal" circular figure. To the extent that this is true, the performance of circular-loop antennas is representative of the performance of the entire family.

Another reason is that theoretical analysis of the circular loop is far more advanced than analysis for other shapes. Extensive tables of calculated characteristics have been published, some with comparisons of measured performance. In contrast, while there are theories of square quad loop and array performance, their complexity makes them impractical for calculation—even on mainframe computers.

Theoretical Basis of Circular-loop Analysis

As shown in Figure 1, only two quantities are necessary to specify the circular-loop antenna: the conductor size, usually given as its radius; and the loop size, also described by radius. It's often more convenient to use two derived descriptive quantities in theoretical discussion. The first is the normalized circumference of the loop at the specific frequency of interest, given by the quantity kb, where k is defined as:

$$k = 2\pi/\lambda,$$

λ being the wavelength. The quantity kb is therefore the circumference of the loop in wavelengths.

Conductor size is usually given by the relationship:

$$\Omega = 2\ln \left(2\pi b\right) a$$

where LN is the natural logarithm, equal to 2.3 times the more common $\log_{10}$ value. The value of omega is given in Figure 2 as a function of the ratio $2\pi b/a$, or loop circumference to conductor radius. Values less than 10 represent very large conductor diameters; those over 20 represent very small conductors. High-

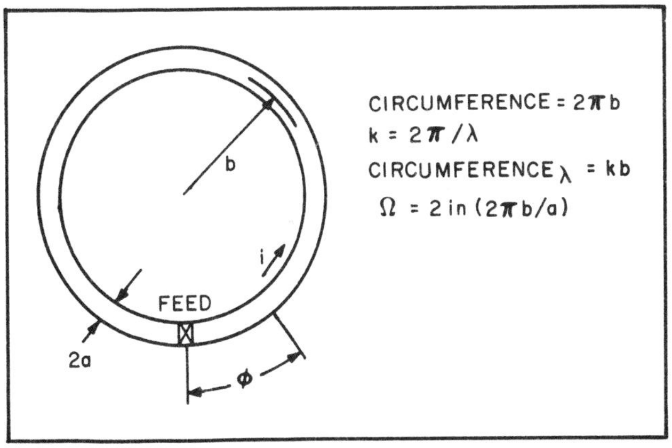

Figure 1. A circular loop can be described by two quantities, the radius of the conductor (a) and the radius of the loop (b). For work, it's convenient to use two derived quantities, a conductor thickness factor (omega) and the circumference in wavelengths at the operating frequency (kb).

frequency antennas will usually have values in the range 20 to 25, and self-supporting ultrahigh frequency antennas values of 10 to 15.

The loop is assumed to be fed at one point, usually taken as the angle reference. This induces a current, *i*, in the loop at angle zero which, in turn, creates a field at the point designated by R,01, for example. The total field at this point is the sum of the fields produced by all points on the loop.

The field components also induce current flow in the loop. When equilibrium is reached (after a few RF cycles), the field close to the conductor must lie only at right angles to it. (If there had been a tangential component, a change in the current would have been induced, so equilibrium would not yet have been attained.) This observation plus standard field equations give the conditions for calculating current distribution, and therefore the drive impedance and radiation pattern.

While the concept is relatively simple, the mathematical operations are difficult. See the references, especially Storer,[1] for details. For our purposes, it's sufficient to note that the current distribution is given by:

$$I(\emptyset) = \frac{V}{(j \cdot \pi \cdot 377)}\left[\frac{1}{A_o} + 2 \cdot sum\,\frac{Cos\,(n \cdot \emptyset)}{A_n}\right]$$

where the sum is for all values of N, from 1 to infinity. This result was derived by Hallen.[2]

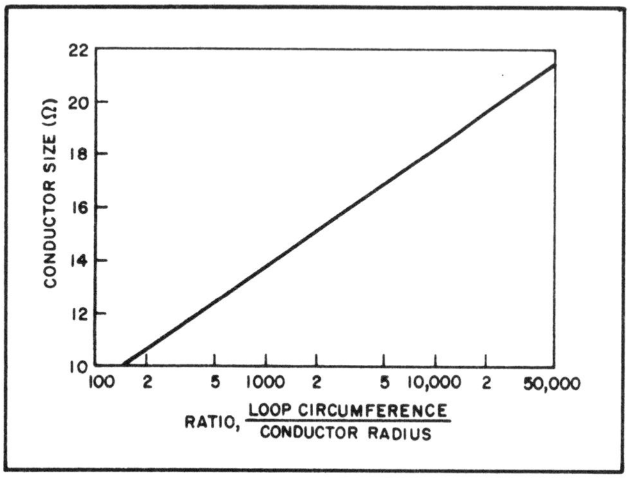

Figure 2. Values of the thickness factor (omega) as a function of the ratio of loop to conductor radius (or diameter). Practical self-supporting loops need an omega around 10 to 12 to have sufficient strength. Wire cage elements may be used to secure low omega factors at low frequencies.

This equation is simple, but its evaluation is complex. The quantities A involve series for which exact solutions are unknown. Even approximate solutions require further assumptions—one being that the conductor diameter is small compared with loop diameter and operating wavelength. This restriction is satisfied by practical antennas.

Additionally, the infinite series in the equation tends to become divergent when the number of terms is reduced for reasonable scale of calculation. Storer[1,3] developed a method of calculation by keeping the first four terms of the series and replacing the remainder with an integral. He also published a set of ten curves giving the real and imaginary components of the series. With these, evaluation of the current distribution and drive impedance is reduced to some simple (but tedious) curve measuring and complex-number algebra. Unfortunately, the curves can't be reduced to simple equation form, so this is unavoidable.

Rather than presenting these curves and usage details here, I'll give only the results of the examination of some specific loop designs. Subsequent analyses have yielded a table of drive impedances which is more accurate for most work. These values are covered later.

To find values outside the range given here, or to obtain the current distribution, we'll need the Storer curves. (The version of the curves in Reference 3 is best, and is available at reasonable cost from Cruft Laboratory Library, Harvard University, Cambridge, Massachusetts.)

A simpler solution for small circular loops is available.[4] Here "small" covers the range from 0 to 0.3 wavelength circumference. This restriction allows simplification of the general equation for current distribution, with only three terms giving adequate accuracy. Approximate equations for each are given, and are easily handled by small computers. Note that the reactance for these small antennas is determined almost entirely by the loop circumference, with almost no effect on conductor size. In contrast, the input resistance varies with both.

Current Distribution on a Circular Loop

Figure 3 shows the magnitude of the current on a one-wavelength circular loop as calculated by Storer, together with the current magnitude on an ideal shorted transmission line made from one wavelength of con-

Figure 3. Current magnitude and phase on a large conductor circular loop one wavelength in circumference as derived by Storer.[1,3] The loop is below resonance. The current on a conductor of essentially 0 radius is shown for comparison. Derived from transmission line theory, this is the first approximation to the current on any loop of the same circumference.

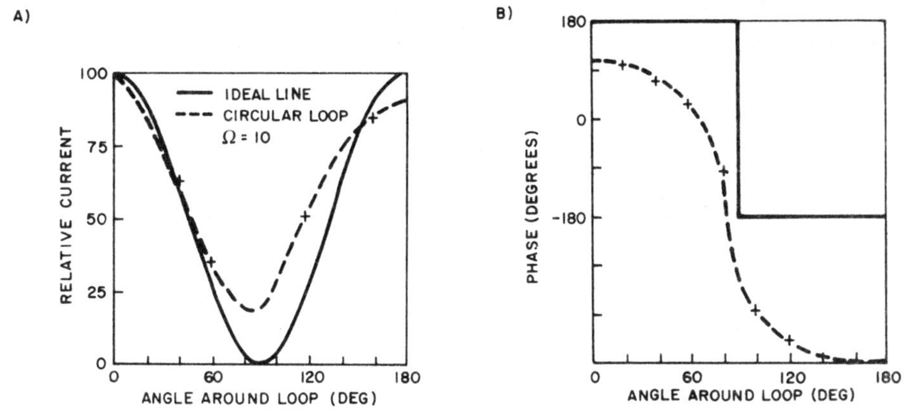

ductor. Figure 3B shows the phase with respect to the driving voltage for both cases.

A number of important factors concerning the entire quad family are illustrated by these curves. The first is that a one-wavelength loop isn't resonant, as indicated by the fact that the angle of the drive impedance is not zero. Because the loop appears as a capacitance, it's below resonance. Unlike dipoles, loops must be longer than a wavelength to be resonant.

A second factor is that there's no point on the loop where the current goes to zero. Associated with this is the fact that the current at 180 degrees from the feed point is appreciably less than at the driving point. Similarly, these two currents are not 180 degrees out of phase, but somewhat more. Consequently, the point of phase reversal is not at right angles to the line of symmetry through the feed point.

One reason for the differences between the loop and the ideal shorted line is the greater separation of the sides. Currents aren't constrained to be equal because of tight coupling, as in the ideal shorted line. Further, power is being radiated, causing a reduction in current when moving away from the feed point. (In a practical antenna the current differences would be somewhat greater, as the analysis assumes zero ohmic loss.)

These current curves show that the usual evaluation of a quad—two separated dipoles with ends bent to touch—can't fully describe the performance. This simple concept is useful in verbal descriptions and can be a valuable tool in approximate analysis. But it must be remembered that numerical results are probably in error by a factor *at least as large as the current error*, or at least 20 percent. The effect of the error should be smallest for pattern calculations and drive resistance, but is likely to be sizeable for drive reactance and resonant frequency.

These detailed current calculations are for loops with relatively large conductors, omega = 10. Other studies, plus the tables presented in a later chapter, show that the current magnitude and angle move progressively toward the curve for the ideal transmission line as the conductor radius becomes smaller. This means that the two-dipole approximation is likely to be better at high frequency than at ultrahigh frequency where omega is large for practical conductor sizes. The current distribution for some other loops is also given in References 1, 3, 5, and 6.

As a first approximation, the current distribution can always be assumed to be that of an ideal transmission line of the same conductor length. A second approximation can be sketched by "rounding all sharp corners," and decreasing the current away from the feed point. Greater accuracy will require tedious evaluation using Storer's curves or the MININEC technique, which I will discuss later.

Drive-point Impedance

We find the drive-point impedance by dividing the drive-point voltage by the drive current calculated in the preceding section. Storer[1,3] gives tables of this impedance for loops from 0.05 to 2.5 wavelengths in circumference, and for omega values from 8 to 12 (large diameter conductors).

In considering the general problem of loop antennas, Wu[7] developed another method for solving Hallen's equation. This method gives the same values for resistance components as Storer, but the reactance values are quite different and agree more closely with

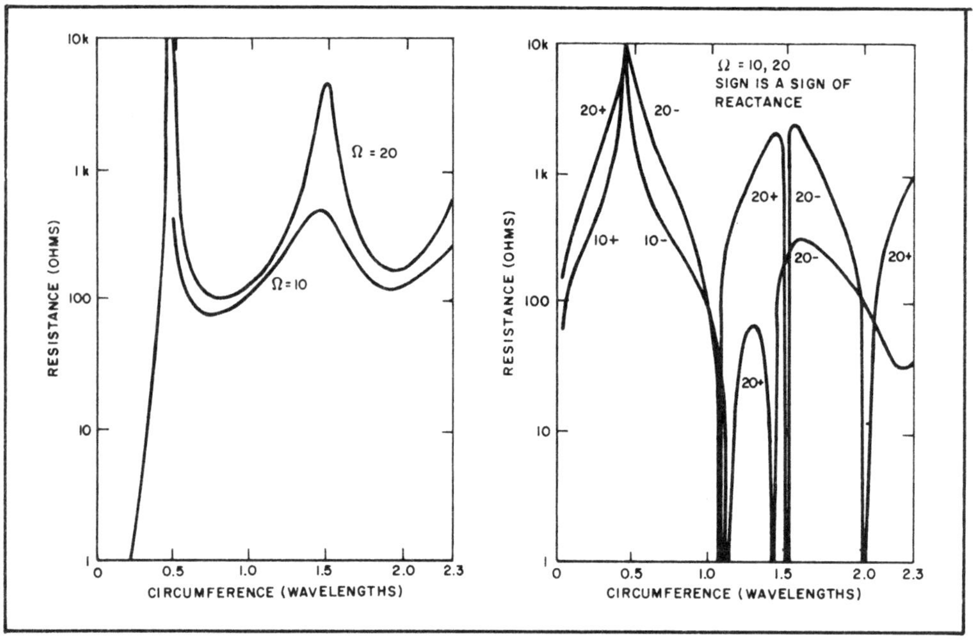

Figure 4. (A) Feed point or self-resistance of one-wavelength circumference loops—one with a large conductor radius, the other small. The drive resistance is sensitive to conductor size around the points of parallel resonance, at 0.5, 1.5, 2.5, etc. wavelengths circumference. These curves may be used as approximations for any loop of equal circumference and radius. (B) Feed point or self-reactance of one wave-length circumference loops. High reactance denotes the parallel resonances, low reactance the series resonances. Both series and parallel resonances are sensitive to conductor size. Series resonance is generally at wavelengths greater than integral values. Parallel resonances are at wavelengths less than odd multiples of 0.5 wavelength.

measurements on real antennas. King, Harrison, and Tingley[8,9] have used Wu's theory to calculate loop-drive admittances for sizes from 0.05 to 2.5 wavelengths and for omega from 10 to 20. For those who have forgotten, or never had occasion to work with admittances,

$$Z = R + jX$$
$$Y = 1/Z$$
$$Y = G + jB$$

(See also Reference 10.)

Figures 4A and 4B show feed resistance (R) and reactance (X), respectively, for loops from 0.05 to 2.5 wavelengths circumference. The two values of omega shown, 10 and 20, are representative of very high frequency and high-frequency loops.

Over this range of loop diameters, three high-impedance or parallel resonance points are noted, corresponding to 0.5, 1.5, and 2.5 wavelengths circumference. For omega = 20, the low-impedance, zero-reactance points correspond to serial resonances at about 1.0 and 2.0 wavelengths circumference. For omega = 10, there's also a serial resonance at 1.0 wavelength circumference. There is no true serial resonance for a circumference of 2.0 wavelengths. Instead, the reactance becomes low and remains low and capacitive. This is the case for all values of omega less than 11, and for all but the first serial resonance.

The resistance at the serial resonance point changes markedly with the value of omega. For a large omega value (20 or so), the resistance also varies markedly with loop diameter. But for omega = 10, the resistance change is much smaller. (For omega = 8, the change is less than 2:1 for any circumference greater than 0.6 wavelength.)

These characteristics mean that thick loops are inherently broadband antennas and relatively easy to match. However, the pattern changes described in Chapter 1 affect the desirability of this broadband operation, as discussed later.

Figures 5A and 5B show the feed resistance and reactance for frequencies of greatest interest (those close to the first series resonance) for loops from 0.8 to 1.2 wavelengths circumference. Figure 6 is derived from Figure 5B, and shows the resonant frequency as a function of conductor size. These three curves give the information needed to design practical loop antennas and arrays, and to calculate their performance. Their use is covered further on.

Gain of Loop Antennas

In considering loop gain, I noted that gain should increase as loop size increases and that there are pat-

Figure 5. (A) Feed point or self-resistance of three loops around one-wavelength circumference. (B) Feed point or self-reactance of five loops around one wavelength circumference. The reactance for all conductor sizes is equal to −95 ohms for a circumference of one wavelength. This may be compared with dipole reactance which shows the same characteristic, with a reactance of 35 ohms at 0.94 wavelength as the common point.

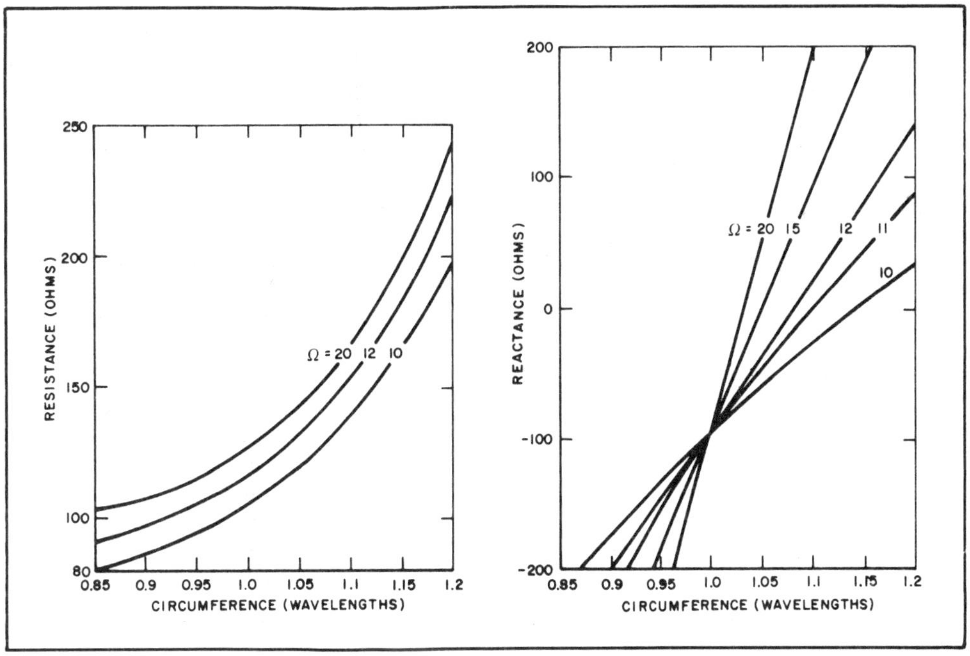

tern changes as size increases. Specifically, gain on the axis of symmetry becomes zero for all loops with 2, 3, or more wavelengths in conductor length.

These two effects are shown in the calculated gain curve of Figure 7 (see Reference 11). For a circumference of one wavelength, the on-axis gain is 3.4 dB above isotropic, or about 1.4 dB above a dipole. (Based on measurements, Lindsay[12] quotes approximately 4.0 dB, and Appel-Hansen[13] quotes 3.4 dB above isotropic.)

Figure 7 shows that larger loops are superior from a gain standpoint. A calculated gain value of 4.5 dB occurs for a loop circumference of 1.5 wavelengths.

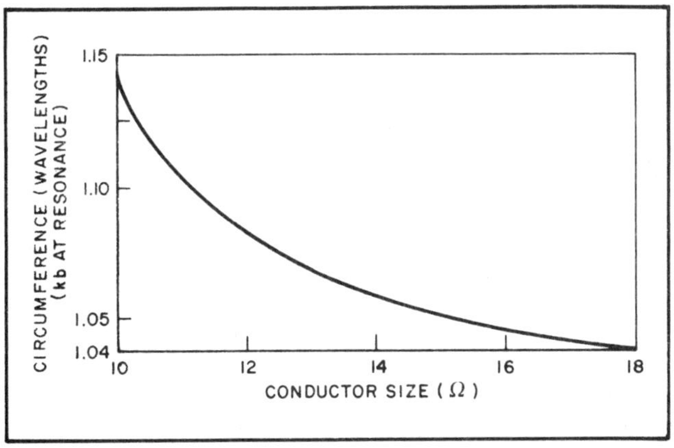

Figure 6. Resonant frequency of a circular loop derived from Figure 5B, as the frequency of 0 reactance. A simple linear relationship of reactance versus wavelength (or frequency) can be developed using two figures. However, the curves are usually less time consuming to use.

The gain decreases above this circumference, as the on-axis lobe splits. Remember, there is no radiation on axis for a circumference of 2 wavelengths.

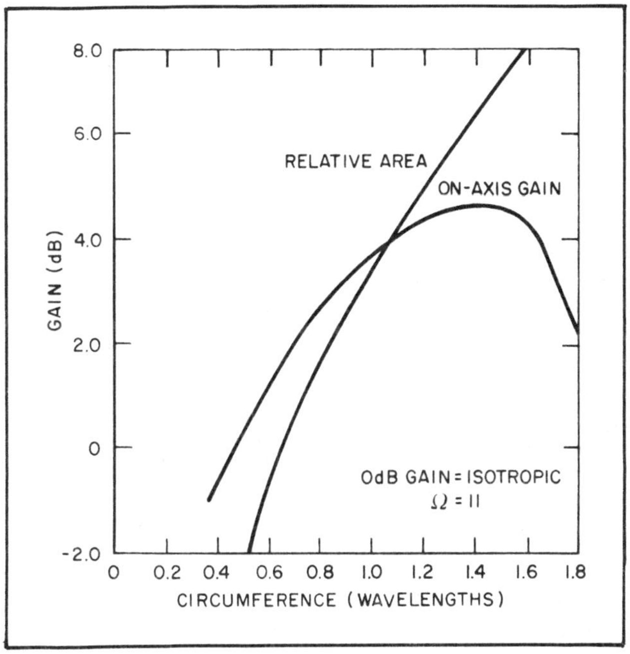

Figure 7. Gain of a circular loop in dB above isotropic, as derived by Ito et al.[11] Maximum gain occurs for a loop circumference of 1.4 wavelengths. Gain is less for smaller sizes due to less radiating area. On-axis gain, the value shown, is less at larger sizes because of lobe splitting. However, the gain of the main lobe increases as circumference increases. The associated curve shows the relative area of the loop.

In smaller sizes, the on-axis gain decreases as the circumference becomes less than a wavelength, and as the pattern changes toward the doughnut pattern of a small loop. The gain is about equal to that of a dipole at a circumference of about 0.65 wavelength, and about 1.5 dB poorer than that of a dipole for a circumference of 0.5 wavelength.

This brings up two important points. First, from the standpoint of gain, the circular loop should be designed to operate away from resonance. Considering the factors of gain, lobe shape, and feed impedance, Ito et al.[11] recommend a design circumference of 1.2 wavelengths, for a gain of 4.2 dB.

Second, a loop has good gain performance over a wide range of frequencies. For example, a 14-MHz loop would be slightly better than a dipole on 10 MHz, and about 1.5 dB poorer on 7 MHz. The gain would be near maximum on 21 MHz, better than a dipole on 18 MHz, and about as good on 24 MHz. The on-axis gain would be poor on 28 MHz, but there would be usable radiation in two lobes.

The acceptability of operation away from resonance is affected by the difficulty encountered in feeding the loop. If we look at Figure 4, it doesn't appear that feeding a large loop on high frequency would present an unusual problem. The feed resistance would increase, requiring the use of a matching section or transformer. There would be an inductive component, easily compensated for by a stub. Depending on conductor size, a very high frequency loop might require only a matching transformer.

Wideband operation would be a greater problem. You'd need a complex matching section at the antenna, or your transmission line would have appreciable VSWR. The combination of open-wire line and a wide-range "match-box" would give excellent performance. This is especially interesting as the basis of a 3.5/7/10-MHz fixed-loop design.

I have taken advantage of this wideband capability on many occasions. I use a 14-MHz loop on bands from 7 to 28 MHz on a regular basis. Much of my 10, 18, and 24-MHz experimental work (KK2XJM) used this antenna. Other than using Teflon™ insulated 75-ohm feedline and a pi-section tuner, I took no special precautions.

Circular-loop Radiation Patterns

Published data on radiation patterns of loop antennas varies from sparse to nonexistent. The equations needed to calculate the currents and patterns are solvable on a small computer only with a lot of programming and computing time. It has been necessary to approximate further to develop the pattern data which follows.

The MININEC Antenna Program

The chosen calculation technique (sometimes called "the method of moments") is the public domain program "MININEC," from Logan and Rockway,[14] currently in its third version. It uses essentially the same initial assumptions as the Hallen[2] approach. But instead of applying the geometry exactly and then making simplifying assumptions, it assumes that the radiator is composed of a series of straight-line sections carrying constant current.

A solution's accuracy increases as the number of sections for a given geometry is increased. While the complexity of the solution isn't greatly affected, the time required to obtain the solution increases as the square of the number of segments. An IBM PC may take several hours to run the program; a very small computer like the Commodore 64 may need 12 to 24. Even so, this approach is the best generally available, and is quite practical.

The original MININEC[15] was written for the Apple computer. I have translated it for the C-64 and the Amiga; KA4WDK has done the same for the PC. The third version was written for the PC, and I translated it for the Amiga; it also runs on the Macintosh.

MININEC's originators used various conditions to examine calculation accuracy, including analysis of loops. One series used a ten-sided polygon approximation of a circle, with two current segments per side plus three for the feed side. Agreement with theory was within 10 percent.

A second series approximated the circle using the circumscribed polygon. This shows good agreement (±6 percent) for susceptance, down to four sides. Equally good agreement for conductance required 16 sides. With 22 sides, agreement was within 6 percent for sizes from 0.1 to 2.0 wavelengths. MININEC solutions are found to be unreliable for circumferences less than about 0.01 wavelength.

The inscribed polygon approximation isn't an ideal check of solution accuracy as the number of segments decreases. It introduces two added factors which change the results. One is the loss of area. (For eight sides this amounts to approximately 3 percent.) This reduces the gain by about the same amount. More important, the total conductor length decreases as the

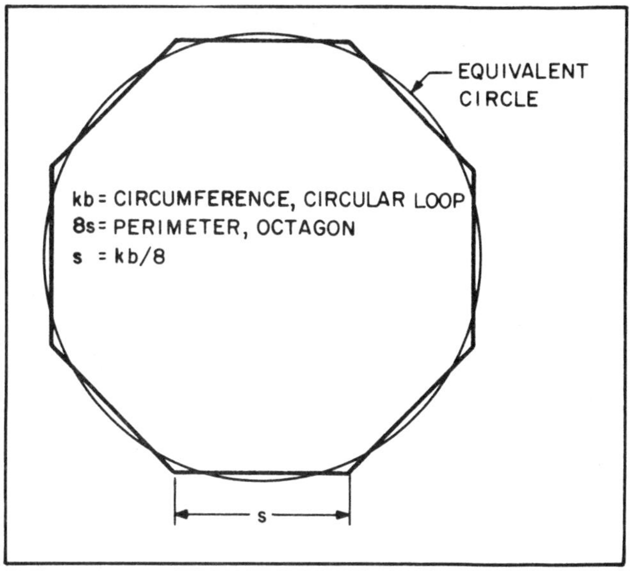

Figure 8. Relative geometry of a circle and an octagon. The difference in area for the same conductor length is very small.

number of segments decreases, by about the same percentage. This change in wire length introduces a change in reactance near the series-resonant points, and a change in resistance near parallel-resonant points. Both may be relatively large.

A better approximation occurs when the conductor length is kept constant. It's not easy to evaluate the precise error this causes, but it appears that it's no greater than the sum of the inherent error of MININEC plus the area error in the approximation.

It also appears that the inherent error will be around 5 to 10 percent if two simple rules are followed in setting up a MININEC analysis:

- Use a minimum of two segments per section of conductor.
- Use a minimum of four segments per halfwave of conductor.

In the practical sense, MININEC is extremely accurate. An antenna can easily depart from its ideal value by 20 percent or more because of often-neglected factors like supports, tower location, and feed-antenna interaction.

Octagonal Loop

Calculated values for an octagonal loop should be a good approximation of circular-loop values. The octagon is also a useful antenna by itself. The best-known type is probably the "Army Loop." This antenna is really a loop operating well below resonance, incorporating both a matching system and low-loss construction. Figure 8 shows the basic factors involved and the small area difference from the circular loop.

The properties of the octagon for conductor lengths close to one wavelength are summarized by Figures 9A and 9B for drive resistance and reactance. Compare these with Figures 4A and 4B to see the validity

Figure 9. (A)Feed or self-resistance of an octagonal loop for $\Omega = 17.9$, calculated by MININEC 3.[14] Values are within a few percent of those for the corresponding circular loop. (B)Feed or self-reactance or an octagonal loop. Differences between these and circular loop values are somewhat greater than for resistance. They are due to different area, and the geometry of current interaction. It's likely that there are also differences due to the calculation approximations, since the expected calculation error may reach 10 percent.

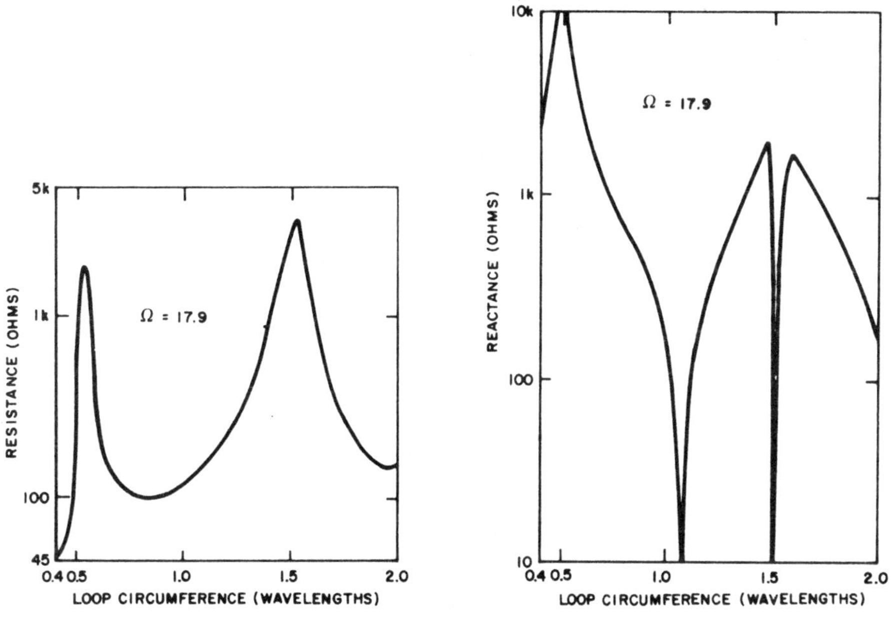

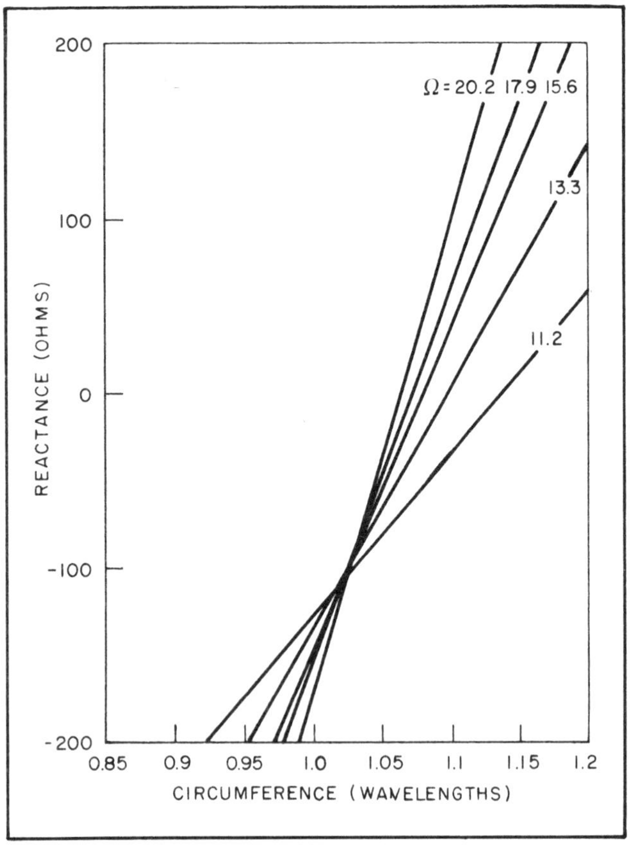

Figure 10. Feed or self-reactance of an octagon for circumferences around one wavelength. The point of equal values occurs at the same reactance as the circular loop but at a greater circumference. Curve slopes are almost identical.

of the octagon approximation to a circular loop. As might be expected from previous discussions, the agreement for resistance is good—no more than a few percent difference. The agreement for reactance isn't

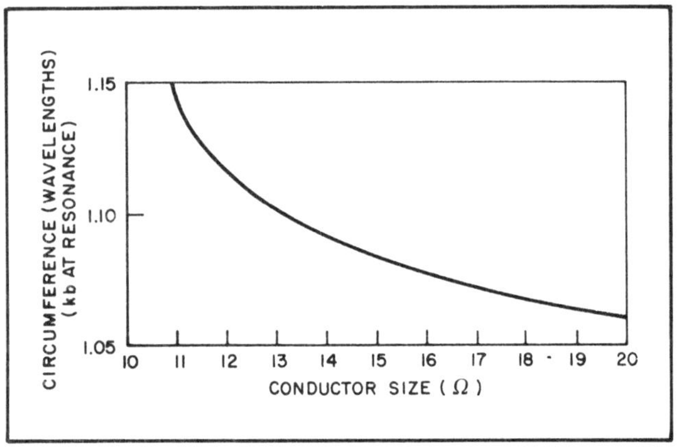

Figure 11. Resonant wavelength of an octagonal loop as a function of conductor thickness. The curve is similar to that of a circular loop, but differs by as much as ±6 percent.

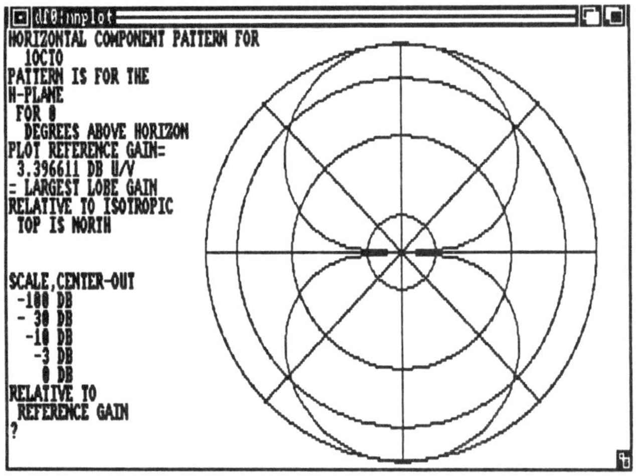

Figure 12. Horizontal radiation pattern for an octagon lying in a vertical plane and fed at the bottom. Only the horizontally polarized component is shown. The pattern is very similar to that of a dipole, but the gain is higher by about 1.25 dB.

quite as good, but this is the usual finding. However, neither error is beyond the expected range.

If you build a practical octagonal antenna, which is to be resonant or nearly so, use Figures 10 and 11. If it is away from resonance, the values from Table 1 give a good approximation. For best accuracy, the table values should be interpolated for the actual conductor diameter and equivalent radius. Alternatively, MININEC can be used to obtain values within 5 to 10 percent error.

Patterns of Circular/Octagonal Loops

Because the octagon has been found to be a good approximation of the circle, we can use its pattern as a close approximation for those of circular loops.

Figure 12 shows the MININEC-calculated horizontal plane pattern for a 1-wavelength octagon fed at the center of the bottom segment; that is, with horizontal polarization. The lobes are very nearly the same as those of a dipole, but are slightly narrower to produce gain. The gain is 3.4 dB—essentially that of a circular quad.

Figure 13 shows the vertical plane pattern in the plane of the loop, for the total radiation. Figure 14 shows the horizontal plane pattern for vertical polarization. There is a major difference from dipole patterns in these two figures. The loop has an appreciable vertically polarized component, zero on axis and maximum at right angles to this. This component isn't present with dipoles and its importance isn't clear. It may

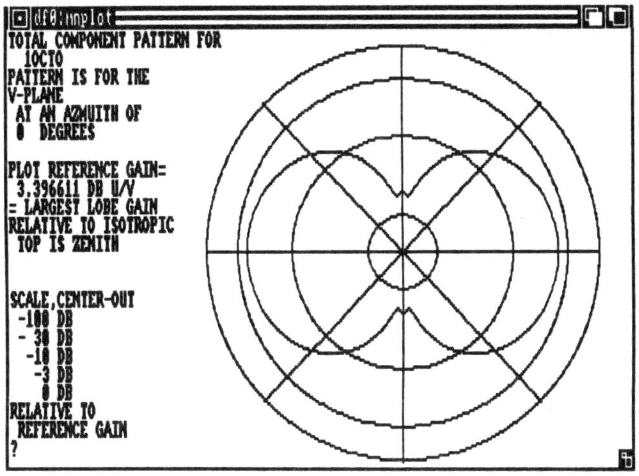

Figure 13. Vertical radiation pattern for the total radiation from an octagon in the plane of the loop. Maxima are along the line of maximum current. The minima represent the expected side radiation (kb = 1.0, Ω = 17.9).

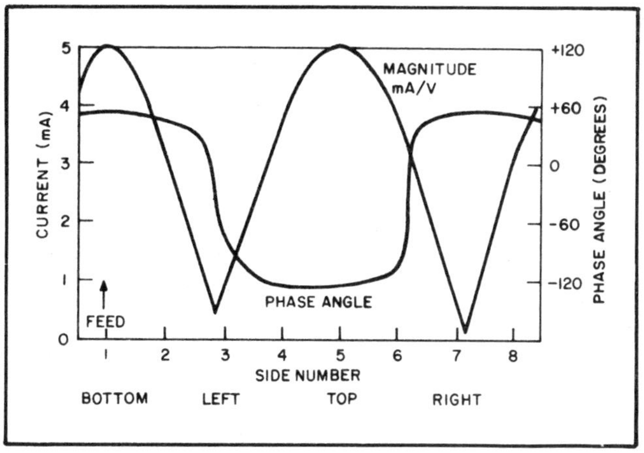

Figure 15. Current distribution on an octagon. Because of the small conductor size, the distribution approaches that of an ideal transmission line.

have an adverse effect because of interference received on the vertically polarized side lobes (and back lobe in beams), or it may tend to reduce fading due to the variation of incoming signals and path splitting. The component may also be a factor in the reputation of quad loops as good performers at low height.

Figure 15, provided for comparison, shows the calculated current on an octagon. Because the example is for a relatively thin conductor, there is close resemblance to the distribution for a transmission line.

Figure 16 compares the calculated gain of an octagon and a circular loop. Some of the variation is due to the difference in areas. The rest appears to be related to distinctions in the calculation method. Practically, the difference is negligible.

General conclusions from this comparison of circular and octagonal loops are that both have good performance, and that the data from one can be used for the other with little error.

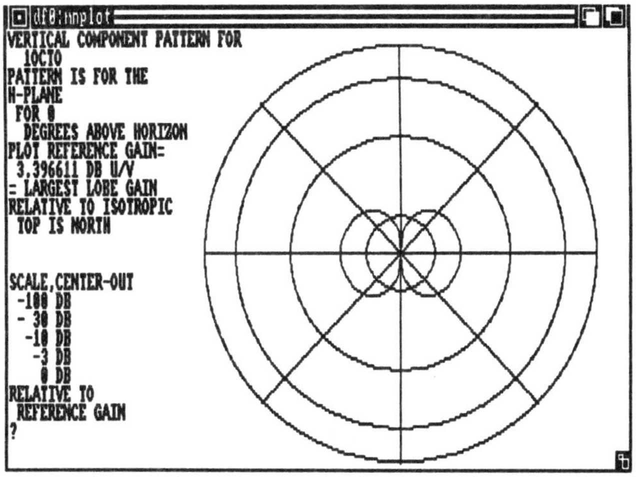

Figure 14. Horizontal radiation pattern for the vertically polarized component of an octagon lying in a vertical plane and fed at the bottom. This component is about 20 dB below the main lobe because of the currents on the vertical and inclined parts of the octagon.

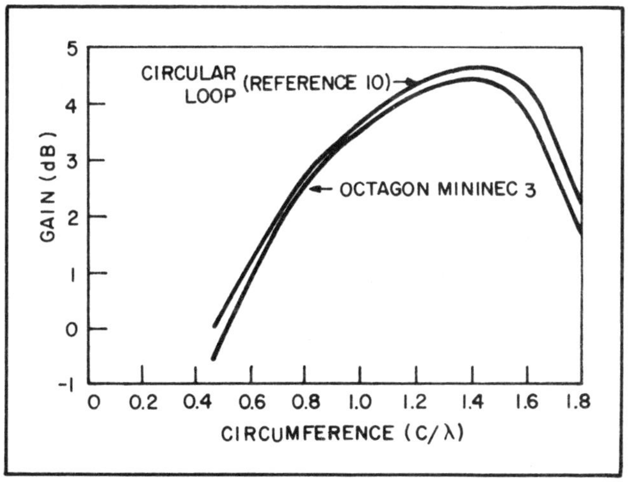

Figure 16. Calculated gain of a circular loop and an octagon. The difference is small—partly real and partly due to differences in the calculation technique. The curves do confirm the extra gain of a loop as compared with a dipole, and also indicate the added gain possible with off-resonance operation.

NOTES

Chapter 3

Circular-loop and Octagonal Arrays

Analysis of multi-element arrays and an all-driven-element design

An array of two or more circular loops (like any other antenna) can have all elements driven, or some driven and the others self-excited or parasitic. The special case of one driven element is sometimes called the Yagi-Uda configuration. This really applies only to the configuration of one parasitic reflector, a driven element, and one or more parasitic directors—all elements being dipoles.

Theory of Dual Circular-loop Arrays

The relationships between two circular loops, parallel and on a common axis, are shown in Figure 1. Compare this with the single-loop drawing in Chapter 2, Figure 1. The important difference is that two loops contribute to the field and, in turn, to the induced currents. They have the same requirement that the component of the field along the wire be zero.

Computational complexity can be reduced if the current on each loop is assumed to be the sum of two currents, of the form:

$$I1 = Ia + Ib$$
$$I2 = Ia - Ib$$

Those with power-line engineering training will recognize this as a form of "symmetrical component" analysis used for multiphase power lines.

The assumed currents Ia and Ib are the same on both loops. This lets us solve for the currents using Hallen's[2] method. Unfortunately, this simplification doesn't help much in practical calculation. There's no easy way to use the ten curves prepared by Storer[1] to derive the currents on the two elements. Theory application is beyond practical small computer use; a large computer is needed. However, a number of published results of this theoretical analysis can be applied to practical antennas. These will serve as the basis for some further analyses, for comparison, and

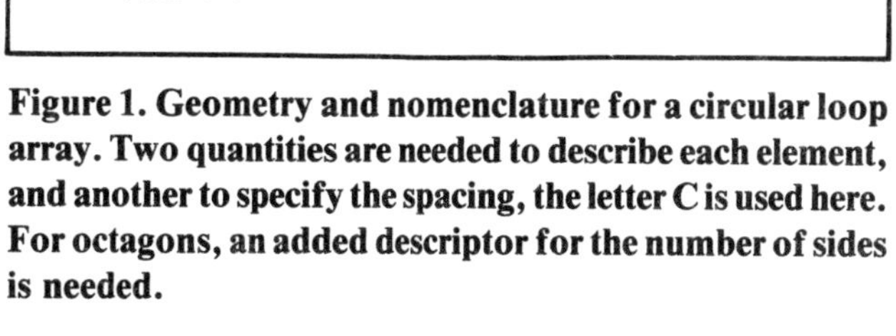

Figure 1. Geometry and nomenclature for a circular loop array. Two quantities are needed to describe each element, and another to specify the spacing, the letter C is used here. For octagons, an added descriptor for the number of sides is needed.

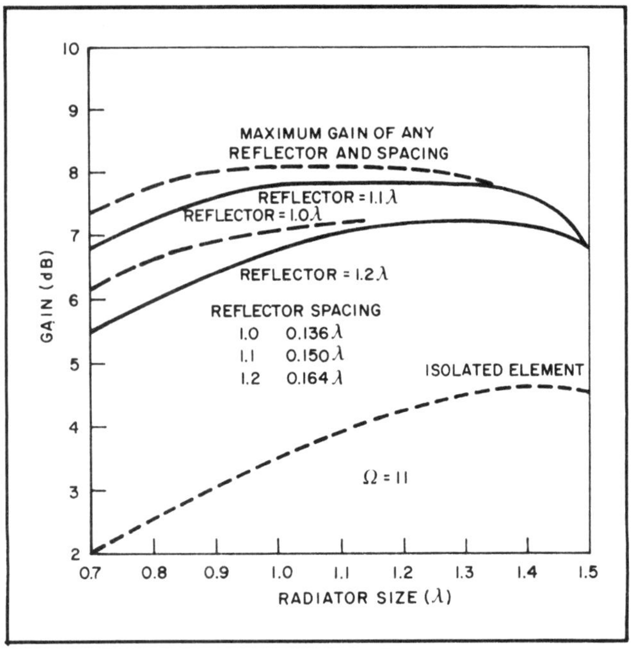

Figure 2. Forward-lobe gain for a reflector-radiator combination, plotted against size of radiator. Separate curves are shown for reflectors of 1.0, 1.1, and 1.2 wavelengths circumference. The top curve shows the maximum gain possible by choice of reflector size and spacing. Gain of an isolated element is shown for comparison. Data from Ito et al.[11]

also provide some data on the performance of practical designs.

The Basic Two-loop Array

As with the Yagi, the two-loop array with one loop parasitically excited is both a useful antenna and the basis for further array expansion. Practically, these two-element arrays give the largest increase in performance for a given investment in size and weight.

All the performance data on the designs described here is derived from Ito, Inagaki, and Sekiguchi.[11] Unless otherwise noted, gain values are in dB above isotropic.

Gain Performance

Figure 2 shows the forward gain performance with the parasitic element tuned as a reflector. The top curve shows the maximum gain with retuning; the nearly parallel curves show the gain for a specific parasitic element size.

For the two-element beam with reflector, the maximum gain is 8.1 dB. The gain is <3 to >5 dB higher than the gain of an isolated radiator of the same size. Over the range of radiator size from 1.0 to 1.25 wavelengths, the optimum gain varies only by 0.1 dB or so. Also, the gain doesn't vary greatly for changes in reflector size, typically by ±0.6 dB for a change from 1.1 to 1.3 wavelengths loop size. This means that the two-element loop with reflector is a good wideband antenna, with nearly constant forward gain over the wide frequency range of ±15 percent.

Forward gain performance with the parasitic element tuned as a director is shown in Figure 3. The top curve shows the maximum gain that can be developed for a given radiating loop size by tuning the director and choosing the best spacing. The nearly parallel curves show the gain for given director sizes. Also shown is the gain of a single isolated element.

The maximum two-element gain is 7.3 dB. This represents almost exactly a 3-dB increase over a single loop of the same size. This 3-dB increase is essentially independent of radiator size, but is sensitive to director dimension.

The combination of nearly maximum gain and good gain stability led the developers of this data to recommend that the radiator loop be larger than resonant size. Their specific recommendation is 1.2 wavelengths circumference. Some consequences of this are discussed later.

The curves for a 1.2-wavelengths radiator shown in Figures 4 and 5 are also useful. The variable for these curves is radiator-parasitic spacing in wavelengths.

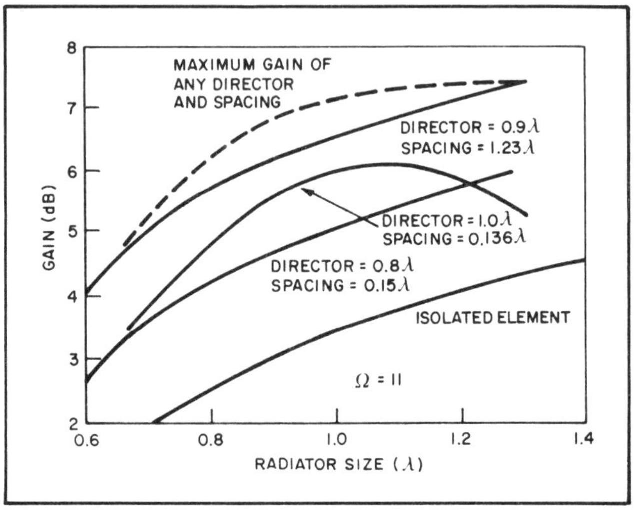

Figure 3. Forward-lobe gain for a director-radiator combination, plotted against radiator size. Separate curves are shown for directors 0.8, 0.9, and 1.0 wavelengths circumference. The top curve shows the maximum gain possible by choice of director size and spacing. Gain of an isolated element is shown for comparison. Data from Ito et al.[11]

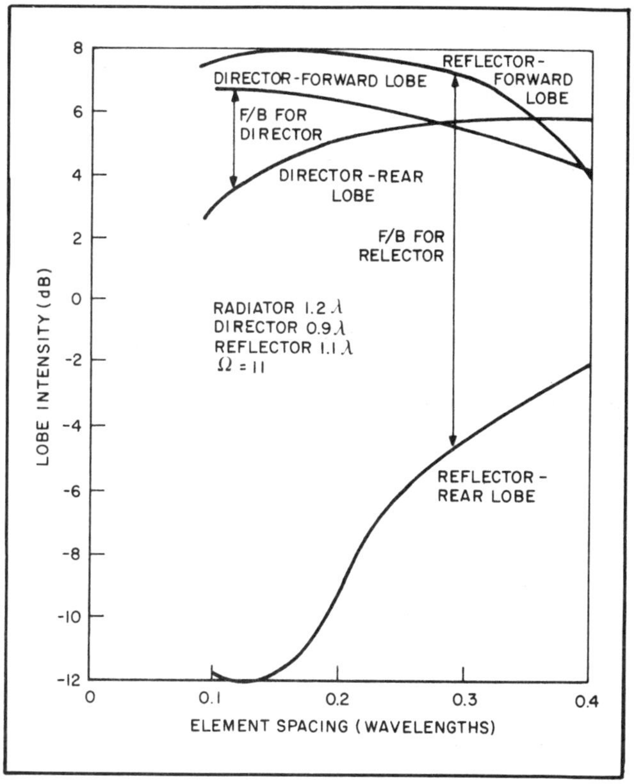

Figure 4. Forward and back lobe intensities for typical radiator-parasitic combinations, plotted against parasitic spacing. As usual for parasitic elements, maximum gain and maximum front-to-back ratio occur for different conditions. For two-element antennas, the reflector at 0.12 wavelength spacing is a good choice. Data from Ito et al.[11]

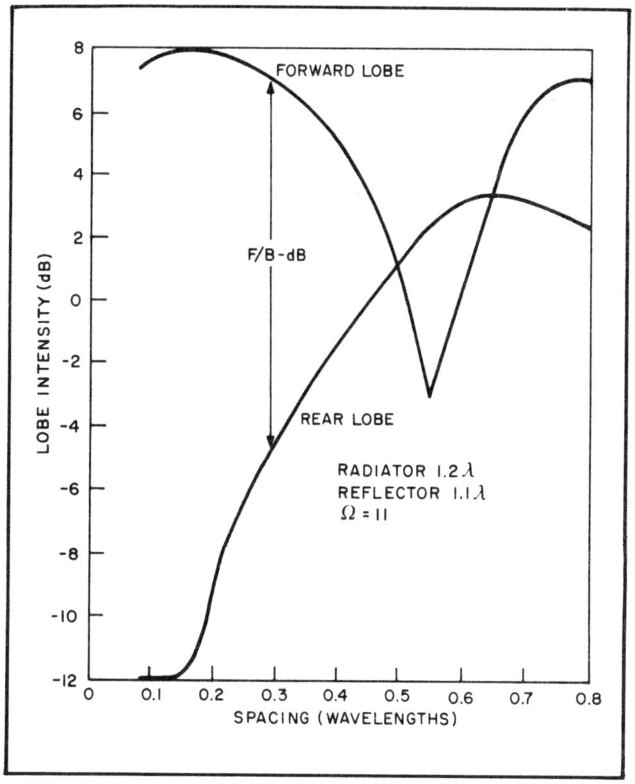

Figure 5. Forward and back-lobe intensities for a reflector-radiator combination over a wide range of spacings. With no change in element size, relative lobe intensities reverse around 0.6 wavelength spacing. Wider spacings can give fair forward lobe performance, but the back lobe remains large. Data from Ito et al.[11]

As we have seen, maximum gain with the parasitic element as a reflector is 8.1 dB; for the parasitic element as director it is 7.3 dB. These values show on the curves. Maximum reflector gain occurs at 1.08 wavelengths loop circumference at a spacing of 0.15 wavelength. Maximum director gain occurs at 0.95 loop circumference and a spacing of 0.1 wavelength.

Minimum back radiation with a reflector is −12.1 dB. This is nearly the same over the range of reflector sizes from 1.1 to 1.15 wavelengths circumference, and the range of spacing from 0.1 to 0.25 wavelength. Maximum front-to-back ratio occurs at a circumference of 1.1 wavelengths and a spacing of around 0.1 wavelength. It's about 20 dB.

The data in Chapter 1 shows that the optimum director size for minimum back radiation varies markedly with spacing. Minimum back radiation with the parasitic as a director is −3 dB, but this occurs for spacings around 0.6 wavelength—beyond the range of practical use. The forward gain is only 2 dB at this point. The practical minimum back lobe level is essentially 3 dB, occurring over the range 0.8 to 0.96 wavelength director circumference and a spacing of 0.1 wavelength. The front-to-back ratio is only 4 dB. A director of 0.9 wavelength circumference and 0.15 wavelength spacing will give nearly as good gain and front-to-back ratio.

These data show that the best design values for a two-element circular loop array, and for the radiator-reflector combination for larger arrays, are the same. Maximum performance is obtained from:

- Radiator circumference, 1.2 wavelengths.
- Reflector circumference, 1.1 wavelengths.
- Element spacing, 0.15 wavelength.

Radiator circumferences from 1.1 to 1.3 wavelengths give nearly the same performance.

Drive Impedance

The drive impedance of an isolated loop is a complex function of loop and conductor diameter. These factors still apply with a parasitic element present and the added complexity of the changes introduced by the currents on that element. Because of this, there is very little data on feed impedances in scientific or engineering literature. It's necessary to use the limited amount available to develop trends and then depend on measurements to design the antenna feed. Fortunately this isn't difficult, given the wideband characteristics of loop antennas.

The paper by Ito, et al.[11] includes a single set of curves that give useful information about the radiator-reflector combination. Data is for a reflector of 1.1 wavelengths circumference, with an element omega of 11, corresponding to a loop-to-conductor radius of 39. This is typical of a self-supporting UHF antenna.

The following values are developed from the curves; resonance, of course, means that the drive reactance is zero:

- Isolated resonant loop.
  Radiator circumference = 1.1 wavelengths.
  Drive resistance = 153 ohms.
- Resonant radiator with reflector.
  Radiator circumference = 1.08 wavelengths.
  Drive resistance = 139 ohms.
- Resonant radiator with director.
  Radiator circumference = 0.87 wavelength.
  Drive resistance = 250 ohms.
- Isolated 1.2-wavelengths loop.
  Drive resistance = 215 ohms.
  Drive reactance = 84 ohms.
- A 1.2-wavelengths radiator with reflector.
  Drive resistance = 158 ohms.
  Drive reactance = 365 ohms.
- A 1.2-wavelengths radiator with director.
  Drive resistance = 33.5 ohms.
  Drive reactance = −132 ohms.

The change in impedance is greater for the director and this behavior is much the same as for the two-element Yagi. The reason for the increase in drive resistance for the resonant element with director lies partly in the marked change in length needed to cancel the induced and self-reactances.

Remember that a standard matching method is to change the length of the driven element, then add a stub to cancel the reactance, leaving the resistance at the value that matches the line impedance. You can also do this with loops, but it may require some loss in loop area and, therefore, loop gain. A gamma match, or stub-transformer sections, would be a better choice.

For most two-element arrays, a good feed is a 4:1 balun at the antenna, used with 50-ohm line. Solid-state transmitters will probably require a better match to obtain rated output, say, by use of a transmatch. A possibility is a gamma match, used by at least one commercial design with good results. The simplest alternative is to use low-loss line, say, 75-ohm Teflon™ or foam, or 300-ohm twin lead, plus an appropriate transmatch. A transmatch is a necessity if you want the full advantage of the wideband loop characteristic.

Patterns

There's a limited amount of theoretically derived pattern data in current literature. Figure 6A-C is replotted from the patterns in Ito et al.,[11] which are similar to the corresponding patterns of two-element Yagi antennas.[16] The front-to-back ratio is reasonably good for reflectors in the 1.1 to 1.2 wavelengths circumference range. Directors of any size give poor front to back, although the forward gain is good, as was mentioned previously.

None of the theoretical analyses found in the literature show patterns in other planes, or for cross-polarization. These will be covered later by approximate analysis.

Multi-element Arrays

The theory for two-element antennas discussed in the preceding sections has been used to examine the properties of multi-element circular-loop arrays, by Ito et al.[11] and by Shoamanesh and Shafai.[17,18] Experimental work has been reported by Appel-Hansen.[13] Their work is summarized in Figures 7 and 8 and shows the gain versus array length. Also shown is the theoretical gain of Yagi antennas derived by Lawson.[19]

The gain for the usual Yagi is determined primarily by array length. The first few directors add about 1 dB per director. Thereafter, the gain increases approximately 3 dB for each doubling of array length.

Information in Table 1 is taken from Shoamanesh and Shafai's[17] theoretical table. All of the entries are for a director spacing of 0.3 wavelength. This gives maximum gain for the number of elements used, and reasonable front-to-back performance. Gain, expected drive impedance, and minor lobe characteristics plus front-to-back ratio are tabulated.

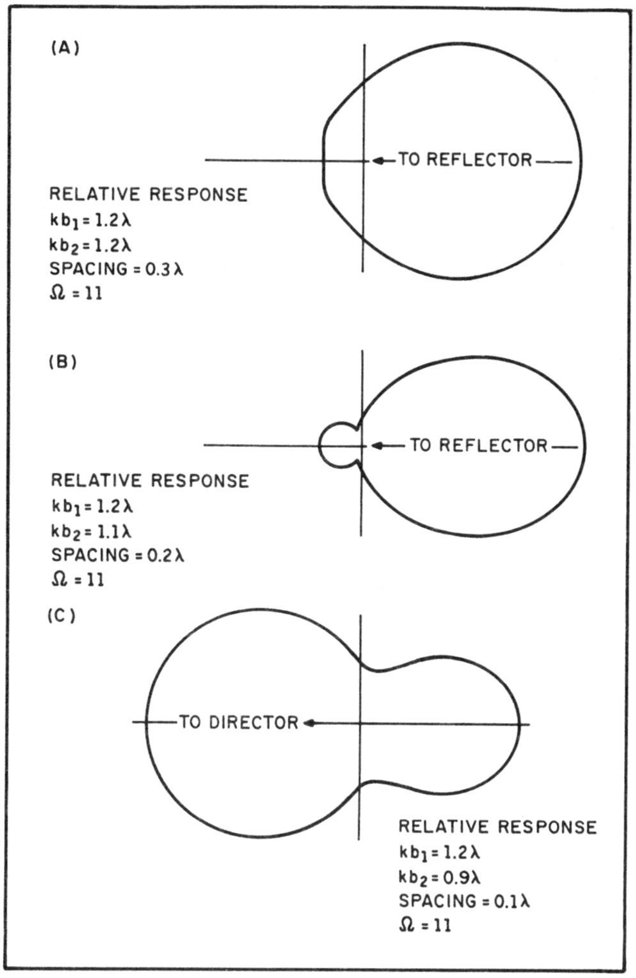

Figure 6. Horizontal plane patterns with horizontal polarization for three selected radiator-parasitic combinations. A and B show patterns with reflectors; C shows a pattern using a director. The front-to-back ratio of a reflector is good for a fairly wide range of conditions. While the director forward gain is also good over a fairly wide range, the front-to-back ratio is always poor. Patterns are scaled from a number plotted by Ito et al.[11]

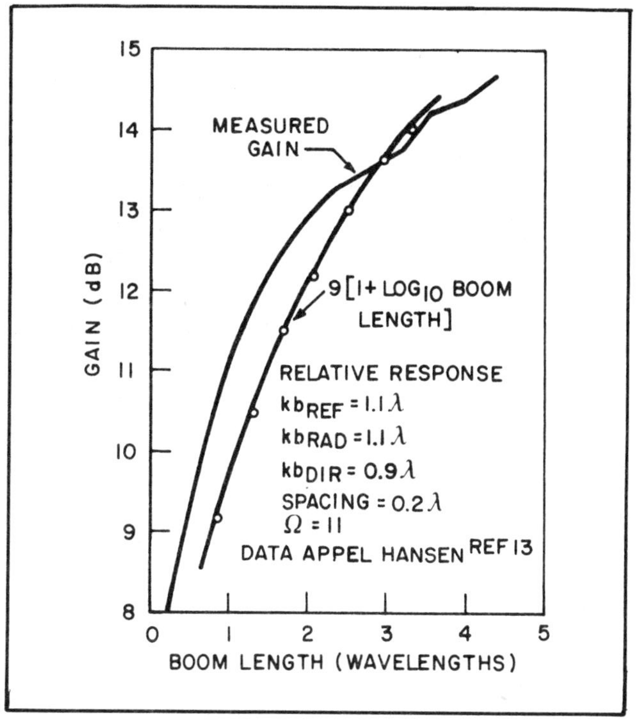

Figure 7. Measured gain of multi-element circular loop arrays, plotted against total boom length. Gain shown is not necessarily the maximum for the boom length given. The arbitrary curve is shown for comparison. Data from Appel-Hansen.[13] See also Lindsay.[12]

Arrays constructed from the table values should be easy to build in self-supporting form for frequencies in the VHF, UHF, and SHF ranges. By adopting the principle of the bicycle wheel, discussed in a later chapter, you can build HF designs.

The design values in Table 1 are known to be near optimum for maximum performance. A small increase in gain may be obtained by making some changes in element size and spacing. Unfortunately, there may be appreciable change in minor lobe structure if much retuning is attempted.

The only theoretical data relating to the gain of multi-element circular-loop arrays comes from the same source,[17,18] and is summarized in Figure 8. This again shows gain versus boom length, but with individual curves for director spacing. A small but definite increase in gain is usually possible by adding directors, but the limit of improvement isn't known. Retuning must be done carefully. Front-to-back ratios vary from about 10 to over 25 dB from one set of spacings to another, and there are probably changes in bandwidth as spacing changes. Shoamanesh and Shafai[17] give some guidelines for design optimization.

The data in Chapters 2 and 3 of Lawson[19] gives more information on results obtainable by optimization. Conversion of this data to loop design conditions is somewhat tedious. First, we calculate the mutual and self-reactances for the length and spacing data given by Lawson.[19] Then we transform these values to loop size and spacing, using the curves and tables given in this book or the references. The basic computational process described in Lawson's Chapter 1[19] or in Kraus[20] can be modified for loop analysis.

Octagonal Arrays and Loop Approximations

Arrays of polygons aren't common; circular elements are usually just as easy to build. There's also the matter

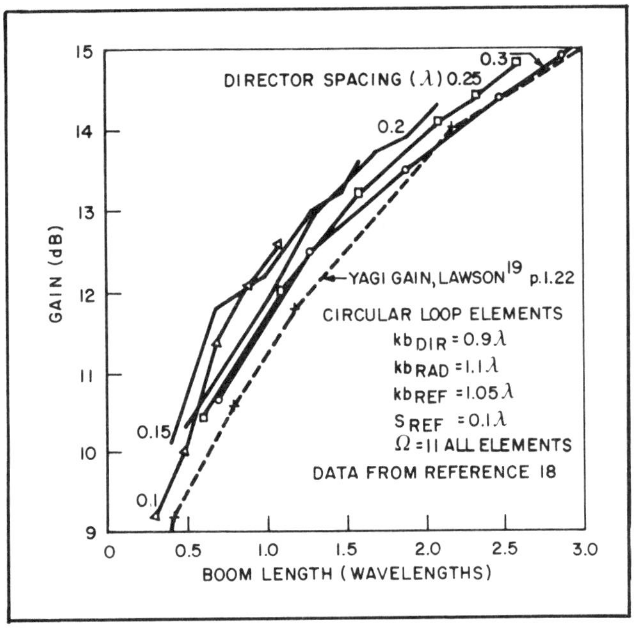

Figure 8. Calculated gain of multi-element circular loop arrays, plotted against boom length. Separate curves are shown for director spacings of 0.1, 0.15, 0.2, 0.25, and 0.3 wavelength spacing. For most boom lengths, a small increase in gain can be attained by using close spacing. There are penalties for this: for example, greater wind load, and usually, smaller bandwidth. Data from Shoamanesh and Shafai.[18]

of added resistance at the joints between segments, if present. The octagonal array data included here is primarily for use in approximating the performance of circular-loop arrays.

Figure 9 is included for comparison and gives the gain and front-to-back ratio for a two-element octagonal array. It's very nearly the same as that for the two-element circular array of Figure 2.

Figure 10 gives the mutual admittances between two 1.0-wavelength elements. Combined with the self-admittance values of Table 1, these curves allow calculation of the currents in any multi-element array and, in turn, the pattern described by Kraus[20] and used by Lawson.[29] As mentioned previously, these curves make the performance data calculated by Lawson[29] useful as an approximation to circular-loop (and other shape) arrays.

Patterns of Octagonal Arrays

Figure 11 shows the MININEC-calculated horizontal-plane pattern of a two-element, bottom-fed array with a parasitic director—an array with nominally horizontal polarization. Radiator and director circumferences are 1.0 and 0.9 wavelengths; spacing is 0.15 wavelength. The pattern resembles that of an equivalent two-element Yagi, which also shows poor front-to-back ratio. Somewhat more gain could be obtained with a larger radiator and closer spacing.

Figure 12 shows the same calculation for a 1.1-wavelengths reflector. Gain is nearly maximum from this element combination, and the back lobe is reasonably small. This appears to be a good choice for a two-element beam or for the exciter section of a large array.

Figure 13 shows the same calculation for a three-element octagonal array, with 0.9, 1.2, and 1.1 director, radiator, and reflector circumferences, with the director spaced at 0.2 and the reflector at 0.15 wavelength. The gain is close to, but below, the maximum attainable with this boom length. The front-to-back ratio is reasonable. Experience with 2, 10, 15, and

No. directors	Boom wave-lengths	Gain (dB)	F/B (dB)	H-B/W (Deg)	E-B/W (Deg)	Drive:R (Ohms)	Drive:jx (Ohms)
2	0.70	10.70	15.20	59.00	54.00	58.50	156.70
4	1.30	12.50	25.80	48.00	44.50	44.20	168.00
6	1.90	13.50	14.20	40.50	38.50	56.00	179.80
8	2.50	14.40	14.20	35.00	32.50	63.60	163.30
9	2.80	14.80	16.40	32.50	30.50	52.00	17.50
10	3.10	15.20	23.00	30.50	30.00	45.60	163.10

Directors L = 0.9λ S = 0.3λ.
Radiator L = 1.2λ.
Reflector L = 1.1λ S = 0.15λ.

Reflector = 1.1, radiator = 1.2, and all directors = 0.9 wavelength circumference. Reflector is spaced 0.15 and directors 0.3 wavelength. The nine-director array needs no matching, the others need only a capacitive stub across the feed point. A balun isn't necessary, but is probably helpful in avoiding feedline and tower radiation.

Table 1. Calculated performance of multi- and circular-loop arrays.

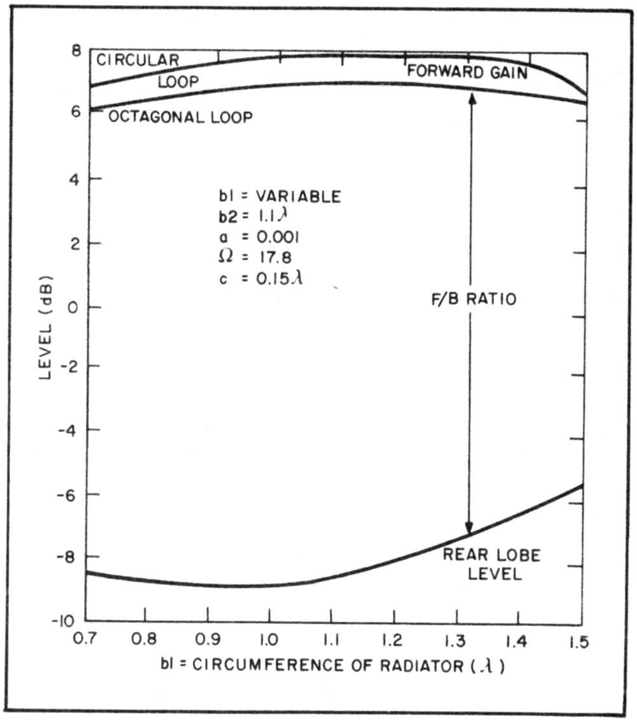

Figure 9. Calculated front and back lobe levels of an octagonal array, plotted against radiator circumference. Radiator circumference is 1.2 and spacing is 0.15 wavelength. The gain of a two-element circular loop is shown for comparison. The variation in gain would be more marked if the size of the reflector were also varied, but this shows again that loop antennas are broadband. The difference in circular and octagonal gain appears to be largely due to calculation technique differences.

20-meter versions of this combination indicates that better front to back, up to about 30 dB, is possible with element tuning. Figure 8 shows that the gain could be increased about 1 dB by a second director at 0.1-wavelength spacing.

Figure 14 shows another pattern for the same beam with the total radiation component. Because of the small cross-polarized component present in loops, the sides of the pattern are filled in, so the front-to-side ratio is poor.

It's my experience that the filled-in pattern is correct for high-angle or E-layer radiation. However, low-angle radiation, typical of F2-layer DX, shows the high front-to-side discrimination of Figure 13. Under good conditions, an S9+ signal can be dropped to the noise level if the signal is placed on the 90-degree null. This results in a reduction of 40 dB or more. The same characteristic is noted on 15 meters when E-layer skip is present. Published theoretical analyses shed no light on this performance difference. It appears to be associated with ground reflection.

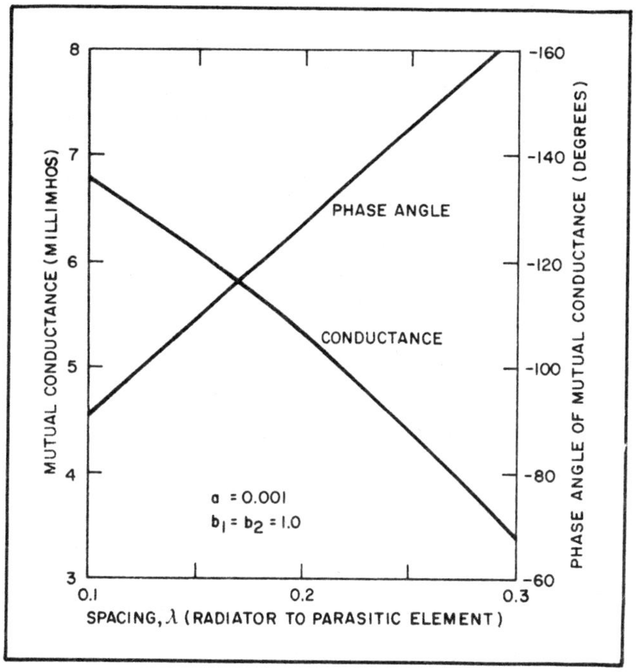

Figure 10. Mutual admittance of two octagonal loops, one wavelength in diameter. For wider separation, the magnitude of the admittance decreases nearly inversely with separation, and the phase linearly. These curves plus the self-admittance data given in Chapter 2 allow determination of array patterns by use of pattern multiplication techniques.

All-driven Arrays

As with dipole elements, loops can be used in driven arrays. Driven arrays aren't common in amateur use because of space requirements. The few occasional-

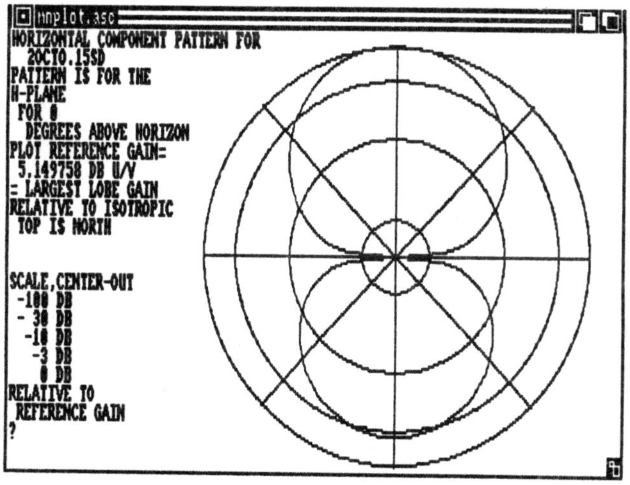

Figure 11. Horizontal plane pattern of the h-polarized component from a bottom-fed director-radiator combination. Circumferences are 0.9 and 1.0 wavelength. Spacing is 0.15 wavelength. Forward gain is reasonable, but the back lobe is large.

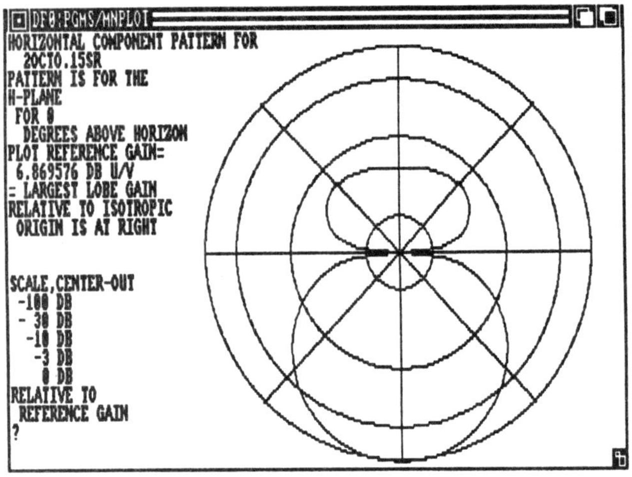

Figure 12. Horizontal plane pattern of a 1.1 wavelength reflector-radiator combination. Other conditions are the same as in Figure 11. Forward gain is excellent for a two-element antenna. The front-to-back ratio of about 13 dB is good.

ly encountered are typically two-element designs. These are also used in place of a single radiator element in some high-performance arrays, partly for gain and front-to-back improvement, but also to improve the effective bandwidth.

Figure 15 shows the MININEC-calculated gain for two 1.0-wavelength loops spaced 0.12 wavelength and fed 135 degrees out of phase, like the "ZL-Special" antenna. The gain performance is good, but the back lobe is rather large.

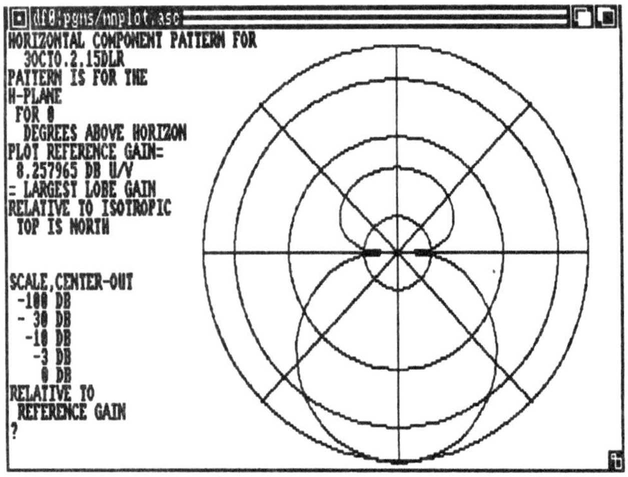

Figure 13. Horizontal plane pattern of a bottom-fed three-element octagonal array. Circumferences are director = 0.9, radiator = 1.2, and reflector = 1.1 wavelengths. Director spacing is 0.2 and reflector 0.15 wavelength.

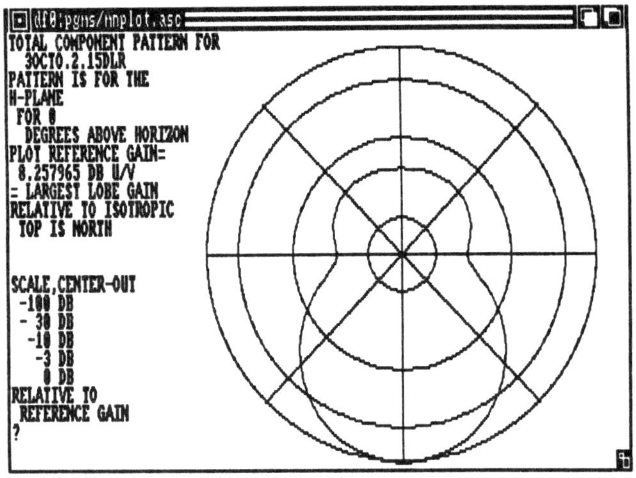

Figure 14. Total component radiation pattern for the array of Figure 13. The vertical component fills in the side-lobe area. See text for practical experience with an antenna of this type.

Super-wideband Operation

Figure 16 shows the calculated performance of a 20-meter two-element array from 8.5 to 20 MHz. The gain curve shows that this parameter exceeds that of a dipole over the entire range. In fact the performance is reasonable over 7 to 30 MHz, although there is lobe splitting on the higher bands and main-lobe reversal on the lower ones. However, front-to-back ratio is high only near the design frequency.

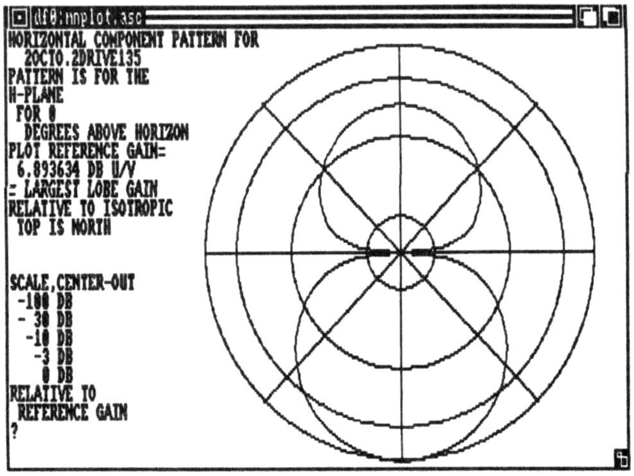

Figure 15. Horizontal component pattern for two one-wavelength loops separated 0.12 wavelength and bottom fed 135 degrees out of phase. The antenna is related to the "ZL-Special," and is useful by itself. The technique is also useful as the feed of large arrays, to give better front-to-back and gain.

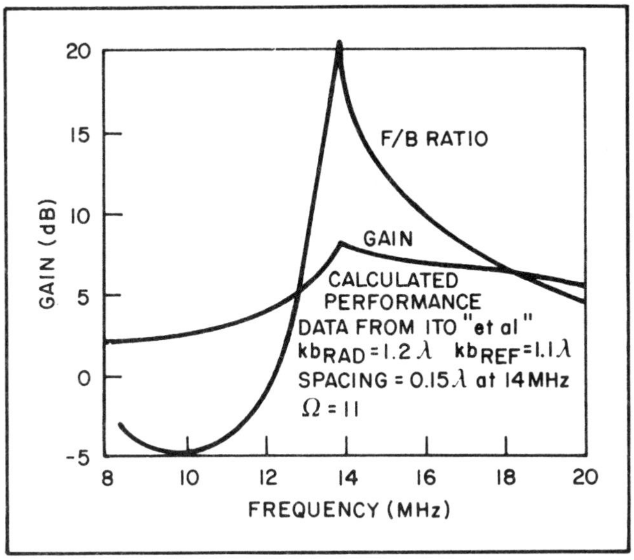

Figure 16. Calculated gain and front-to-back ratio of a two-element array designed for 14 MHz, but operated over the range 8 to 22 MHz. Forward lobe gain exceeds that of a dipole over an entire range. The back lobe can even be larger than the main lobe, but effective interference reduction is usually possible by rotating the antenna to place the interference in a null. Any antenna of the quad family is a useful wide-band antenna for occasional operation on other than designated bands. Based on data from Ito et al.[11]

The secret of using this wide bandwidth lies in the transmission-line matching technique used. I prefer Teflon-insulated 75-ohm cable. Open-wire line or twin lead is also good. A match-box is a necessity. (I think that a match-box should always be employed to simplify using the entire band and to reduce harmonic radiation.) One fact seems clear: If you are using a quad, you can have multiband operation without a lot of real estate for an antenna farm.

NOTES

Chapter 4

Rectangular and Square Loops

Design and performance data for any square, rectangular, or diamond loop

The classic quad is made from a pair of square loops. The usual feed point for these is at for these is at the bottom center, and occasionally the side center. Some variations change from a square to a rectangular shape. A more common variation rotates the square by 45 degrees, making a diamond. Again, the feed is usually at the bottom point of the diamond, less commonly at the side. "Squashed" diamonds are often advertised, but most are really fat dipoles rather than quad loops.

Notes on the Quad Literature

There's a lot of Amateur literature on quad loops used both alone and in arrays. In contrast, scientific and engineering material is sparse. The only theoretical studies I've found are King[21] and Prasad.[27] King's study places a generator at each corner of the loop, allowing a set of four independent currents using symmetrical components. The theory is complex; the only practical usable result is that a first approximation of the current distribution is given by its distribution on a transmission line, as for the circular loop. As a consequence, the tables of circular loop admittance of Chapter 2, or Storer's curves,[1] can be used as approximations for all quad loops. The approximation will be better for square loops than for markedly rectangular ones. Diamond and square loops are identical to this degree of approximation.

Kennedy[5] has reported measurements for thick conductors, typically omega = 11. Feed impedances and current distributions are given for a limited number of conditions. The measurements confirm the close relationship between circular and square loops.

Orr[22] gives the best summary of amateur findings. His work mainly concerns the quad array of loops and gives estimates of size, performance, feed impedance, and construction details. Almost all other amateur material relates to construction, with occasional reports of measurements of drive impedances and performance.

Quad loop literature reflects this lack of useful theoretical analysis. The material which follows is intended to correct this, at least to the extent that full design data is available for all common quad types and for a number of the less popular quad designs.

I used the MININEC analysis program to obtain data for selected loop families. The resulting material was organized into a format I believe to be both useful and explanatory.

MININEC gives approximate answers. The accuracy of the data in this section should be sufficient for any practical purpose. Analyses have been made with a total of sixteen segment divisions of the loop perimeter. For the most important data, this is also sixteen segments per wavelength. The smallest segmentation is eight per wavelength—usually sufficient to give 10 percent accuracy.

Quad Loop Geometry

The general geometry and nomenclature for this loop family is shown in Figure 1. The previous nomenclature of a = conductor radius is retained. The perimeter (analogous to circumference and denoted by b) is used to define the conductor thickness factor, omega. New quantities introduced are: s = side length and sf = shape factor (defined as twice the length of a vertical side divided by the perimeter). This factor is 0.5 for a square loop, zero for a horizontal folded dipole, and 1.0 for a vertical transmission line, all fed at the bottom center.

Square Loop Drive Impedance

The self- or drive resistance of a typical square loop is shown in Figure 2 for wavelengths from 0.05 to 2.0 The theoretical curve for a circular loop is shown for comparison.

The general pattern of these two types is the same,

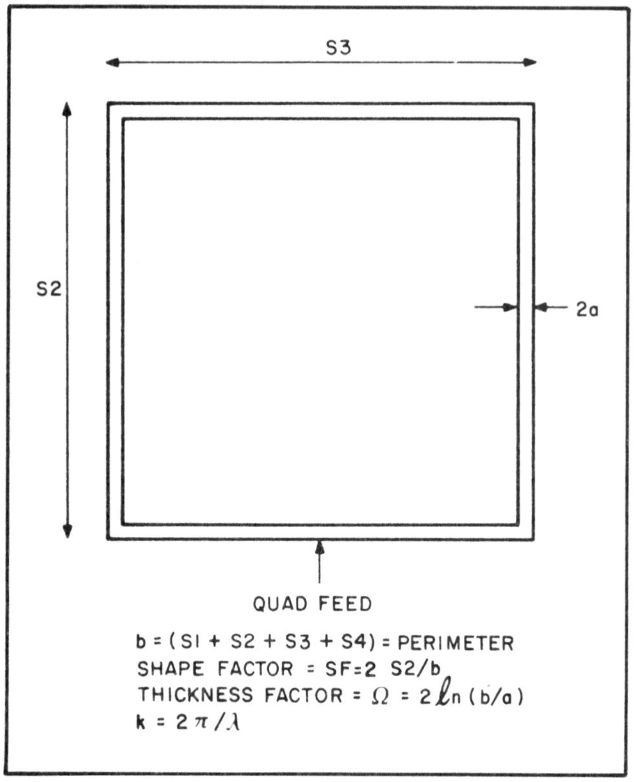

Figure 1. Geometry and nomenclature of a quad loop. The circular loop circumference is redefined to be the perimeter of the loop. The quantity kb is the perimeter in wavelengths, and is also, more generally, the conductor length. Shape factor is defined in terms of a rectangle, but can be used for any regular figure.

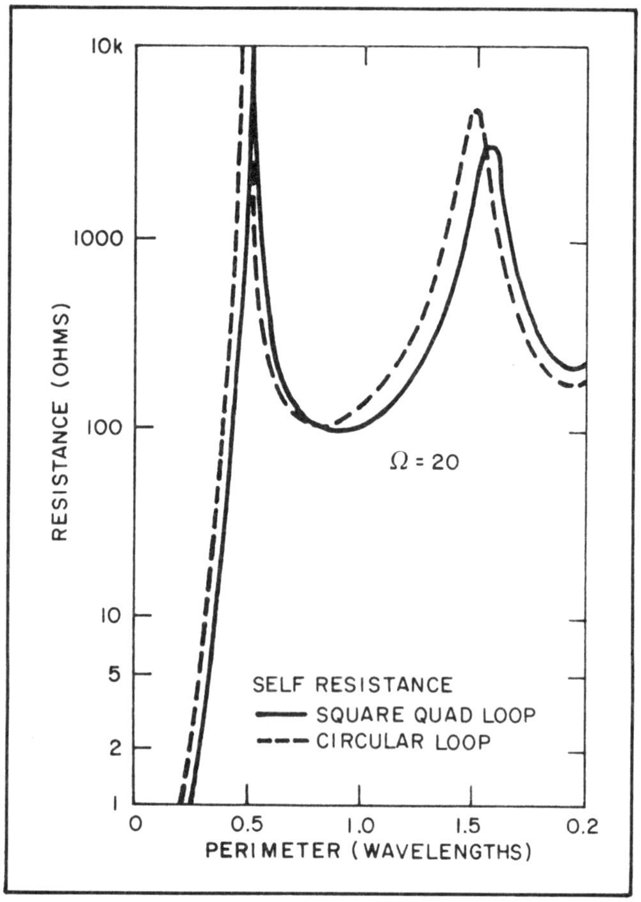

Figure 2. Self-resistance of a square quad loop. The self- or driving-point resistance is shown as a function of perimeter for a thickness factor of 20. The circular loop curve is for comparison. See also Figures 4A and 9A of Chapter 2.

with parallel resonance points at 0.5 and 1.5 wavelengths, and series resonant minima at 1.0 and 2.0 wavelengths. There's appreciable difference in the resonant frequencies and some difference in the magnitudes.

Figure 3 shows the corresponding reactance curve; the comparison curve has been omitted for the sake of clarity. Circular data was given in Chapter 2. Comparison shows nearly identical curve shape and close correspondence in magnitude, but appreciable shift in resonant frequency with the square loop resonance occurring at a larger loop size.

Remember that this pair of calculated values results from two different theories, so some of the difference may be due to approximation errors. This factor appears to be negligible for practical purposes. The shift in resonant frequency is close to that measured by Kennedy,[5] and in the same direction but greater than that implied by Orr.[22] However, Orr's values are to

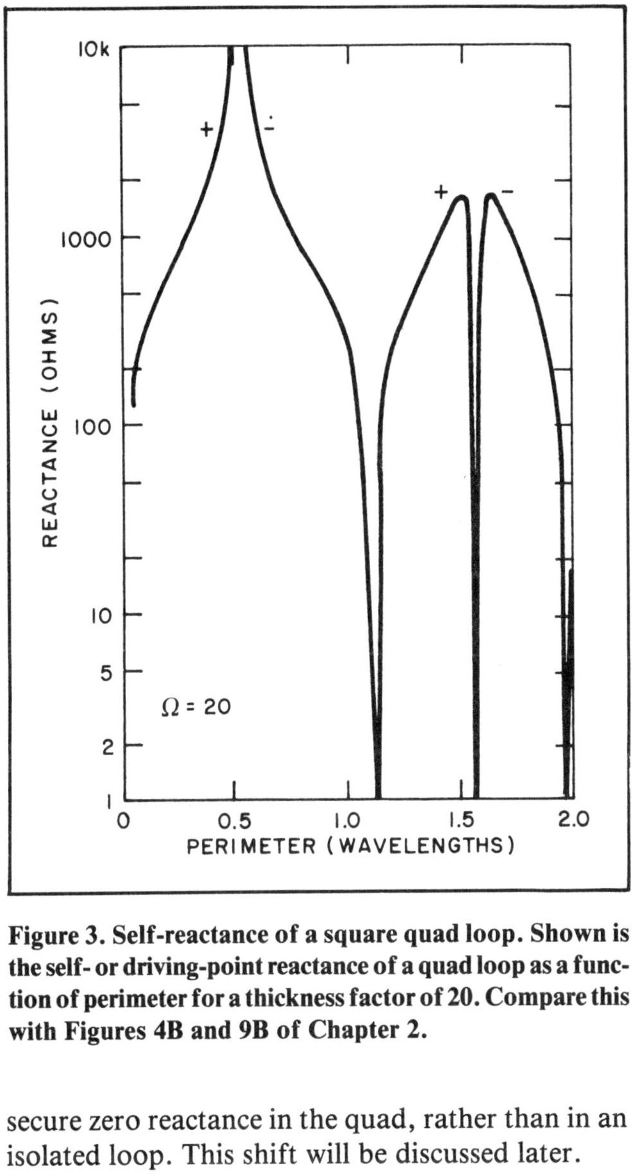

Figure 3. Self-reactance of a square quad loop. Shown is the self- or driving-point reactance of a quad loop as a function of perimeter for a thickness factor of 20. Compare this with Figures 4B and 9B of Chapter 2.

secure zero reactance in the quad, rather than in an isolated loop. This shift will be discussed later.

Table 1 summarizes the source data for the entire family of square and quad loops for the size range from kb = 0.05 to kb = 2.0, and for selected thickness factors of omega = 10 to omega = 22.5. (Similar data for rectangular and diamond loops are discussed and given in Table 2.) Use the nearest table value for estimations. Interpolate for exact perimeter and thickness for accuracy within a few percent. You'll need the values of this table for calculation of array performance.

I've mentioned using Storer's curves[11] as approximations for any loop. You can also use Table 1 from this chapter. If you need better accuracy, I suggest you measure the resonant frequency of the design loop, or a scale model. Use the percentage offset of the desired frequency to enter the table as a percent change from its corresponding resonance. If you use a scale model, don't forget to correct the resonant frequency for the change in wire thickness factor.

Drive Impedance Near Resonance

The normal use of loops is at or close to the first series-resonance point, with the circumference close to 1 wavelength. Figure 4 shows the self-resistance for kb between 0.85 and 1.2, with the thickness factor as the parameter. In Figure 5 the data is plotted against the thickness factor, with perimeter as the parameter.

Typical self-supporting VHF/UHF loops have thickness factors in the range of 10 to 12. For these, the drive resistance can be taken as 100 ohms. High frequency loops with wire conductors have a drive resistance close to 110 ohms.

Figure 6 shows the drive reactance versus perimeter, with thickness factor (omega) as the parameter. As with the other loops and with dipoles, there's a point where reactance is independent of thickness factor (in this case 1.063). The reactance at this point of crossover is essentially the same as for the circular and octagonal loops, but the common point occurs at a perimeter which is about 5 and 3 percent larger, respectively. The zero-reactance point on these curves gives the resonant frequency of an isolated loop. This is plotted against the thickness factor in Figure 7. Note that the resonant perimeter approaches one wavelength as the conductor becomes smaller, as with a dipole. However, the direction of change is just opposite the direction of change with dipoles.

When dealing with arrays of antennas, it's often convenient to use numerical values from equations rather than from graphs. The data in Figures 5 through 7 can be approximated by the following relationships:

resistance
  thick loop = 100 ohms
  thin loop = 110 ohms

perimeter for resonance, RP
  $RP = 1.057 + 0.24/OM + 13.57/OM^2$

reactance

$$X = 100\left(\frac{kb - RP}{RP - 1.063}\right)$$

where OM is the thickness factor, omega, RP is the perimeter at resonance, and the loop perimeter, kb, is within 20 percent of unity.

ADMITTANCE OF SQUARE QUAD LOOPS
Values in Millimhos

kB	Omega=10	Omega=11	Omega=12	Omega=15	Omega=17	Omega=20	Omega=22.5
.05	.0003-j19.3072	.0002-j16.6672	.0002-j14.6462	.0001-j10.7208	.0001-j 9.1118	.0001-j 7.4230	.0001-j 6.4038
.10	.0012-j 9.4046	.0009-j 8.1273	.0007-j 7.1448	.0004-j 5.2355	.0003-j 4.4492	.0002-j 3.6260	.0001-j 3.1286
.15	.0028-j 5.9916	.0021-j 5.1859	.0016-j 4.5637	.0008-j 3.3496	.0006-j 2.8481	.0004-j 2.2222	.0003-j 2.0044
.20	.0053-j 4.1984	.0039-j 3.6421	.0031-j 3.2100	.0016-j 2.3586	.0012-j 2.0101	.0008-j 1.6403	.0006-j 1.4164
.25	.0089-j 3.0489	.0066-j 2.6542	.0052-j 2.3446	.0028-j 1.7290	.0020-j 1.4754	.0013-j 1.2053	.0010-j 1.0416
.30	.0139-j 2.2174	.0105-j 1.9405	.0081-j 1.7200	.0044-j 1.2753	.0032-j 1.0902	.0021-j .8922	,0016-j .7718
.35	.0211-j 1.5629	.0159-j 1.3794	.0123-j 1.2294	.0067-j .9194	.0049-j .7880	.0033-j .6467	.0024-j .5602
.40	.0310-j 1.0138	.0234-j .9090	.0182-j .8182	.0010-j .6210	.0073-j .5349	.0049-j .4410	.0037-j .3830
.45	.0450-j .5289	.0341-j .4936	.0267-j .4550	.0146-j .3574	.0107-j .3111	.0072-j .2590	.0054-j .2261
.50	.0648-j .0819	.0492-j .1103	.0386-j .1195	.0212-j .1134	.0159-j .1046	.0104-j .0902	.0078-j .0805
.55	.0929+j .3458	.0708+j .2574	.0556+j .2029	.0308+j .1222	.0232+j .0986	.0152+j .0733	.0114+j .0606
.60	.1330+j .7689	.1018+j .6225	.0802+j .5241	.0447+j .3584	.0338+j .3030	.0222+j .2381	.0167+j .2032
.65	.1910+j 1.1999	.1469+j .9969	.1162+j .8552	.0653+j .6043	.0495+j .5167	.0326+j .4112	.0246+j .3534
.70	.2758+j 1.6509	.2136+j 1.3925	.1699+j 1.2077	.0965+j .8701	.0734+j .7488	.0487+j .6005	.0368+j .5182
.75	.4013+j 2.1338	.3138+j 1.8221	.2516+j 1.5947	.1450+j 1.1683	.1111+j 1.0114	.0742+j .8168	.0563+j .7075
.80	.5901+j 2.6601	.4678+j 2.3001	.3791+j 2.0323	.2234+j 1.5163	.1725+j 1.3215	.1164+j 1.0758	.0890+j .9361
.85	.8785+j 3.2381	.7100+j 2.8410	.5846+j 2.5395	.3559+j 1.9394	.2784+j 1.7053	.1910+j 1.4036	.1473+j 1.2288
.90	1.3249+j 3.8645	1.1014+j 3.4539	.9284-j 3.1355	.5943+j 2.4746	.4745+j 2.2045	.3341+j 1.8451	.2616+j 1.6308
.95	2.0181+j 4.5023	1.7474+j 4.1211	1.5258+j 3.8212	1.0584+j 3.1683	.8743+j 2.8833	.6444+j 2.4832	.5184+j 2.2324
1.00	3.0704+j 5.0313	2.8132+j 4.7336	2.5874+j 4.5091	2.0445+j 4.0206	1.7962+j 3.7937	1.4464+j 3.4464	1.2312+j 3.2071
1.05	4.5374+j 5.1725	4.4529+j 4.9448	4.3822+j 4.8006	4.2016+j 4.5795	4.1049+j 4.5118	3.9412+j 4.4329	3.8166+j 4.3878
1.10	6.1843+j 4.5134	6.4142+j 4.0894	6.6583+j 3.7226	7.3606+j 2.6607	7.6932+j 1.9784	8.0230+j .6364	7.9836-j .4714
1.15	7.3145+j 2.9334	7.5535+j 1.9410	7.6588+j .9708	7.0230-j 1.5839	6.1725-j 2.5952	4.5238-j 3.4388	3.4412-j 3.5373
1.20	7.3452+j 1.0751	7.0571-j .2716	6.5152-j 1.3599	4.3811-j 2.9724	3.3298-j 3.1343	2.0751-j 2.9086	1.4795-j 2.6118
1.25	6.5262-j .2650	5.7191-j 1.4586	4.8229-j 2.1875	2.6905-j 2.6998	1.9396-j 2.5441	1.1612-j 2.1500	.8204-j 1.8685
1.30	5.4653-j .9012	4.4419-j 1.7651	3.5288-j 2.1600	1.7991-j 2.1665	1.2745-j 1.9559	.7563-j 1.5976	.5348-j 1.3732
1.35	4.5151-j 1.0379	3.4819-j 1.6303	2.6700-j 1.8329	1.3048-j 1.6718	.9194-j 1.4804	.5456-j 1.1911	.3871-j 1.0189
1.40	3.7679-j .8948	2.8066-j 1.3150	2.1086-j 1.4271	1.0107-j 1.2456	.7119-j 1.0927	.4239-j .8723	.3019-j .7442
1.45	3.2117-j .6066	2.3396-j .9320	1.7378-j 1.0115	.8274-j .8726	.5840-j .7622	.3496-j .6053	.2500-j .5151
1.50	2.8138-j .2421	2.0226-j .5234	1.4945-j .6021	.7128-j .5331	.5052-j .4651	.3044-j .3670	.2187-j .3107
1.55	2.5479+j .1664	1.8206-j .1019	1.3449-j .1967	.6473-j .2100	.4615-j .1835	.2804-j .1413	.2026-j .1168
1.60	2.4007+j .6033	1.7174+j .3313	1.2748+j .2122	.6242+j .1123	.4488+j .0978	.2757+j .0855	.2006+j .0789
1.65	2.3726+j 1.0583	1.7132+j .7775	1.2853+j .6330	.6468+j .4483	.4706+j .3937	.2934+j .3270	.2153+j .2889
1.70	2.4766+j 1.5156	1.8235+j 1.2334	1.3932+j 1.0704	.7306+j .8129	.5404+j .7207	.3439+j .5998	.2553+j .5292
1.75	2.7346+j 1.9425	2.0791+j 1.6804	1.6340+j 1.5170	.9109+j 1.2186	.6900+j 1.0965	.4522+j .9262	.3412+j .8226
1.80	3.1633+j 2.2748	2.5200+j 2.0651	2.0633+j 1.9333	1.2586+j 1.6630	.9883+j 1.5336	.6779+j 1.3345	.5245+j 1.2040
1.85	3.7399+j 2.4113	3.1634+j 2.2750	2.7357+j 2.2081	1.9013+j 2.0776	1.5810+j 1.9955	1.1719+j 1.8360	.9485+j 1.7109
1.90	4.3543+j 2.2502	3.9263+j 2.1469	3.6126+j 2.1275	2.9731+j 2.1845	2.6958+j 2.2163	2.2861+j 2.2362	2.0213+j 2.2206
1.95	4.8107+j 1.7858	4.5569+j 1.5931	4.4037+j 1.4994	4.1962+j 1.4126	4.1467+j 1.4018	4.1058+j 1.3963	4.0945+j 1.3938
2.00	4.9481+j 1.1725	4.7710+j .7859	4.6703+j .4989	4.5064-j .1617	4.4116-j .4981.	4.1701-j 1.0440	3.9178-j 1.4038

Table 1. Source data for the entire family of square quad loops for the size range from kb = 0.05 to kb = 2.0.

Admittances of Rectangular and Diamond Loops
Millimhos

kb	SF=.875	SF=.75	SF=.625		SF=.25	DIAMOND
.50	.0019-j .1672	.0057-j .0911	.0099-j .0910	.0162-j .1176	.0174-j .1563	.0125-j .0690
.55	.0025+j .0343	.0078+j .1021	.0138+j .0963	.0244+j .0720	.0276+j .0399	.0181+j .1170
.60	.0032+j .2340	.0106+j .2950	.0194+j .2839	.0371+j .2646	.0438+j .2414	.0261+j .3028
.65	.0042+j .4398	.0145+j .4948	.0275+j .4790	.0568+j .4685	.0705+j .4581	.0381+j .4954
.70	.0056+j .6602	.0200+j .7101	.0394+j .6898	.0882+j .6937	.1155+j .7023	.0563+j .7029
.75	.0075+j .9063	.0283+j .9517	.0575+j .9270	.1403+j .9536	.1948+j .9911	.0851+j .9358
.80	.1167+j .9418	.0402+j 1.2768	.0917+j 1.2432	.2373+j 1.2897	.3467+j 1.3619	.1388+j 1.2413
.85	.1433+j 1.3919	.0668+j 1.6466	.1431+j 1.6029	.4055+j 1.6855	.6385+j 1.8071	.2246+j 1.5856
.90	.1860+j .9827	.1079+j 2.1433	.2367+j 2.0810	.7357+j 2.1917	1.2495+j 2.3195	.3854+j 2.0338
.95	.2645+j 2.8217	.1918+j 2.8785	.4281+j 2.2772	1.4343+j 2.7996	2.4942+j 2.5910	.7198+j 2.6491
1.00	.4381+j 4.1590	.4040+j 4.1423	.8976+j 3.8906	2.9217+j 3.1807	4.1113+j 1.5678	1.5173+j 3.4988
1.05	.9504+j 6.7131	1.2119+j 6.9670	2.4643+j 5.9364	5.0234+j 1.9466	4.1101-j .5127	3.6399+j 4.2287
1.10	3.6996+j13.5819	10.7764+j17.3337	10.1740+j 7.3369	4.9892-j .8090	2.9074-j 1.4354	7.0262+j 1.7086
1.15	63.7356+j13.8577	6,6962-j13.8007	9.2204-j 6.5827	3.3438-j 1.9172	1.9756-j 1.4644	5.1970-j 2.4017
1.20	3.8913-j14.7580	1.1239-j 5.7745	2.6748-j 5.0409	2.1680-j 1.8629	1.4227-j 1.2545	2.6759-j 2.5939
1.25	1.2454-j 8.0592	.4519-j 3.3796	1.1838-j 3.2802	1.5090-j 1.5570	1.0922-j 1.0140	1.5434-j 1.9966
1.30	.6766-j 5.5879	.2506-j 2.2318	.6769-j 2.2656	1.1296-j 1.2434	.8842-j .7887	1.0138-j 1.4692
1.35	.4644-j 4.2971	.1658-j 1.5307	.4499-j 1.6085	.8988-j .9631	.7466-j .5834	.7320-j 1.0599
1.40	.3621-j 3.4947	.1241-j 1.0342	.3314-j 1.1318	.7465-j .7142	.6520-j .3936	.5652-j .7353
1.45	.3051-j 2.9405	.1030-j .6444	.2653-j .7527	.6462-j .4871	.5857-j .2132	.4588-j .4668
1.50	.2702-j 2.5298	.0942-j .3117	.2296-j .4272	.5797-j .2723	.5394-j .0361	.3871-j .2352

Table 2. Source data for rectangular and diamond loops.

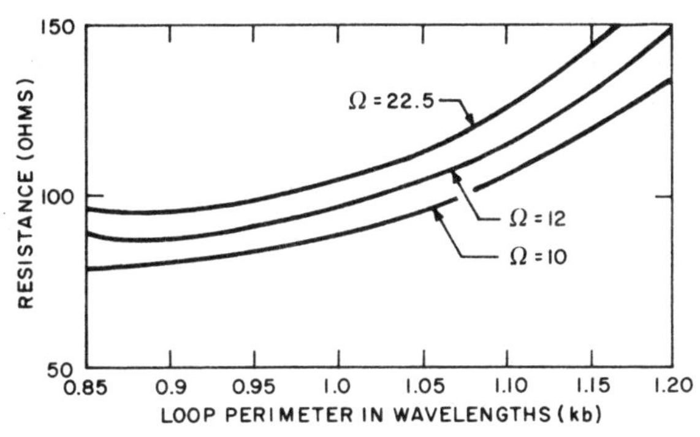

Figure 4. Self-resistance of a square quad loop near resonance. Self-resistance is plotted against perimeter for three values of thickness factor. See also Figure 5.

Gain of the Square Loop

The MININEC-calculated gain of the square loop is shown in Figure 8. This is the magnitude of the lobe at 90 degrees to the plane of the loop, on the axis of symmetry. The comparison curve shows the gain to be below that of a circular loop, as expected from area considerations. The gain increases with size, up to about 1.4 wavelength circumference. The gain is greater than for the corresponding dipole, by about 1.0 dB at kb = 1 and about 1.5 dB at kb = 1.4. Conductor size has a small effect on gain, as shown in Figure 9. This is not of practical significance.

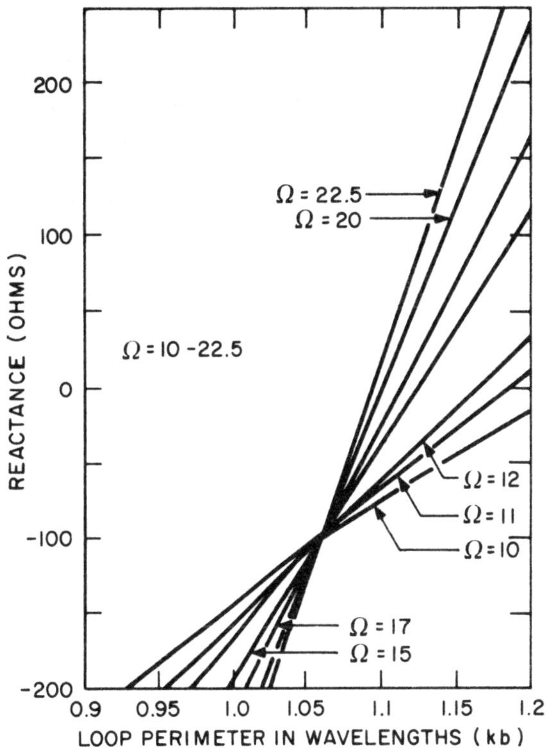

Figure 6. Self-reactance of a square quad loop near resonance. Self-reactance is plotted against perimeter, with selected values of thickness factor as a parameter. Resonance is indicated by the zero reactance intercept. Compare with Figures 5B and 10 of Chapter 2.

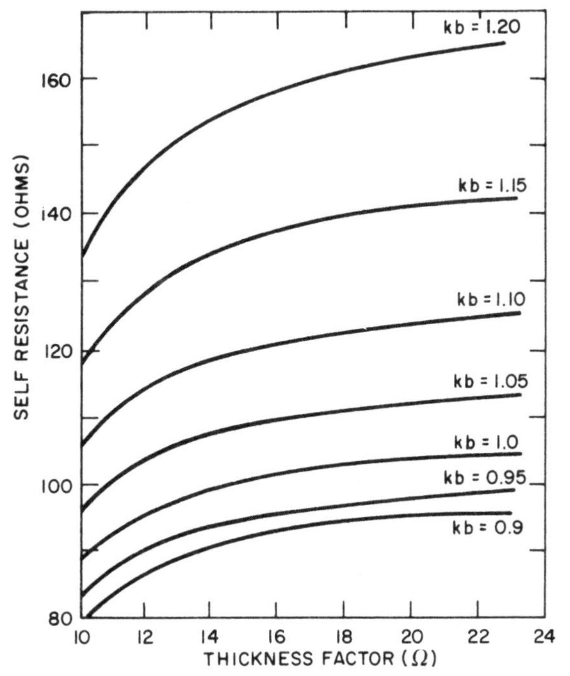

Figure 5. Self-resistance of a square quad loop near resonance. Here self-resistance is plotted against the thickness factor, with selected values of perimeter as the curve parameter. For a typical one-wavelength HF loop, the self-resistance is about 105 ohms.

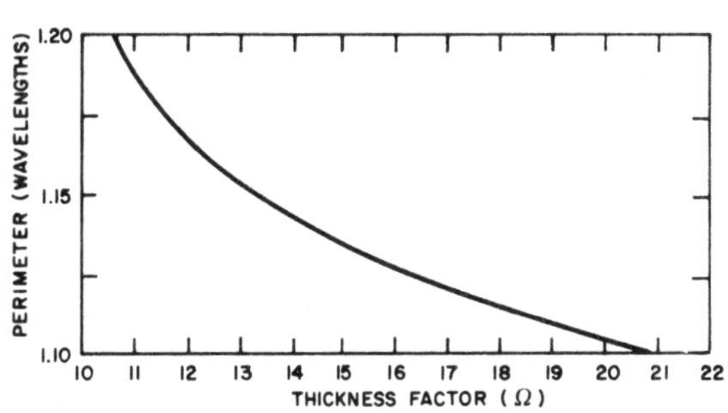

Figure 7. Resonant frequency of a square quad loop. Resonant point data from Figure 6 is plotted against thickness factor. Compare with Figures 6 and 11 of Chapter 2.

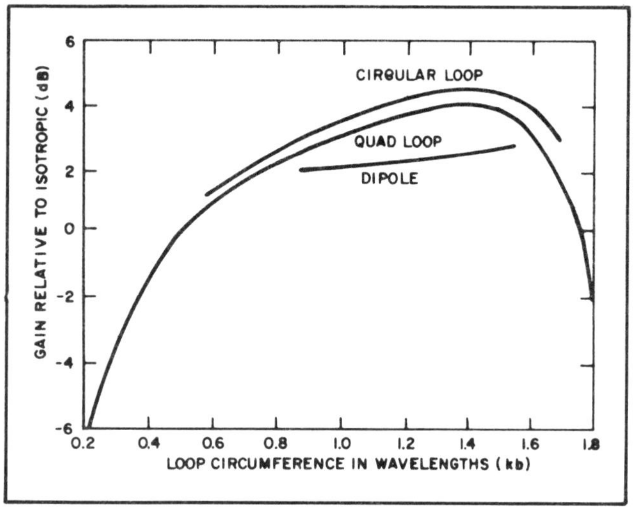

Figure 8. The gain of a square quad loop as a function of circumference. The gain of a circular loop and a folded dipole are shown for comparison. The gain is for the lobe at right angles to the plane of the loop, along the line of symmetry. The differences are partly due to area change and partly to differences in separation of the high-current points.

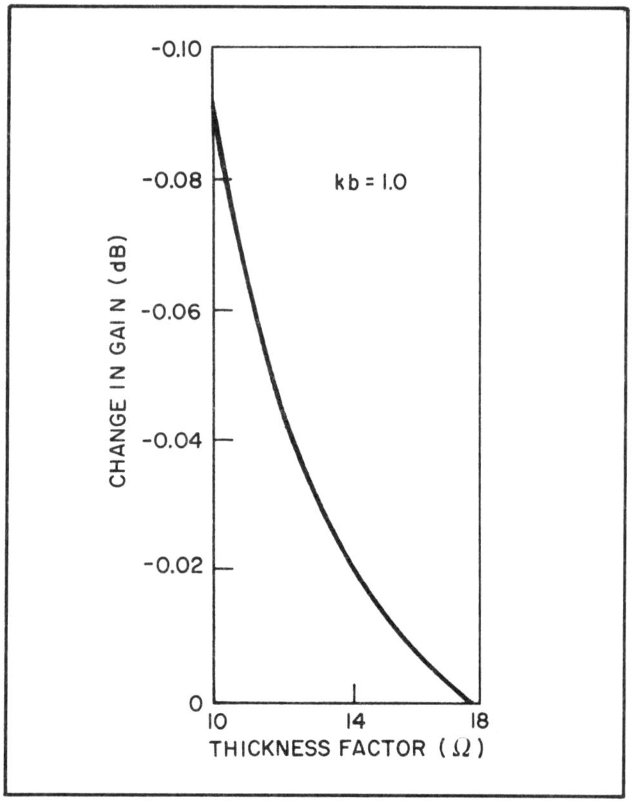

Figure 9. Effect of thickness on loop gain. The small effect appears to be due to the effective reduction in area by conductor blockage. The change is negligible in a practical sense.

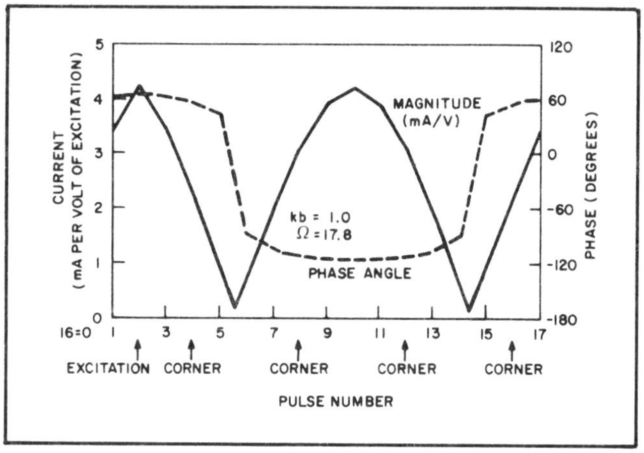

Figure 10. Current distribution on a square quad loop. Comparison with Figure 15 and published data on circular loops shows near identity in distribution, with some change in magnitude, reflecting the difference in drive impedance.

From the viewpoint of gain, better performance is obtained by operating with kb above unity—that is, above resonance. As for the circular loop, kb = 1.2 appears to be a good compromise value. As is also true for the circular loop, wideband operation is feasible for at least the range from one-half to twice the nominal design frequency. You'll need transmission line that will withstand high VSWR without puncture or excessive loss. Open wire line, ladder twin lead, or Teflon™—insulated coax will work. Solid-state transmitters, in particular, will need a match-box.

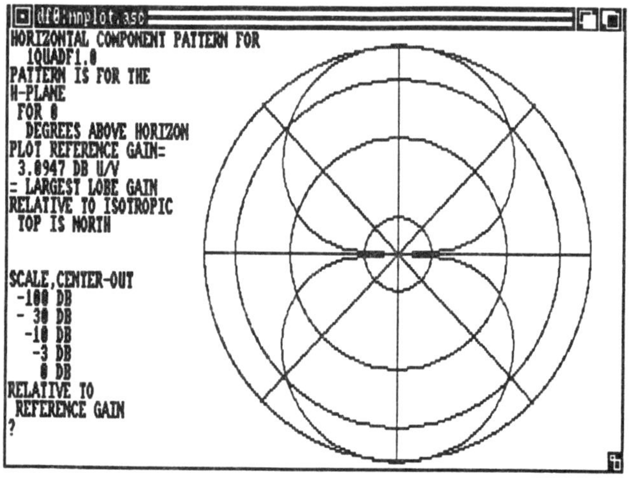

Figure 11. H-H radiation pattern for a quad loop near resonance, showing the horizontally polarized radiation component in the horizontal plane. In this and all following patterns, the loop is vertical, fed at the bottom, and lies in the X-Z plane (from right to left in this figure). The pattern is essentially that of a magnetic dipole. See Figure 12.

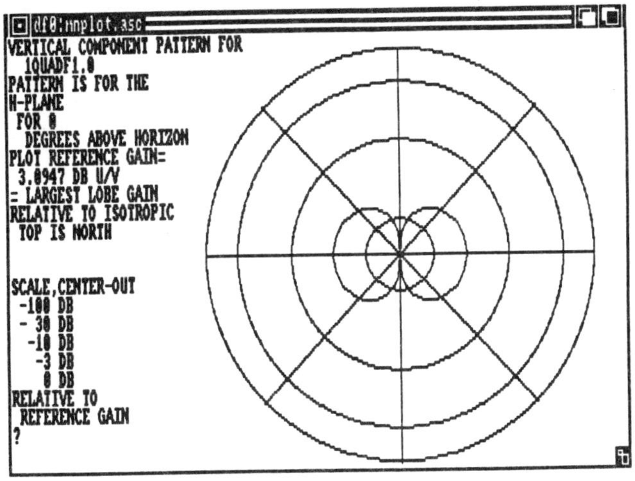

Figure 12. V-H pattern for a quad loop near resonance. Similar to Figure 11, but for the vertically polarized component. The presence of this component is a major difference from a resonant dipole. See Figure 13.

Current Distribution on a Square Loop

Figure 10 shows a typical current distribution for kb = 1.0 and a thickness factor of 17.8. The maximum current is larger for the square loop, reflecting its lower impedance, compared with the distribution for an octagon (see Chapter 2). The general shape of the magnitude and angle curves is the same. See References 1 or 5 for the overall agreement with circular loops.

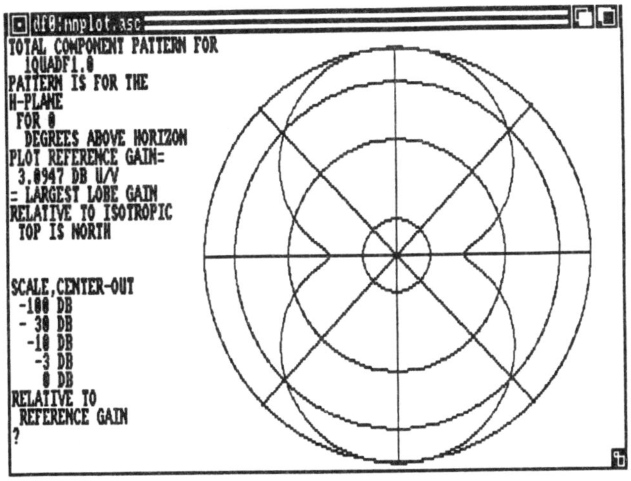

Figure 13. T-H pattern for a quad loop near resonance. Similar to Figure 11, but for the total radiation component. The effect of the vertical component shows in the radiation along the line of the loop, from right to left in the figure.

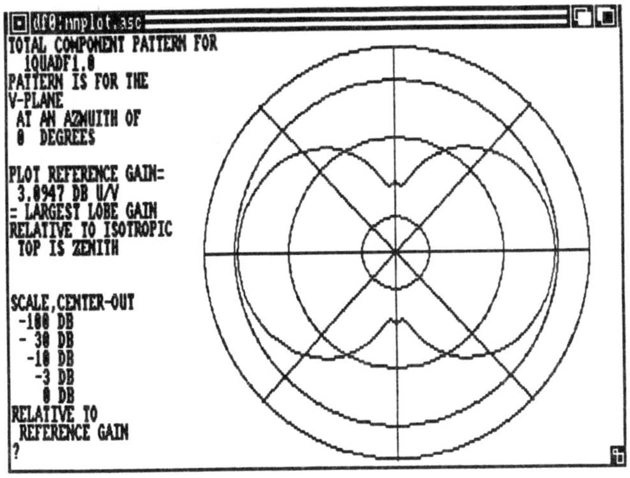

Figure 14. T-V pattern of a quad loop near resonance. The total component is shown for the vertical plane, which contains the loop. Zenith is at the right of the figure.

Radiation Patterns of Square Loops

Figure 11 shows the horizontally polarized radiation pattern of a square loop at kb = 1.0 in the horizontal plane. The beamwidth is just a bit less than 90 degrees. The pattern is identical in shape to that of the corresponding octagon, but the gain is slightly less—3.076 dB compared with 3.397 dB.

Figure 12 shows the vertically polarized component in the horizontal plane. The lobes are about 18 dB smaller than for the main lobe, and are lower than for the octagon, because of the relatively longer wire component in the vertical direction. This component is essentially ignored in the literature; its importance is yet to be determined.

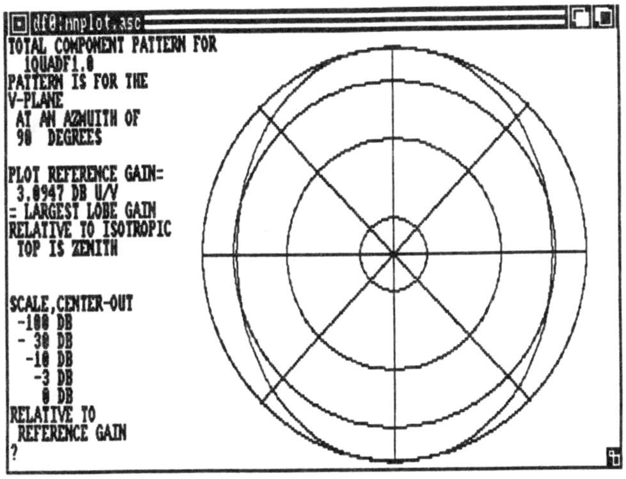

Figure 15. Orthogonal T-V pattern of a quad loop near resonance. The total component is shown for the plane at right angles to the loop. Zenith is at the right of the figure.

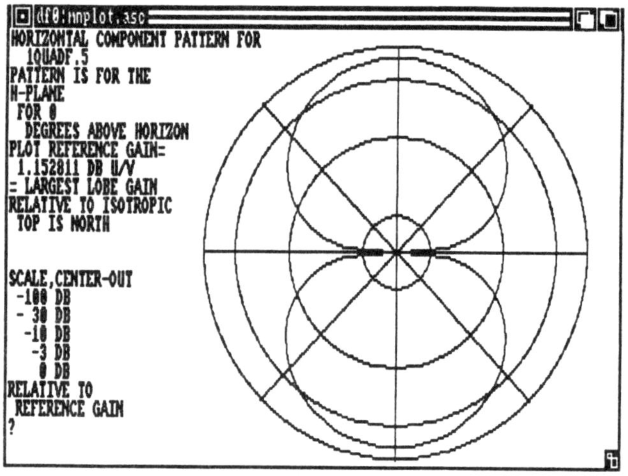

Figure 16. H-H pattern for a 0.5-wavelength quad loop. The horizontally polarized component is shown for the horizontal plane. Compare to Figure 11.

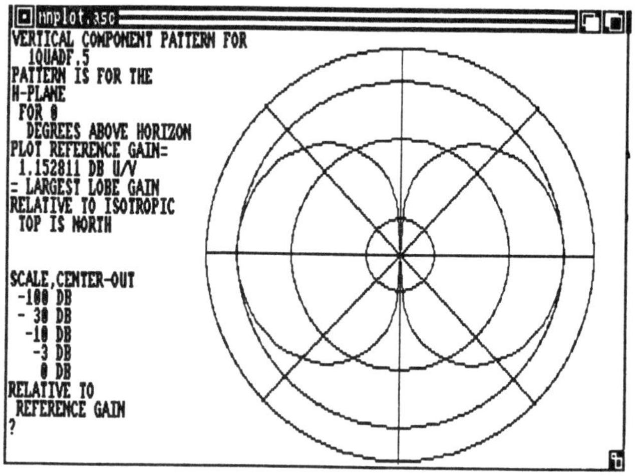

Figure 17. V-H pattern for a 0.5-wavelength quad loop. Note the growth of this component as compared to Figure 12.

Figure 13 shows the total radiation in the H-plane. Because of the vertically polarized component, the front-to-side ratio is only 18 dB, compared with the theoretical infinite value for a dipole (see Chapter 3).

Figure 14 shows the total radiation in the plane of the loop, with zenith at the right. Note that the reference is the maximum lobe size in this plane, in the direction of the feed point. The zenith lobe is a bit smaller, reflecting the slightly smaller current in the upper part of the loop.

Figure 15 shows the pattern at right angles to the loop, in the vertical plane. Front-to-top ratio is very small—about 3 dB. The V-plane beam width is essentially 180 degrees.

There are both similarities and differences in the pattern at other frequencies. Figure 16 shows the H-component, H-plane pattern for kb = 0.5. The direction of maximum gain is not at right angles to the loop; the gain in this direction is down about 2.5 dB. The beamwidth is slightly less than 90 degrees.

The vertical component pattern for this plane is shown in Figure 17. The intensity is markedly increased—only 4.3 dB below maximum gain. This is the lobe used in the usual, but smaller, direction-finding loop.

The total component pattern in the plane of the loop for kb = 0.5 is shown in Figure 18. Compared with Figure 14, the side-directed radiation has increased and the pronounced dip has disappeared. The up-and-down components are nearly the same. The differences are a result of the more uniform current distribution around the loop.

At kb = 1.5, the H-component, H-plane pattern of Figure 19 is nearly the same as that of the one-wavelength loop. The major difference is an increase in gain. The V-component pattern (Figure 20) has increased to about 6 dB below the main lobe. There is also a major difference in the plane of the loop (Figure 21). Because there are now three points of maximum current around the loop, the pattern shows three lobes —one toward the zenith, the others at an angle of about 120 degrees. At a still higher frequency, kb = 2.0, the H-component in the H-plane has become

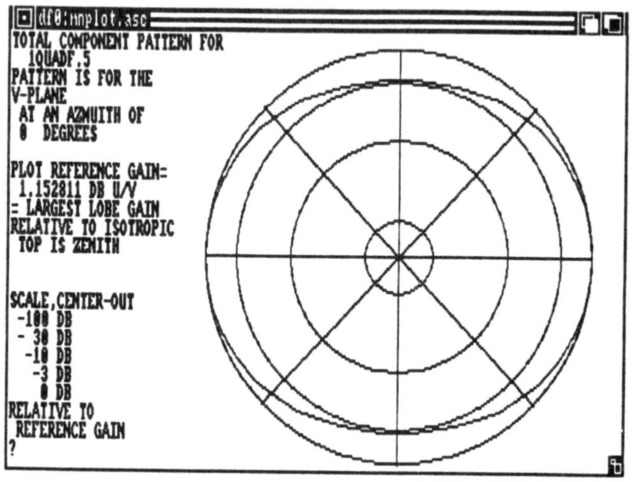

Figure 18. T-V pattern for a 0.5-wavelength quad loop. Total component pattern is in the plane of the loop. Zenith is at the right. Note the growth of the component toward the horizontal as compared to Figure 14.

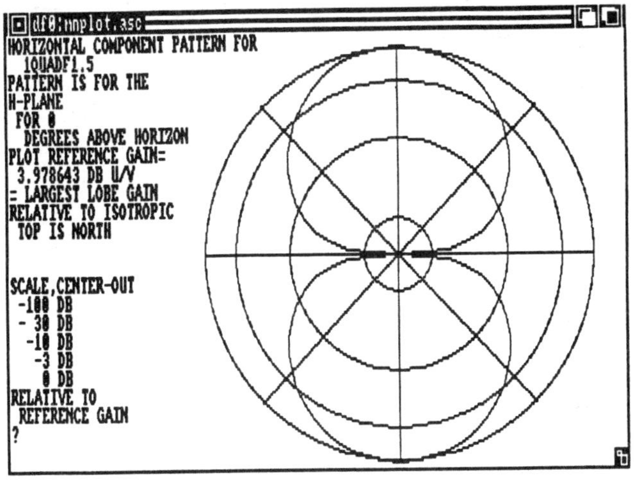

Figure 19. H-H pattern for a quad loop of 1.5-wavelength perimeter. Despite the dimension change, the pattern is essentially that of a magnetic dipole.

small (Figure 22)—about 12 dB below the main lobe. In contrast, the V-component lobe in this plane (Figure 23) has increased, and is only 1 dB below the maximum radiation. Note that the maximum radiation at right angles to the loop has decreased with respect to that for kb = 1.5.

Again, the major difference is in the plane of the loop (Figure 24). There are now four lobes reflecting the four current maxima around the loop. This multilobe pattern persists for all larger values of kb (see Chapter 1).

At right angles to the loop plane, the component toward the horizon becomes small (see Figure 25). At exact resonance, the minimum would occur at the horizon, rather than below it.

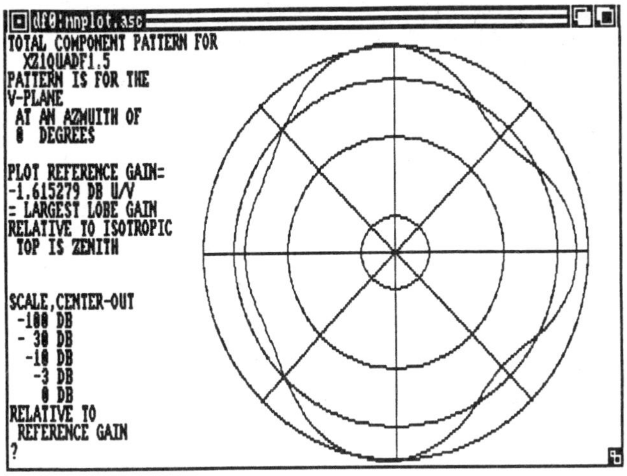

Figure 21. T-V pattern for a 1.5-wavelength loop in the plane of the loop. Zenith is at the right. The three lobes reflect the three high-current points around the loop, one being opposite the feed point, the other approximately 120 degrees away.

Rectangular Loops

Several changes occur as the shape of a square loop becomes rectangular. One is a shift of driving resistance from the 300-ohm value of a folded dipole to the zero resistance of a shorted ideal line. The MINI-NEC-calculated pattern of change is shown in Figure 26. The change is not linear with shape factor, but nearly square law. The associate reactance change is also shown. At the folded dipole end (shape factor less than 0.15 for this conductor size) the reactance is positive, so the loop/dipole is above resonance. Dipoles and loops with small shape factors must be

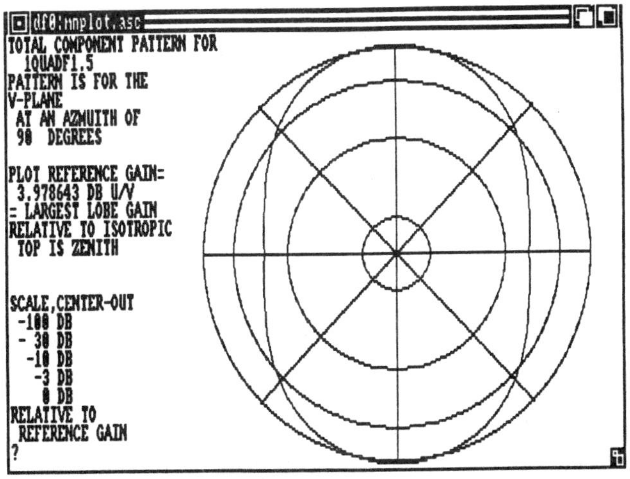

Figure 20. V-H pattern for a 11.5-wavelength loop. Compare to Figures 12 and 17.

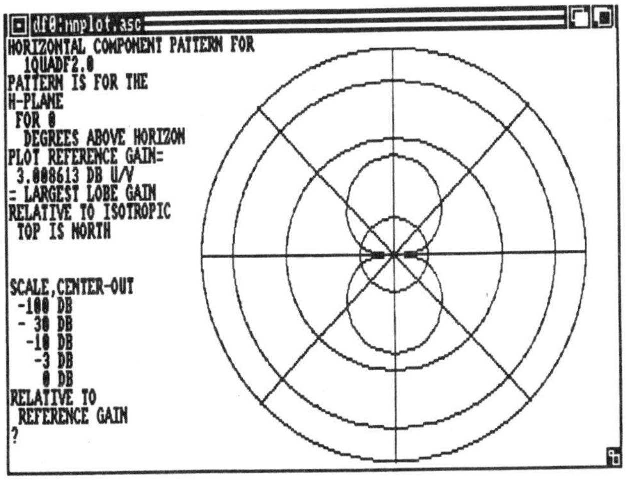

Figure 22. H-H pattern for a 2.0-wavelength loop. Note the marked decrease as compared to Figure 11.

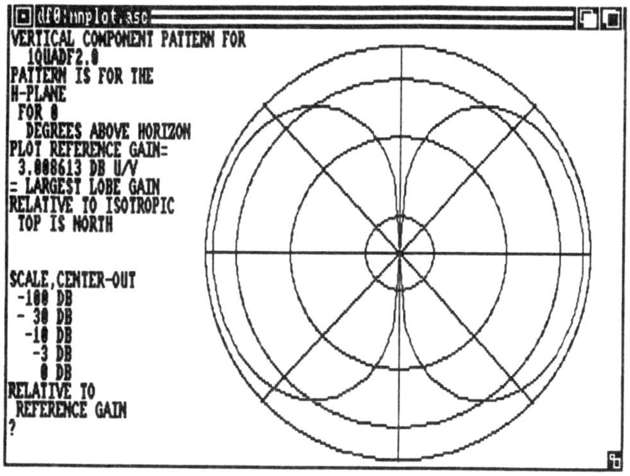

Figure 23. V-H pattern for the 2.0-wavelength loop. Note the marked increase as compared to Figure 12.

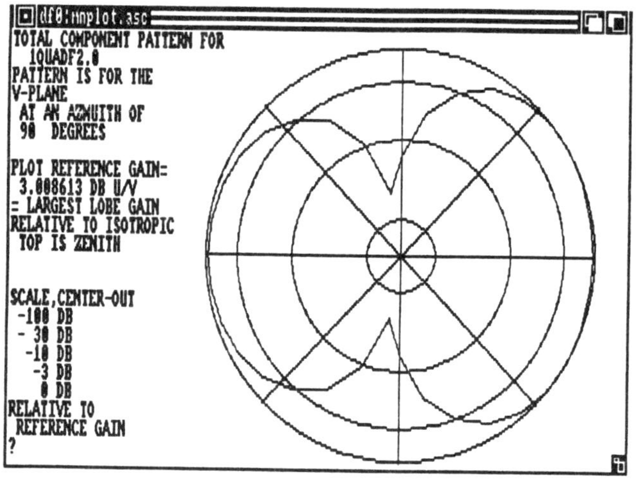

Figure 25. T-V pattern for the 2.0-wavelength loop at right angles to the loop plane. Zenith is at the right. At resonance, the minimum would be at the horizon.

shortened for resonance. For shape factor greater than 0.15, a loop with perimeter kb = 1.0 will be below resonance.

The curves are dotted for shape factors greater than 0.75 because MININEC calculation errors become large if segment lengths are too small. In the extreme, calculated values of radiation resistance may even become negative—a physical impossibility. No attempt has been made to find calculation conditions that give good results because it seems unlikely that very large shape factors will find practical use.

Gain is affected by two factors in these rectangular loops. The lower area implies lower gain, but increased separation of the high current points implies increased gain. Figure 27 indicates that the area loss is the least important, because the gain increases to at least sf = 0.75.

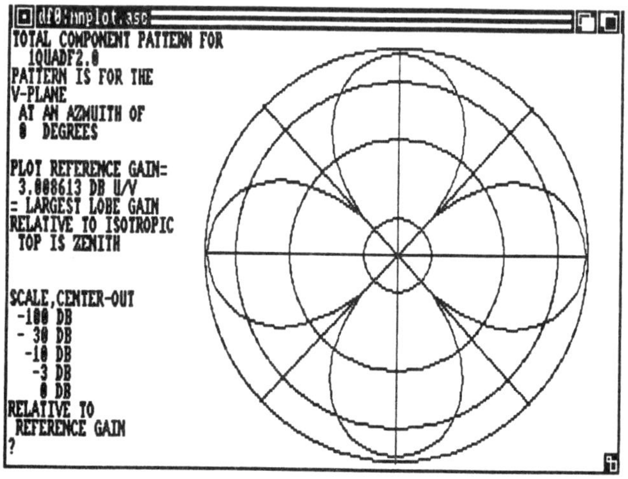

Figure 24. T-V pattern for the 2.0-wavelength loop in the plane of the loop. There are now four high-current points and four lobes. A multi-leaved pattern occurs for all higher integral values of kb.

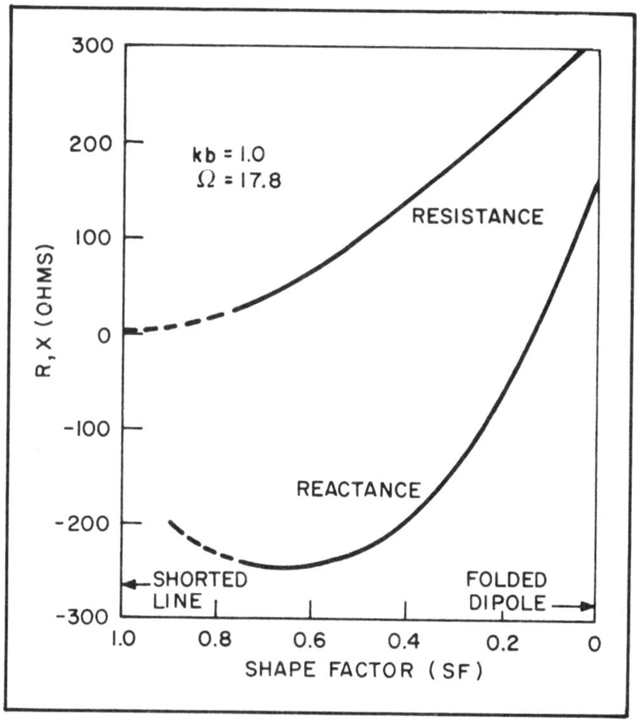

Figure 26. Self-impedance of a typical rectangular quad loop as a function of shape factor. At sf = 1.0, both resistance and reactance must be zero, the values for a perfect shorted transmission line. Change in resistance with conductor size would be relatively small, but reactance change would be sizable. The pattern is related to those of Figures 5 and 6.

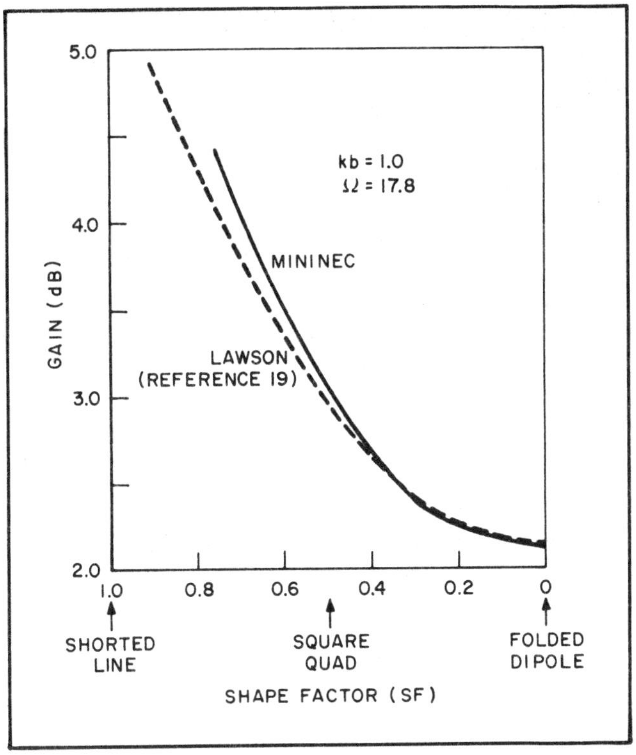

Figure 27. Gain of rectangular loops around 1.0-wavelength perimeter. The loops with high shape factor are related to the skeleton slot loop used in VHF/UHF stacked beams.

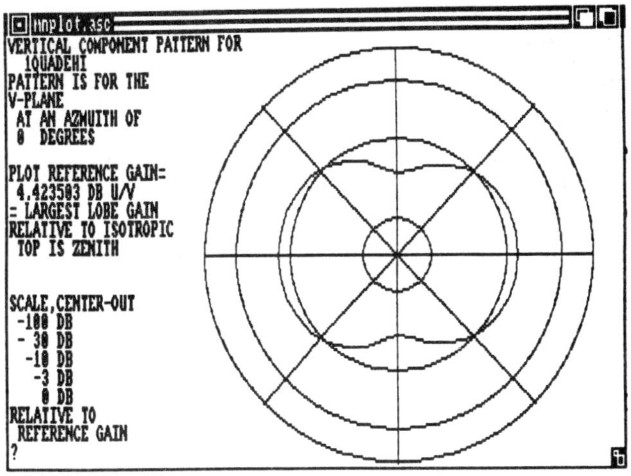

Figure 28. V-V pattern for a rectangular loop at right angles to the loop plane. With sf = 0.75, this is one of the skeleton-slot family. Maximum gain is toward the horizon.

loops will have slightly higher gain, but lower drive resistance. The flattened ones will show reduced gain and higher drive resistance. If maximum gain is needed, the narrow loops operated at kb greater than unity offer interesting possibilities.

Lawson[19] estimated the gain of square loops, considering both the separation of the current points and the length of the high-current part of the loop. His calculated values are also shown in Figure 27. Agreement is reasonably good, but Lawson ignored the vertically polarized component, so his values are smaller than the MININEC results where this component is large.

Lawson[19] shows the gain increasing to sf = 0.9. However, MININEC results at sf = 0.875 imply a marked reduction in gain. As discussed above, the value cannot be considered trustworthy, so it isn't shown on the curve. Practically, the gain must reach a maximum somewhere around sf = 0.8 to 0.9 and fall to zero for a shorted line.

The added gain must have some effect on the radiation pattern. The only plane where this is found to be important is shown in Figure 28. Comparison of this with Figure 15 shows that the up-down directed components have decreased by about 5 dB, not an important change.

The overall conclusion is that rectangular loops are just as useful as square ones. The high and narrow

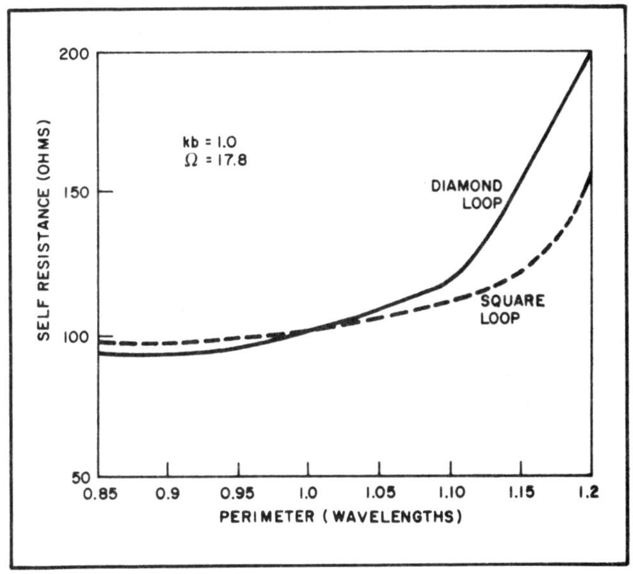

Figure 29. Self-resistance of a diamond loop close to resonance. The curve for a square loop is shown for comparison. While the resistance components are identical at perimeter = 1.0 wavelength, there is appreciable departure above perimeter = 1.1. This is related to the way current maxima change position.

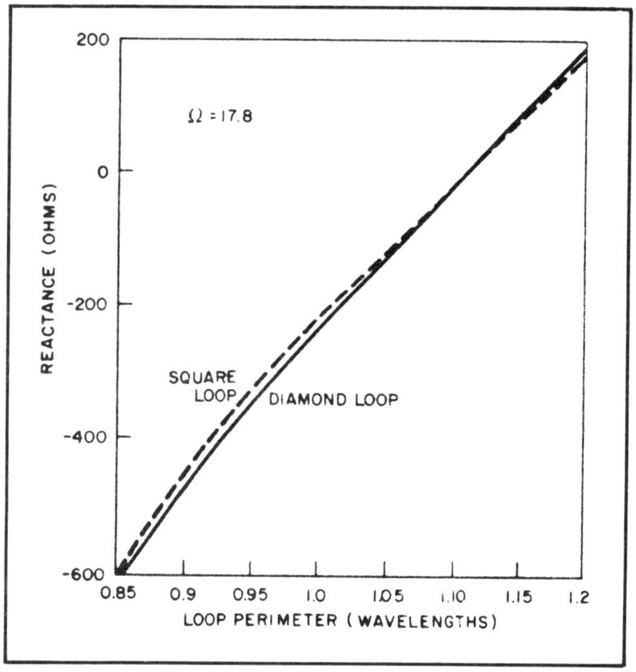

Figure 30. Self-reactance of diamond and square loops near resonance. Practically, the small difference is negligible.

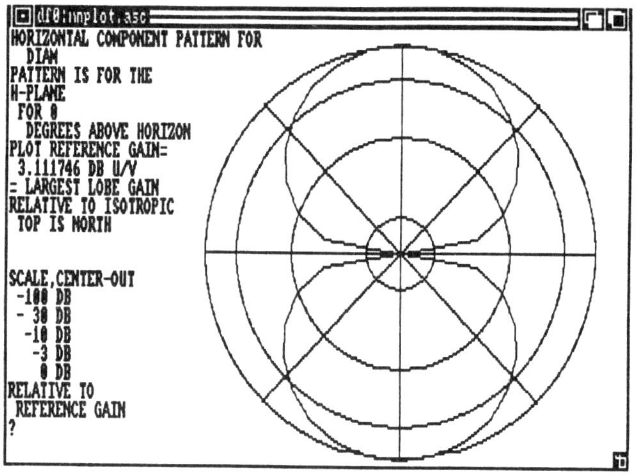

Figure 32. H-H pattern for a diamond. Compare with Figure 11.

The Diamond Loop

The quad literature indicates that the square and diamond loops have identical performance. This is found to be nearly the case, but there are some differences.

Figure 29 shows the major difference—the variation of self- or drive resistance with loop size. At kb = 1.0 the values are identical, but there's some difference at all other values of kb. The difference reaches some 55 percent for kb = 1.15. In contrast, there's only a small difference in self-reactance, as shown in Figure 30. Over the range from kb = 0.9 to 1.02, the difference is about 7 percent.

The current distribution on the diamond is shown in Figure 31. The maximum current is slightly lower than in the square loop (Figure 10); this current difference does induce some small pattern changes. Comparison of Figures 31 and 11 shows a slight increase in gain for the diamond. Other pattern characteristics are essentially identical, as seen by comparing Figures 32, 33, and 34 with figures 11, 12, and 14.

These data indicate that the diamond and square

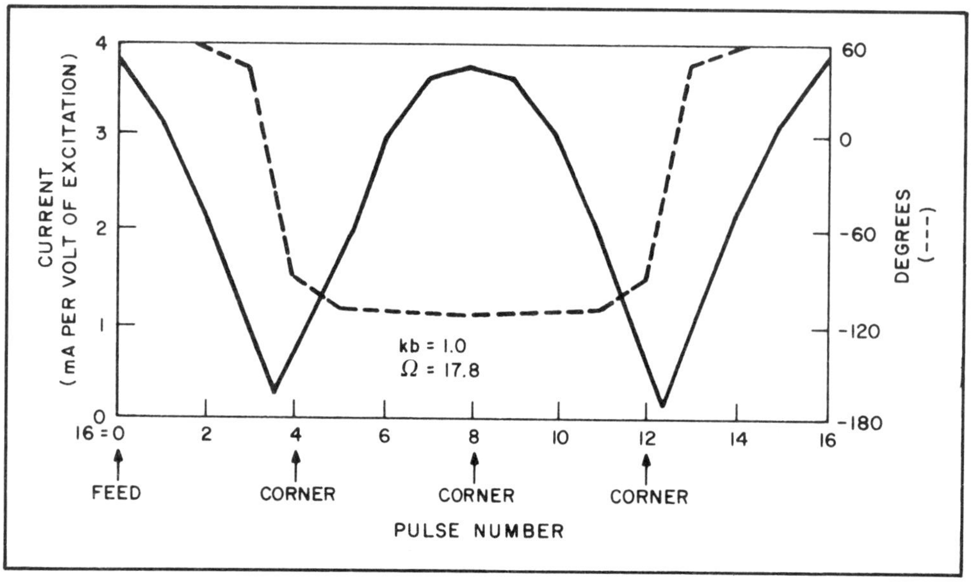

Figure 31. Current distribution on a diamond. Compare with Figure 10.

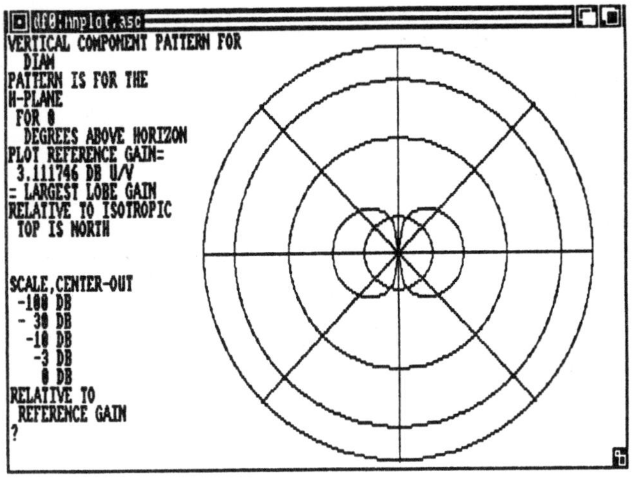

Figure 33. V-H pattern for a diamond. Compare with Figure 12.

loops are fully interchangeable in a practical sense. A small size adjustment is indicated if resonant operation is needed but other differences are negligible, especially if line matching is used. The size adjustment may be important in arrays. This is a factor to be studied. Basically, however, the choice between the diamond and quad loop (and even between these and rectangular ones) can be based on mechanical design and installation considerations.

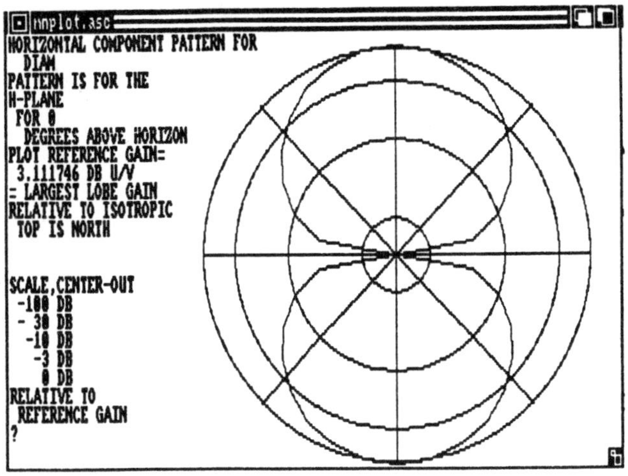

Figure 34. T-V pattern for a diamond in the plane of the loop. Compare to Figure 14 noting the difference in reference level. These patterns confirm the common view that a square and diamond loop have practically identical properties.

Summary

The curves and tables shown here are sufficient to design any practical square, rectangular, or diamond loop, and to estimate performance. The data also provides the basic building blocks for the development of arrays using these loops.

NOTES

Chapter 5

The Two-element Array

General performance data on the most popular quad design

Antenna Theory Simplified

To review, antenna elements in proximity show both self- and mutual impedance, as do other tuned circuits. The voltage and current relationships for the case of exciting voltage applied to only one loop is:

$$V1 = I1 \times Z11 + I2 \times Z12 \qquad (1)$$

$$0 = I1 \times Z21 + I2 \times Z22 \qquad (2)$$

where all quantities have magnitude and phase angle.

Self-impedances Z11 and Z22 are determined by the size of the loops and the conductor size, as described in Chapter 4. Mutual impedances Z12 and Z21 are determined primarily by the spacing between loops, with smaller influence from loop orientation and size. Except in some special cases like ionized gases, Z12 = Z21.

The input or drive impedance of the system is Zd = V1/I1. From Equations 1 and 2, this is:

$$Zd = Z11 - (Z12 \times Z21)/Z22 \qquad (3)$$

With these quantities, the current in each element is easily calculated.

The mutual impedance is often determined experimentally. The input impedance of an isolated element is first determined; this is the self-impedance. (If two different elements are used, they both must be measured.) Then the input impedance, Zd, is measured with both elements present. Equation 3 is then solved for the mutual impedance.

The reason these impedances are important is because the radiation in any direction from any antenna of reasonably similar elements (to a good approximation) can be regarded as the product of the radiation of a single element and an "array factor." This factor is determined by the relative current magnitude and phase angle in the antenna elements. It is given by:

$$F(A1, A2) = 1 + K1 \times \exp(j \times P1) \qquad (4)$$

where:

F() is the array factor

A1, A2 are orthogonal angles from the array center point, usually chosen for symmetry

K1 is the ratio of currents

P1 is the phase difference due to spacing and current phase angle given by:

$$P = S \times \cos(A2v - A2r) \times \sin(A1) + A3 \qquad (5)$$

where:

S is the spacing between the loops

A2v is the angle value being calculated
A2r is the angle between the line of antennas and the angle source
A3 is the phase difference in the currents

A number of these array factors are shown graphically in *The ARRL Antenna Book.*[23] *There is an extensive set in the NAB Engineering Handbook.*[24] These are useful for understanding the general behavior of any type of array.

These relationships are easily extended to more than two elements. Subscript notation is essential for this, to keep from getting lost. Hand calculation of patterns is possible, but tedious. The current practice is to make the calculation with a computer, and there are a number of programs available which can perform this task. The most difficult step is calculating impedances—essentially that of inverting a matrix of complex numbers. (See Reference 25 for a program which does this.) The remainder of the equations are straightforward, and aren't given here.

In working with these equations, data is often given as admittances, rather than impedances. It's easy to interchange the "self" values, since Z11 = 1/Y11 (take the inverse of the magnitude and the negative of the angle). For mutual impedance/admittance, it's necessary to solve as a matrix inversion operation, which is identical to solving Equation 3.

This approach is not the one followed in this book. All of the data presented here are based on MININEC program calculations. These give impedances, currents, and patterns in a single program run. To save work, the MININEC results have been used, rather than the essentially duplicate pattern-factor approach. However, data has been developed and is presented to allow use of pattern factors, if desired. Chapter 4 covers the self-admittances; the following section gives the mutual admittances.

Mutual Impedance and Admittance of Quad Loops

Just as with coupled inductances, the coupling between antennas decreases as the separation between them increases. This means that the mutual impedance, or the mutual admittance, decreases with separation.

Figures 1A and 1B show how admittance varies with

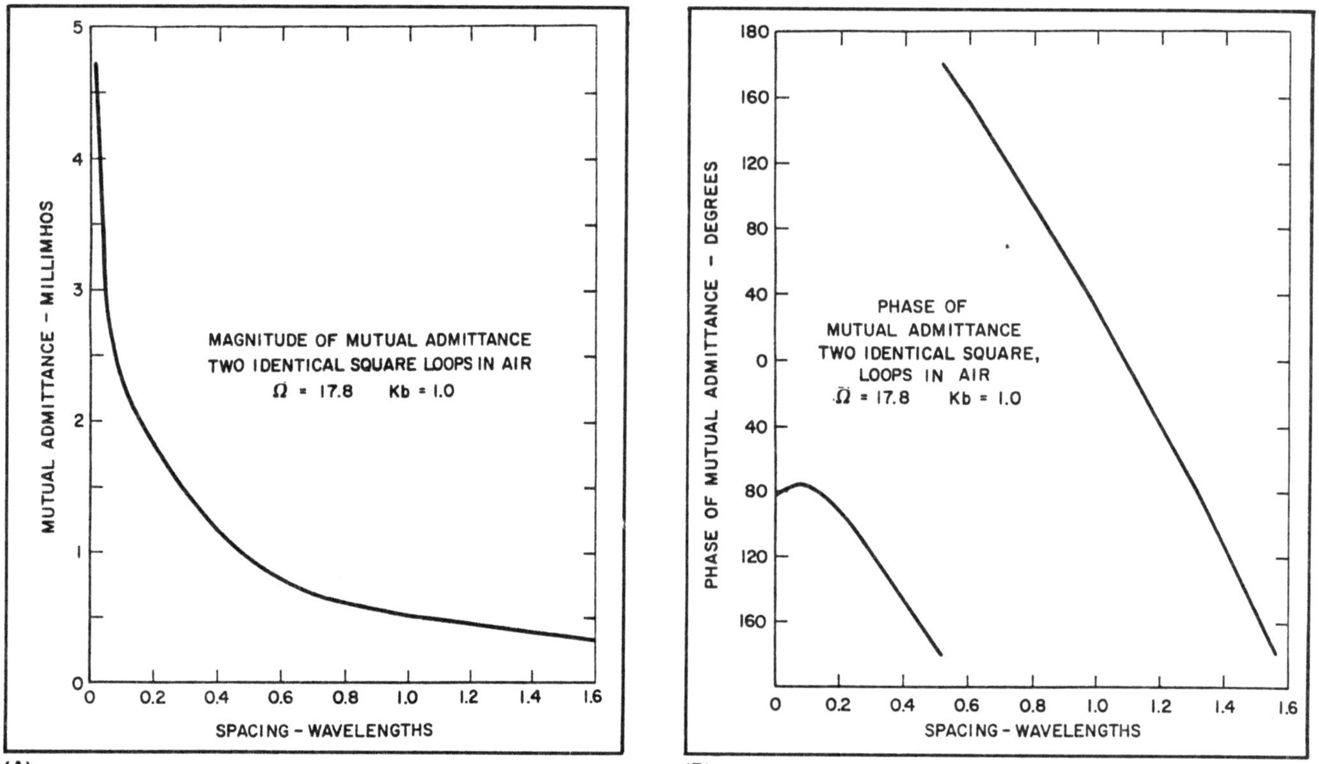

(A) (B)

Figure 1. Mutual admittances of two quad loops. (A) = mutual conductance, (B) = mutual susceptance. These data allow calculation of drive impedances, element currents, and from these, the pattern factors for driven and parasitic arrays. See text for approximations to extend range of spacing.

Spacing (wavelength)	Magnitude (mhos)	Angle (degrees)
0.025	3.74723	− 80.6661
0.050	2.90236	− 77.7098
0.100	2.27445	− 76.5081
0.200	1.84452	− 90.1502
0.400	1.16962	−146.6690
0.600	0.78919	156.6409
0.800	0.65634	92.2287
1.600	0.33556	172.2883
3.200	0.17193	− 29.8165
6.400	0.08579	−109.0183

Table 1. Values of mutual admittance for quad loops of one wavelength perimeter. Data are for a perimeter/radius ratio of 8000, but can be used for other conductor sizes with small error. Remember that the values of mutual impedance aren't obtained by simple inversion of the mutual admittance alone, but by matrix inversion of self- and mutual-admittances.

distance. The data (to greater precision) is also shown in Table 1. These values are for square loops of one wavelength perimeter, and for a specific conductor size. However, the values are very insensitive to conductor size, and reasonably so for perimeter, so the curves can be used for all quads with loops of one wavelength perimeter or close to that measurement.

The values of the curve sections shown don't follow any simple relationship, but simple approximations can be used to extend the admittance beyond the curve points. Very closely, the admittance magnitude decreases inversely with element separation, and the admittance phase angle increases linearly with separation. These approximations allow calculation of currents on very long arrays.

If the admittance or impedance is plotted in rectangular coordinates (X-Y), the values are found to approach zero by a spiral of decreasing radius, terminating at zero. Both the real and imaginary components can take on positive and negative values. One consequence of this appears in multi-element beams. Drive impedances and gains can show cyclic change as separation is varied. This leads to preferred element spacings for a given number of elements. See Lawson[19] for a discussion of this topic.

General Quad Patterns

In most antenna applications, the pattern is by far the most important design factor. Usually, maximum gain of a "forward" lobe is sought. However, because of interference elimination possibilities, the design goal may be maximum F/B ratio, or even minimum rear lobe intensity. Let's examine quad patterns in general terms.

Lobe Intensity With Two Identical Loops

Figure 2 shows the variation of radiation toward and away from the parasitic element. These data are plotted for two identical loops and as a function of loop spacing. The intensity of the two equal lobes of an isolated loop is shown for comparison.

The curve for intensity away from the parasitic element shows that the presence of a parasitic element increases the intensity of this lobe for all spacings shown. There is an attendant narrowing of the lobe width. There is also appreciable variation in lobe size, with the maximum being 5.8 dB at 1.5 wavelengths separation. (Note that the lobe intensity at zero separation would be identical to that of an isolated element.)

In contrast to this behavior, the lobe toward the parasitic element can be greater or less than that of an isolated loop. The parasitic element decreases the gain up to about 0.2-wavelength spacing. It acts as a reflector of fair gain, but not very good F/B ratio. Between 0.2 and about 0.54 wavelength, the lobe toward the parasitic element is stronger than that of an isolated element. The parasitic acts as a director, but gain leaves something to be desired and F/B ratio is very poor.

From 0.54 to about 0.78 wavelength separation, the parasitic element again acts as a reflector, but with

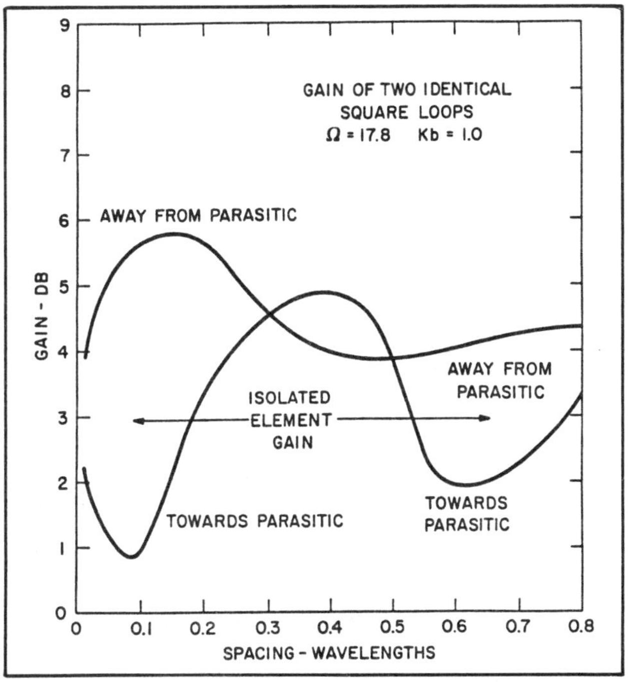

Figure 2. Gains, or lobe intensities, of two identical loops. Both reflector and director action are present.

less gain and poorer F/B ratio. This alternate change repeats approximately each half wavelength of separation, with the lobe intensities decreasing in peak magnitude as separation increases. Two identical loops at 0.1 to 0.15 wavelength separation give fair "forward" lobe gain, but rather poor F/B ratio. As with other parasitic antennas, tuning by changing parasitic element dimensions is necessary to maximize performance. The data in these curves indicate that the range of design interest is for parasitic element spacings of about 0.1 to 0.4 wavelength.

General Radiation Patterns of Tuned Quads

Before attempting to find optimum tuning conditions, it's desirable to look at the way complete patterns vary. Figure 3 shows twenty patterns for spacings from 0.1 to 0.4 wavelength and for parasitic element dimensions from 0.9 to 1.1 wavelengths in steps of 0.05 wavelength, which clearly illustrate the pattern of variation.

The top row is for director tuning, with a perimeter of 0.9 wavelength. Maximum gain with a director is 4.5 dB. Front-to-back ratio is poor in all cases, with a maximum of about 2 dB.

The second row is for a perimeter of 0.95 wavelength, also with director tuning. Maximum gain has increased slightly, to around 5 dB. Front-to-back ratio remains poor, the maximum being about 3 dB. Note that the 0.4-wavelength spacing pattern shows reflector rather than director action.

The center row is for identical elements, corresponding to the curves of Figure 1. Best gain is over 5.5 dB. Except for 0.3-wavelength spacing, director F/B ratio has increased slightly.

The fourth row shows the pattern for a circumference of 1.05 wavelengths perimeter, normally considered reflector tuning. However at close spacing, the parasitic element is still acting as a director, with nearly 7 dB gain, but only 3 dB F/B ratio. At 0.3 and 0.4-wavelength spacing, reflector action is apparent. Best reflector gain is over 6 dB with about 2 dB F/B ratio.

The bottom row is for a 1.1-wavelength parasitic element. Reflector action is present at all spacings. Maximum gain is nearly 7.8 dB, and maximum F/B ratio is nearly 15 dB. Both occur at a spacing of 0.1 wavelength. Note the change in shape of the rear lobe with spacing.

All of these plots are for the horizontal plane and for the horizontally polarized component of radiation. The vertical component found for single loops is present for all of the size and spacing combinations. A typical vertical component pattern is shown in Figure 4. At this size and spacing, the vertical component of radiation is nearly 30 dB below that of the horizontally polarized component. The intensity is not much affected by spacing change, so it isn't shown. There's some increase as the parasitic element spacing decreases. This is the result of the relative current increase in the vertical arms. This vertical component does fill in the lateral nulls, reducing the front-to-side ratio. A later figure shows the relative magnitude of the total and just the horizontal component. As with circular loops, the importance of the vertical component isn't apparent.

Optimum Spacing and Loop Size

To optimize performance, the available variables are spacing and parasitic-element size. Let's look first at the intensity or gain of the lobe away from the parasitic element. Figure 5 shows this, as a function of reflector spacing, for five reflector sizes. Each curve shows an "optimum" spacing, which is quite different for each curve. Maximum gain is obtained for the 1.10-wavelength reflector, at a spacing of about 0.15 wavelength. As this is essentially the same maximum gain found for circular and octagonal loops, it is at, or close to, the maximum possible gain for a two-element quad. We'll look at this again later.

For the smaller or director loops, Figures 6 and 7 show the intensity of both lobes. In the first, there's very little difference in the lobe intensity (the F/B ratio is very low). However, gain is above that of an isolated loop. Two spacings show somewhat more gain the than others. The one at 0.3-wavelength spacing is the more important. This gives the gain for a minimum number of elements in larger beams. The one at the small spacing implies that a small gain increase in large arrays is possible using very close director spacing. This was evident in circular loop arrays (see Figure 8 in Chapter 3).

For 0.95-wavelength directors, Figure 7 shows some gain improvement and greater F/B ratio. Maximum gain occurs at 0.15-wavelength spacing. F/B ratio is constant over the 0.1 to 0.2-wavelength range.

The contrast between director and reflector action

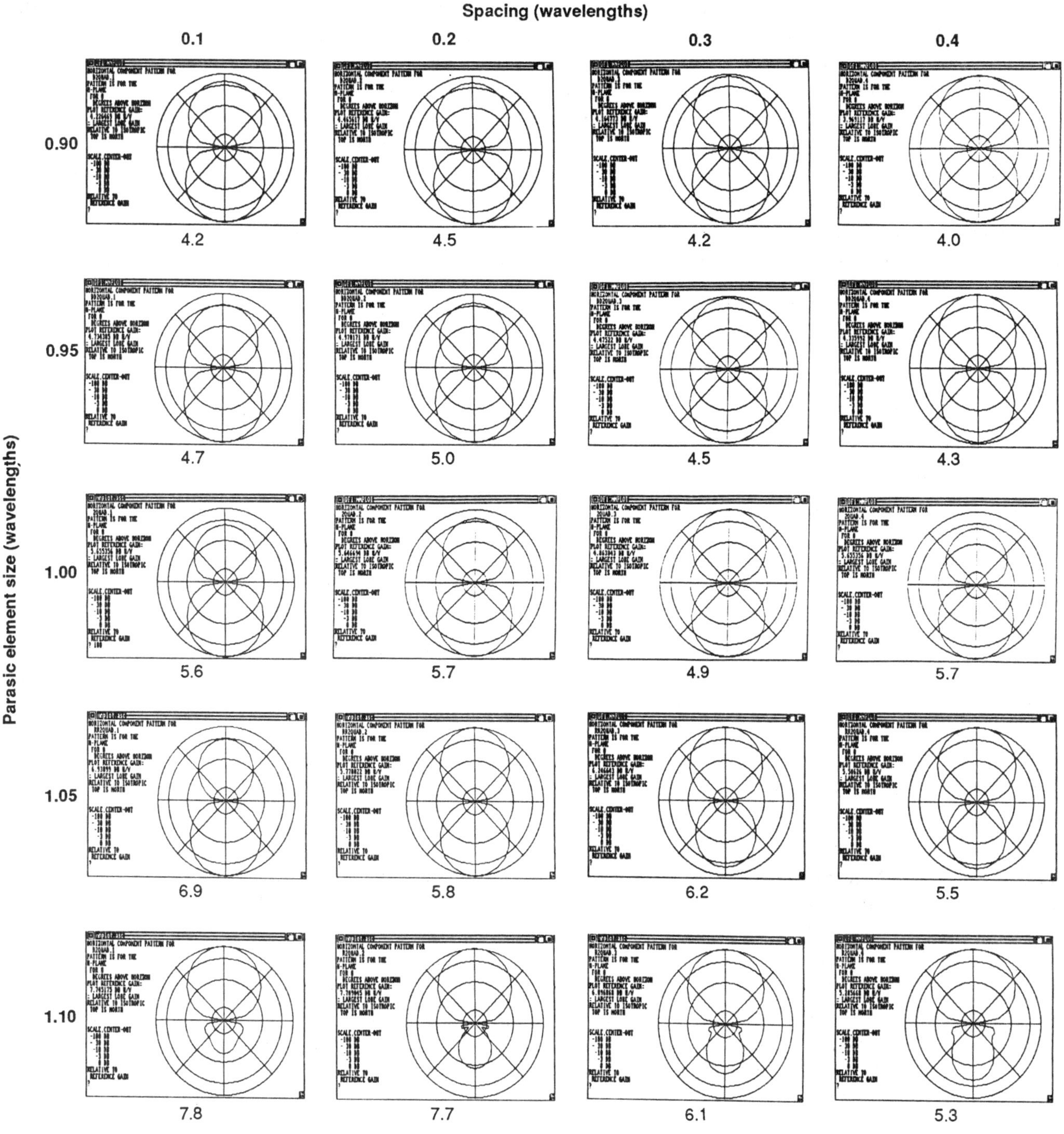

NOTE:

1. Pattern family of two element quads Ω = 17.8.
2. Number by figure is gain of largest lobe.
   Pattern is normalized to this value.

Figure 3. The twenty patterns of this figure show the effect on the pattern and gain by changing element spacing and tuning the parasitic element by changing its size or by adding reactance. Note that each of the patterns is normalized to the gain of its largest lobe, as a way to show pattern detail. All patterns are for the H-component in the H-plane. The lobe away from the parasitic element is toward the top in all patterns.

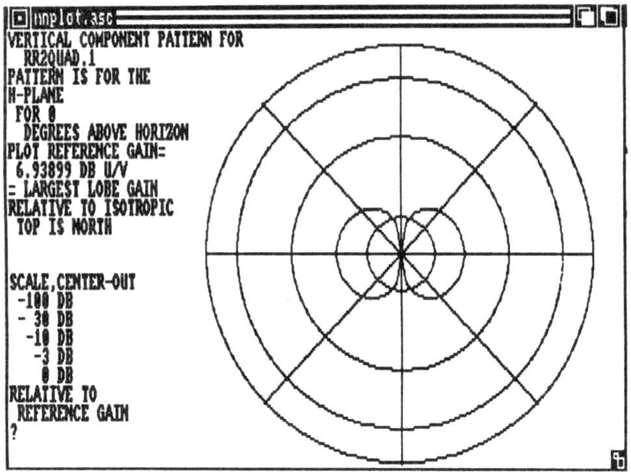

Figure 4. A typical V-component pattern in the H-plane, corresponding to one of the patterns of Figure 3. All V-component patterns have the same shape, with a magnitude variation in the range −12 to −25 dB below the largest lobe.

is very apparent when Figure 8 is considered. As shown by the patterns, there is a marked increase in gain in one lobe and a marked decrease in the other. Maximum gain is at a reflector spacing of 0.15 wavelength. As is common in parasitic arrays, best F/B ratio doesn't occur at the point of maximum gain, but at a lower frequency.

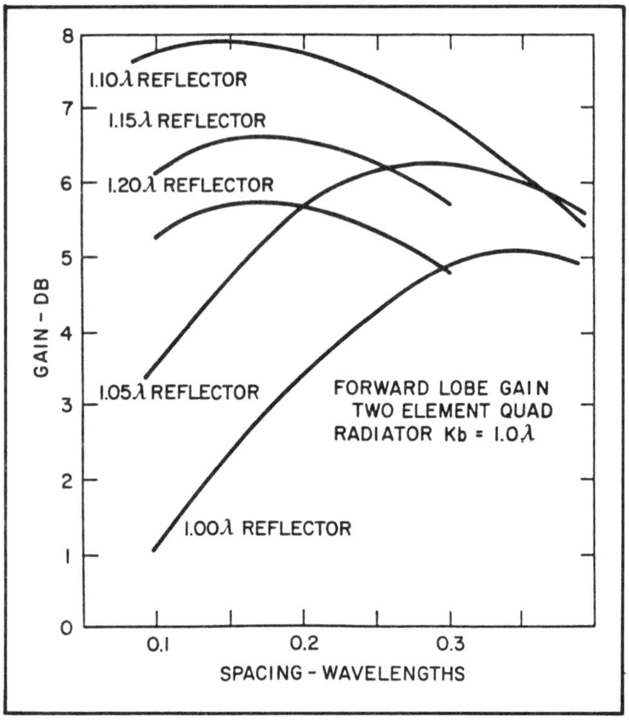

Figure 5. Forward lobe gain with the parasitic element acting as reflector, as a function of spacing and for five parasitic loop sizes.

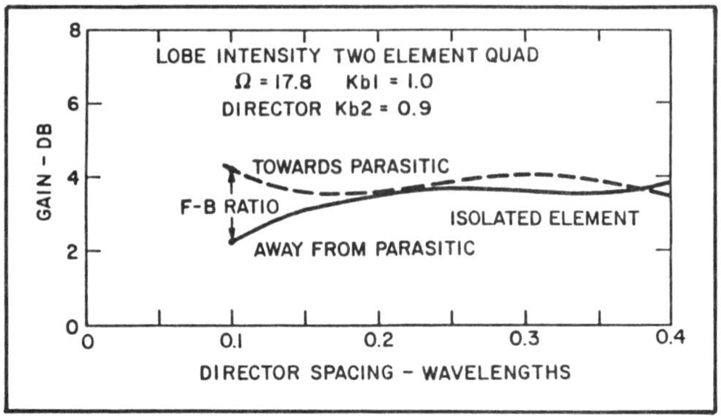

Figure 6. Intensity of front and rear lobes as a function of spacing, for a 0.9-wavelength director.

It must be remembered that there will be a change in the relationship between size, spacing, and lobe structure as the conductor size is changed. The reason for this is the smaller reactance change with a thick conductor, which changes the current phase in the parasitic element.

Figures 9 through 12 are plotted to avoid the effect of this change in conductor size. Gain is plotted as a function of loop reactance rather than against element size. This reactance may be the loop self-reactance, varied by changing the loop circumference. It may also include a tuning reactance, which can be a short section of transmission line, an inductor or capacitor, or a tuned circuit. Strictly speaking, there will be small differences between the behavior of loops with size change as compared to those with added reactance, due to change in current distribution around the loop. However, this difference is minimal for small changes in loop size, and can be ignored.

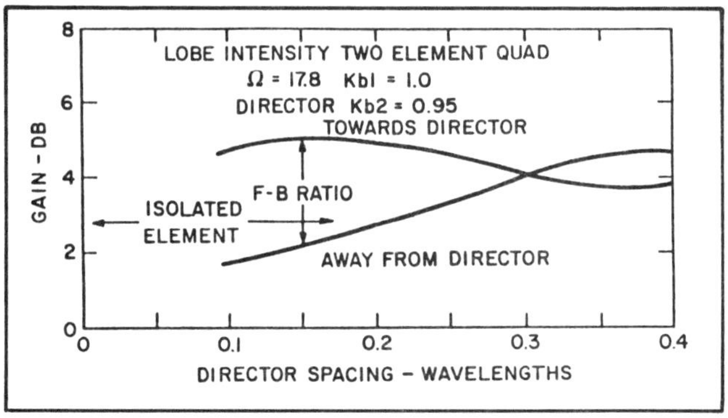

Figure 7. Intensity of front and rear lobes as a function of spacing, for a 0.95-wavelength director.

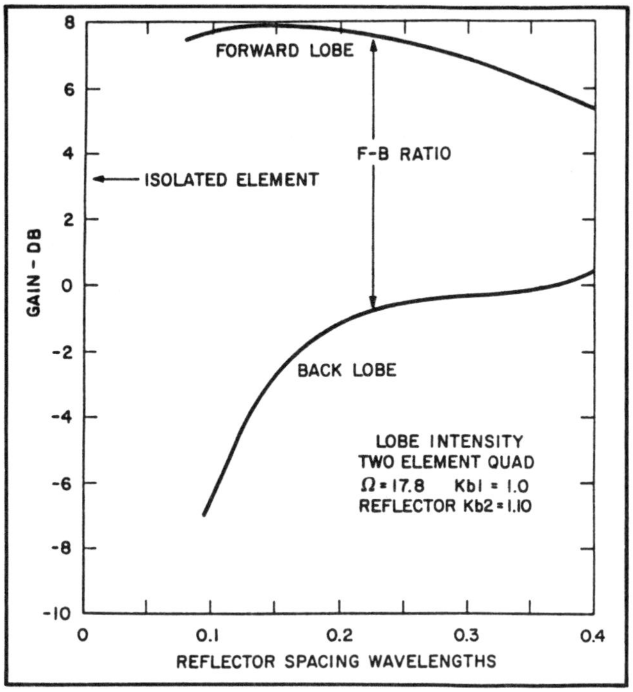

Figure 8. Intensity of front and rear lobes as a function of spacing, for a 1.10-wavelength reflector. The conventional quad with radiator and reflector has better performance than the radiator-director combination. (This does not indicate that directors have no value in multi-element beams.)

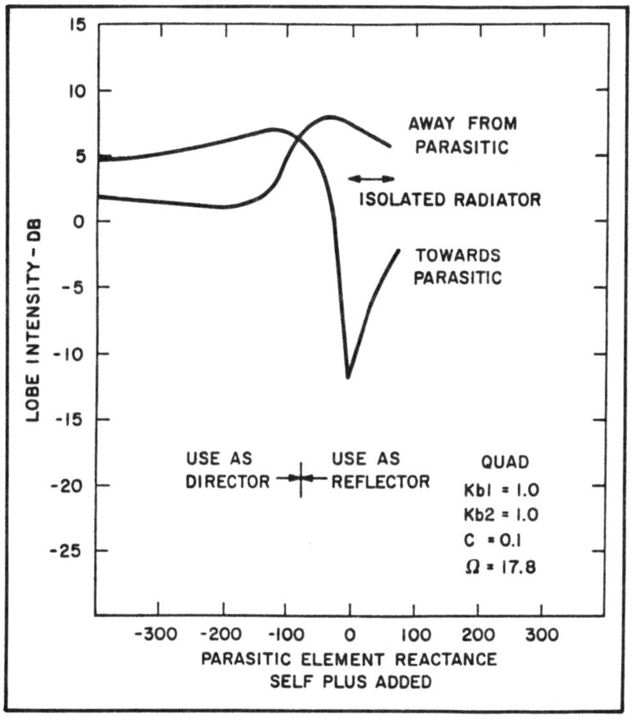

Figure 9. Lobe intensities as a function of parasitic tuning, for 0.1-wavelength spacing. The variable is total parasitic element reactance at a point of maximum current. This may be element self-reactance, or a combination of this and added reactance, as by a "tuning stub."

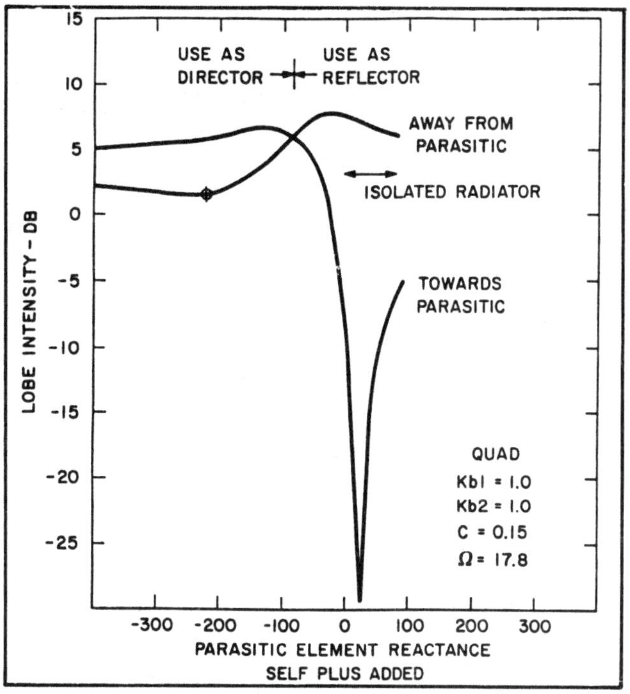

Figure 10. Lobe intensities for a spacing of 0.15 wavelengths. It is not possible to secure maximum gain and maximum F/B ratio with a single tuning condition. A design compromise is necessary, usually based on operating conditions on the design-frequency band.

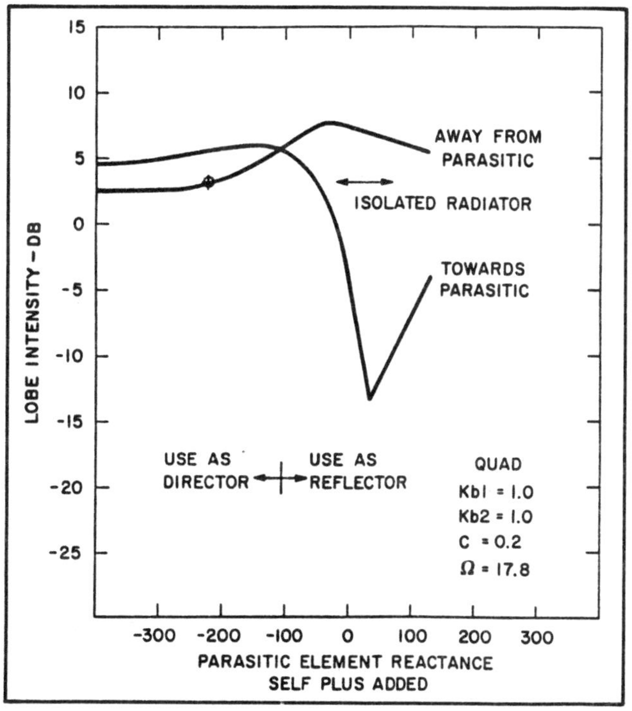

Figure 11. Lobe intensities for a spacing of 0.2 wavelength. The maximum F/B ratio is reduced, but the tuning condition is less critical.

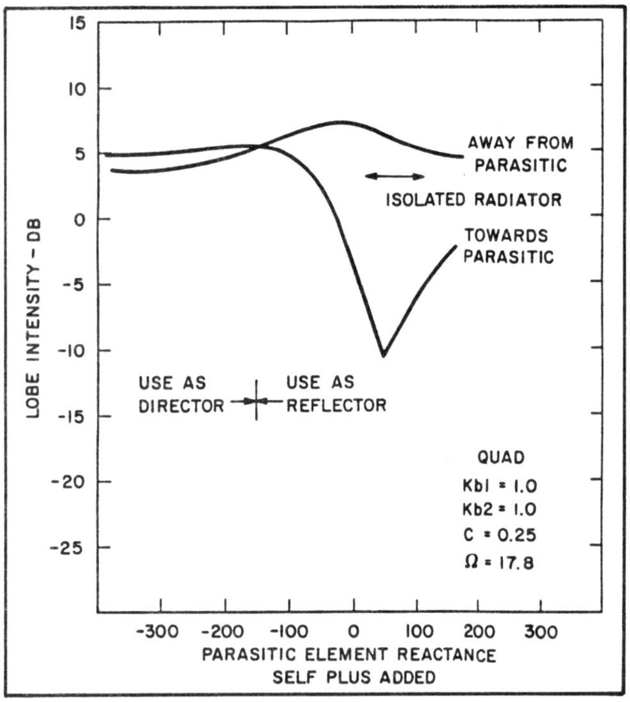

Figure 12. Lobe intensities for a spacing of 0.25 wavelength. The wider notch in the rear lobe indicates that the rear rejection is less affected by frequency shift than for the smaller spacings.

In each figure, the lobe intensity is shown as a function of total parasitic element reactance—self plus added. Both the lobe away from and the lobe toward the parasitic element are shown. Gain and F/B ratio depend on the tuning. The region of director and reflector use is shown. Performance can be evaluated by comparison with the gain of an isolated radiator (also shown).

Largely, these curves confirm the previous conclusions. *Most gain is secured with the reflector-radiator combination.* There is a small improvement in gain at the 0.1-wavelength spacing. However, best F/B ratio occurs for a spacing of 0.2 wavelength. The maximum calculated F/B ratio is very high—over 35 dB.

Unfortunately, the F/B ratio is high only over a narrow range of loop reactance, which also means a narrow range of frequencies. This is common to all parasitic antennas (see, for example, the curves for Yagis given by Lawson[19]). If good F/B performance over a wide range of frequencies is important, the best design would use a larger (more inductive) loop at wide spacing, as illustrated by Figure 12.

These four curves can be used as "universal" quad performance design curves. The procedure is as follows:

- Establish the design criteria, maximum gain, best F/B ratio, best gain bandwidth, or best F/B bandwidth.
- Select the spacing that provides a compromise which seems to best fit the requirements.
- Establish the parasitic tuning and the reactance which seems to be the best fit to the requirements.
- Select the conductor size using mechanical considerations.
- Determine the parasitic size, using the parasitic reactance to enter Figure 6 from Chapter 4.
- If possible, confirm the design choices by inserting the design values into MININEC runs. (Small improvements may be found by varying design choices slightly.)
- Convert the per unit dimensions to physical dimensions at the design operating frequency.

Even without the MININEC recalculation, these steps should produce an antenna very close to calculated performance for main-lobe gain. The maximum F/B ratio will be close to the calculated value, but may be at a different frequency than that calculated. This is sometimes due to calculation approximations, but the practical factors of element support, feed-line interaction, and tower influence are usually far more important. If maximum F/B ratio at a specific frequency is needed, provide an adjustable tuning stub in the reflector, or expect to resize the reflector after measuring the frequency of lowest rear radiation.

Quad Examples

Until recently, quad design data has been empirical, based on scale-model tests or on full-scale "cut and try" attempts. One such set of design data is given by Orr.[22] This work is based on tests of scale models plus a number of full-size antennas. For an antenna without stubs, Orr's data is summarized by the following factors in wavelengths (Chapter 3 of Reference 22, Figure 16):

radiator perimeter = 1.016
reflector perimeter = 1.049
spacing = 0.12

The wire size isn't stated. These values have been run on MININEC, using the wire radius of 1/2500 wavelength mentioned in Reference 22 as typical wire size.

The gain performance of this design is shown in Figure 13. Attainable performance is good: 8.0 dB for-

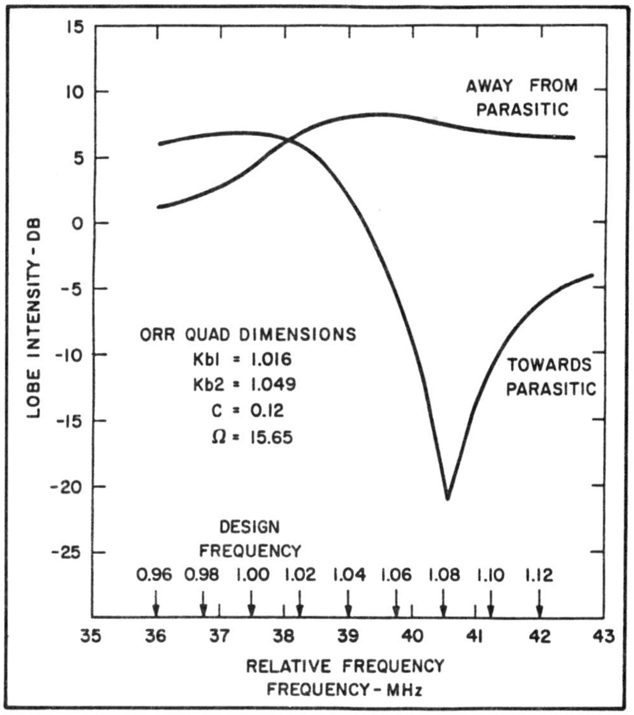

Figure 13. Lobe intensities for a two-element quad based on dimensions given in Reference 22. The data indicates that best gain is given by a reflector of 1.096 wavelengths, and best F/B ratio with a reflector of 1.13 wavelengths, for the conductor size assumed. See text for change in conductor size.

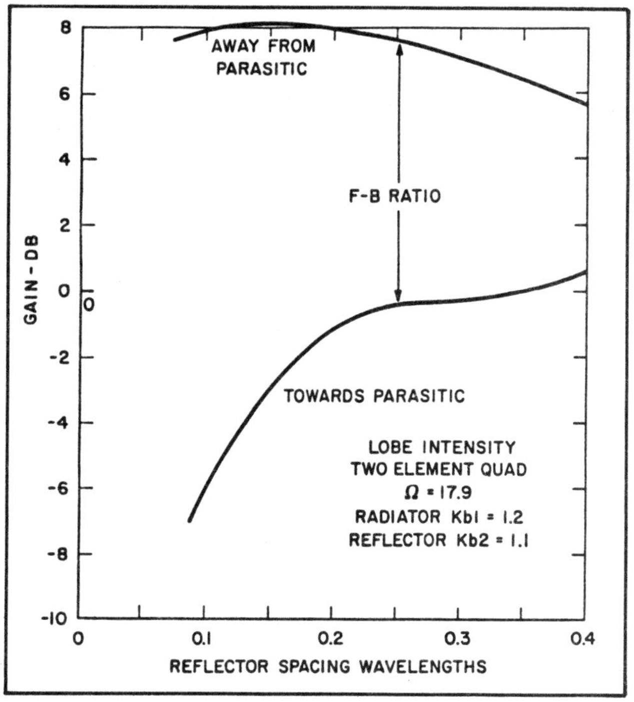

Figure 14. Lobe intensities for a 1.2-wavelength large-radiator quad, following the recommendation in Chapter 3. The larger loop does give a small increase in gain.

ward gain and 24 dB F/B ratio. The gain is slightly higher than that shown by Orr,[22] and the F/B ratio is essentially identical.

Differences occur in the MININEC results for frequencies of maximum gain and minimum F/B ratio. This could be a matter of difference in conductor size, the use of band-edge instead of band-center dimensions, or the influence of tower and feed-line supports on experimental values. In any event, the MININEC calculation shows that the reflector size should be about 1.08 for maximum gain and about 1.12 for maximum F/B ratio. Both these values are for the design frequency, not the band edge. With this change, or with the retuning suggested by Orr,[22] or with elements resized after measurement, this is a good design.

Chapter 3 includes a recommendation that the radiator be sized at 1.2 wavelengths for maximum gain. The effect of this for square loops is summarized in Figure 14 for a specific reflector size. There's a slight increase in forward lobe gain as compared to a 1.0-wavelength radiator (see Figure 8), but not as much as is implied by the increase in radiator area. The F/B ratio is nearly unchanged.

Figures 15 and 16 investigate the large-loop radiator even further. The 0.15-wavelength separation gives excellent gain and F/B ratio (Figure 15). At 0.2-wavelength separation, the gain is nearly as good, but the F/B ratio is lower. However, as for the 1.0-wavelength radiator, the bandwidth for reasonable F/B ratio is better at the wider spacing.

The performance of these large-loop quads can be studied further by looking at the four pattern plots of Figure 17. The forward gain at the closer spacings is impressive. The beamwidth is only 72 degrees—very good for a two-element antenna. The rear lobe is fairly small; it can be reduced at a particular frequency by changing reflector size. In this figure, the pattern for 0.2-wavelength spacing is for the total radiation, not just the horizontally polarized component. Comparison with the 0.1-wavelength spacing shows the fill-in at right angles to the loop. The effect on the main lobe is negligible, and very small on the back lobe.

Practically, having an oversize radiator can be a disadvantage because of increased size. A possibility is to make the radiator and reflector the same size. The effect is shown in Figure 18. These data are for elements of 1.1 wavelengths perimeter, at a spacing

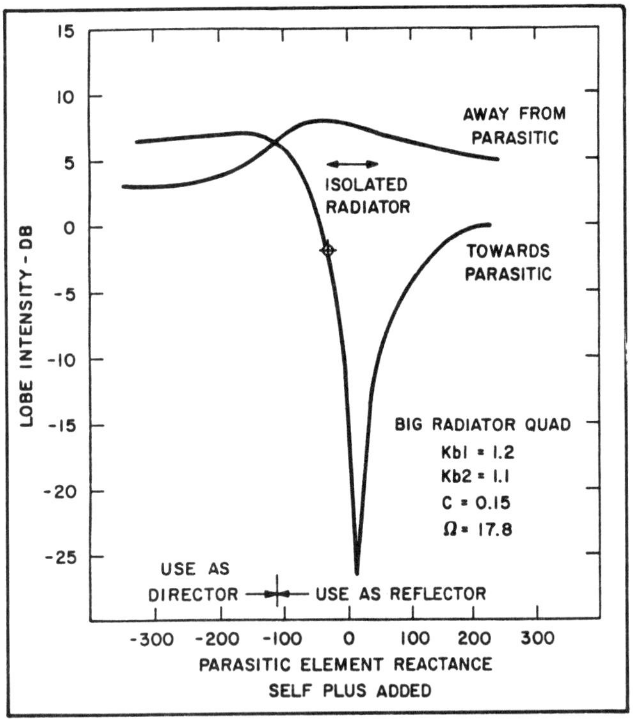

Figure 15. Lobe intensities as functions of parasitic tuning for an element spacing of 0.15 wavelength in a large-radiator quad. A compromise design will give gain of 7.5 dB and a F/B ratio of about four S-units.

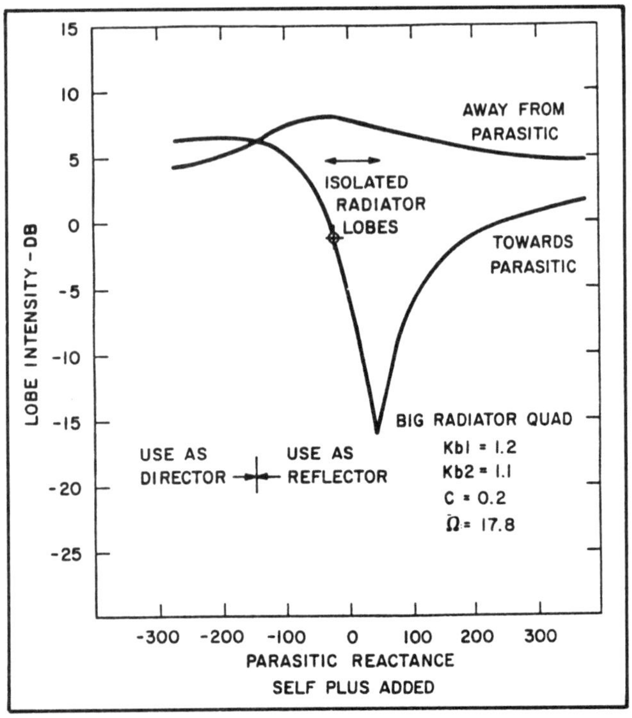

Figure 16. Same as Figure 15, but for a spacing of 0.2 wavelength. Gain is nearly as good, but F/B ratio has decreased: the bandwidth for moderate F/B ratio has increased.

of 0.15 wavelength. Note that this curve is plotted against frequency rather than against parasitic-element reactance, thus giving a good picture of the bandwidth. For gain it is very good, decreasing only 1 dB for a change of ±4 percent in frequency. This is about twice the bandwidth of a two-element Yagi. The F/B bandwidth is also good, a total of about two percent for 20 dB and nearly four per cent for 15 dB. Gain is as good as any other two-element design, and the F/B ratio is better—over 35 dB maximum. Other spacings were tried, but 0.15 wavelength spacing was found to be best.

Figure 19 shows the behavior of this semi-large radiator quad with the reflector tuned by adding reactance. As a demonstration of the validity of the technique of adding self- plus added reactance, three sets of points are shown for three different conductor sizes. The departure of the points from a by-eye average curve is just noticeable over a small range of reactances. Note that a very high F/B ratio can be obtained with any of the wire sizes chosen. Because of its good performance, this combination will be used as the exciting section of large quads, created by adding directors. These are considered in Chapter 6.

Quad Drive Impedances

Power must be fed to the antenna. This is easier if the drive resistance is a reasonable value and the reactance is either inherently low, or easily corrected to zero. The overall picture for quads can be seen in Figures 20A and 20B, and Figures 21A and 21B. The first of these figures shows the change in impedance, as spacing is changed, for a pair of 1-wavelength loops. Very close spacing gives low drive resistance and capacitive reactance—a consequence of the radiator being below resonance. Increasing the spacing increases the resistance, but has little effect on reactance. Both values oscillate about the value of a single isolated loop.

Changes in element size affect both drive resistance and reactance (as well as gain and F/B ratio). Figures 21A and 21B show this for changes in parasitic element size. Change is greatest for close spacings, and is more for reflector tuning than for the smaller (director) loops. (This means that drive impedance won't be changed radically as directors are added to form a higher-gain, multi-element quad). Change in radiator size has very little effect on the drive resistance. The drive reactance does change, practically by the same amount as for an isolated loop.

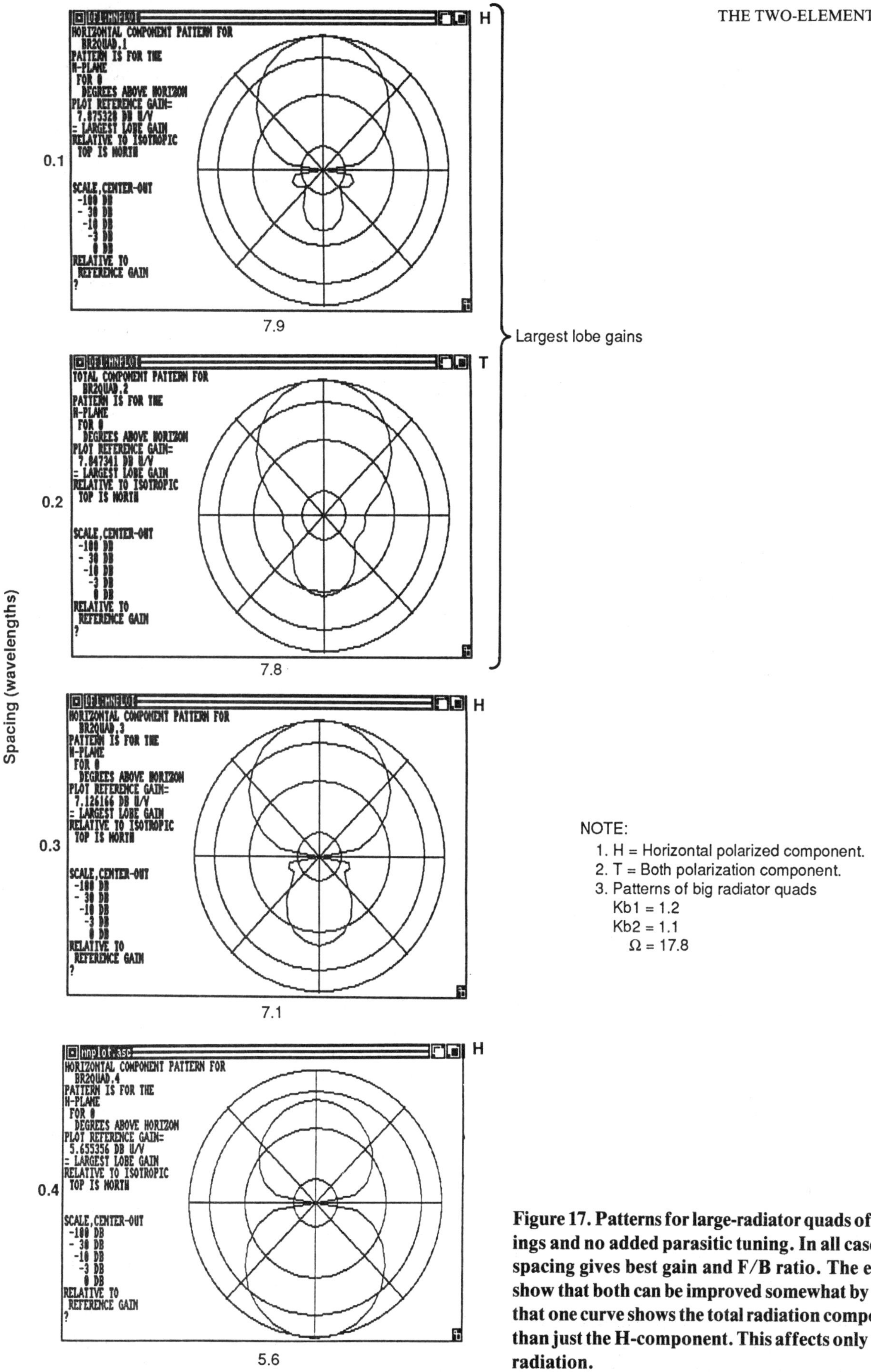

Figure 17. Patterns for large-radiator quads of several spacings and no added parasitic tuning. In all cases, the closer spacing gives best gain and F/B ratio. The earlier curves show that both can be improved somewhat by tuning. Note that one curve shows the total radiation component, rather than just the H-component. This affects only the side-lobe radiation.

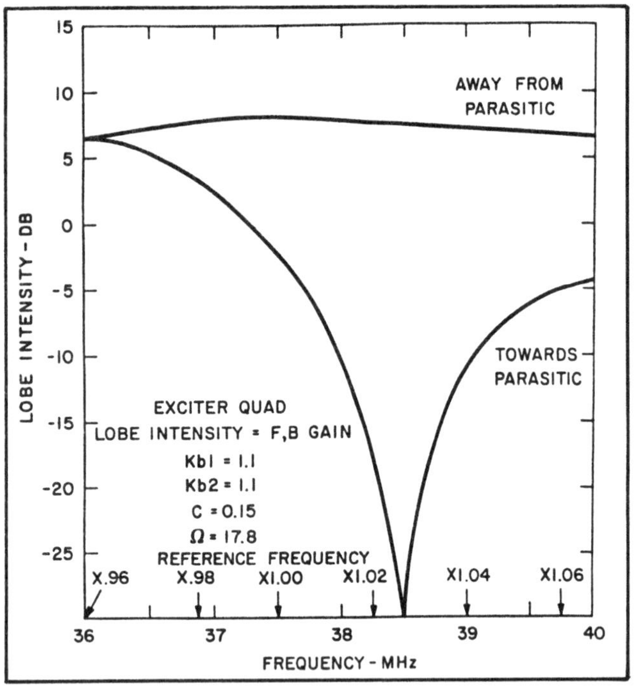

Figure 18. Lobe intensities for a semi-large radiator quad with radiator and reflector sized to 1.10 wavelengths perimeter. Maximum gain occurs at the nominal frequency, but best F/B ratio is about 2.5 percent higher. A good compromise point is 1 percent higher than the nominal frequency. For example, for 20 meters, use 13.9 MHz as the nominal frequency, and obtain maximum gain at MARS/CW frequencies, and maximum F/B ratio at 14.25 MHz.

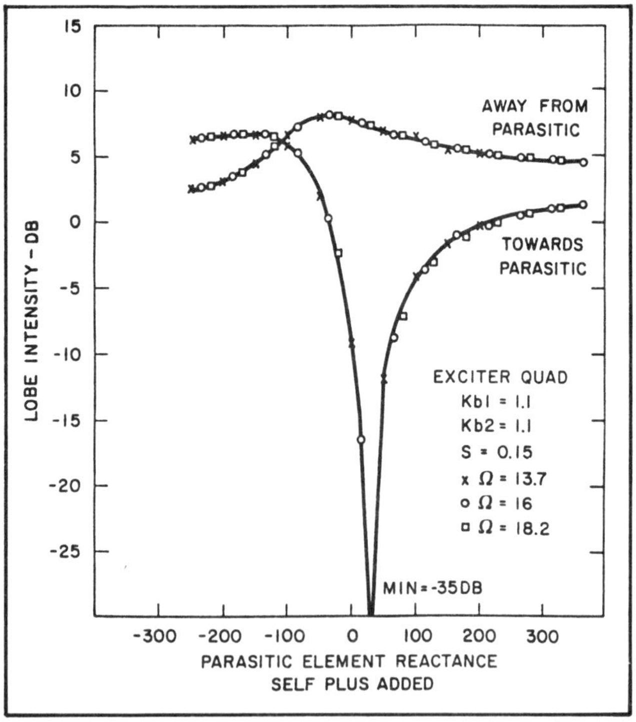

Figure 19. Lobe intensities for a semi-large radiator quad as a function of reflector tuning. A slight increase in gain is possible with a smaller reflector. Maximum F/B ratio requires a larger reflector. The 1.1-wavelength size is a good compromise value, and has been chosen as the excitation section for further work on multi-element beams.

To determine the input impedance of a specific design, read the resistance and reactance from Figure 20, interpolating between curves if needed. Also read the difference between the reactance of a 1.0-wavelength loop and the radiator loop size of the design, using Figure 6 from Chapter 4. Apply this difference as an algebraic correction of the curve reactance. So calculated, the drive resistance will be close to the measured value for a practical antenna. Drive reactance may show variations for the reasons mentioned previously for F/B ratio differences.

Matching techniques will be covered in another chapter. For now, it's sufficient to note that any of the common matching devices can be used.

Specific Design Impedances

The input impedance of the two-element quad sized to the values found in Reference 22 is shown in Figure 22. There's appreciable variation of resistance and reactance across the range of frequencies shown. With a fixed matching device, bandwidth will usually be limited by the allowed VSWR range. (The gain bandwidth can be extended markedly if variable matching elements are used.)

The impedance variation of the large-radiator quad is plotted against element spacing in Figure 23, and against parasitic tuning in Figure 24. A wide variation is again noticeable. It must be remembered that the wide variations of Figures 19, 20, 22, and 23 will be encountered in design. They can be compensated for by other design choices. For example, Figure 24 shows the input impedance versus element tuning; Figure 25 shows the input impedance versus frequency for the exciter quad having 1.1-wavelength elements. A matching section designed for 75 + j110 ohms will have 1.0 VSWR at the operating frequency. At the upper limit of the curve, the feed impedance would be 250 + j200 ohms—corresponding to impedance and VSWR ratios of less than 2.5:1. It's true that this is above the values of 1.5:1 or 2.0:1 often considered acceptable. But with low-loss line and a match-box at the transmitter, this is a perfectly acceptable VSWR. Of course, there would be some variation in gain across this wide frequency range, and the rear lobe

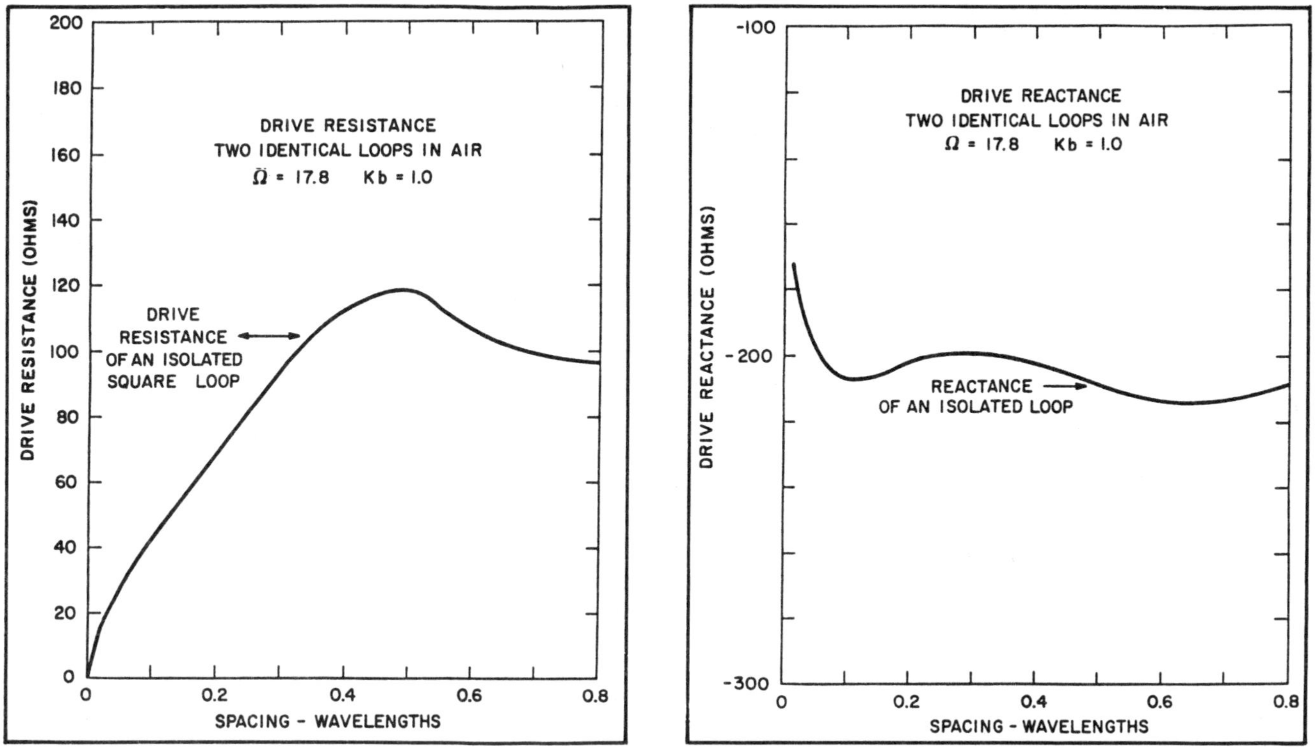

(A) (B)

Figure 20. Drive impedance for two identical loops in air, as a function of loop spacing, (A) = resistance, (B) = reactance. At larger spacings, the drive impedance oscillates about the impedance of an isolated loop.

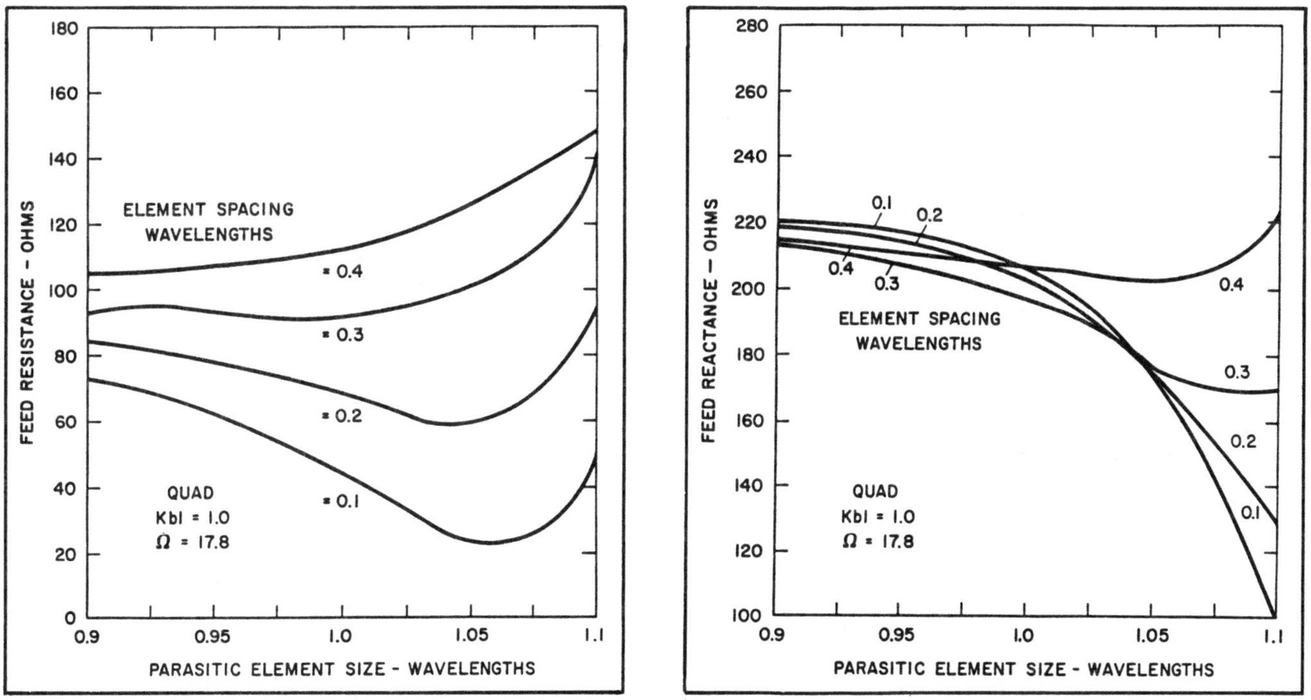

(A) (B)

Figure 21. Drive impedances of a one-wavelength radiator as functions of parasitic element size, for several element spacings. (A) = resistance, (B) = reactance. See text for application.

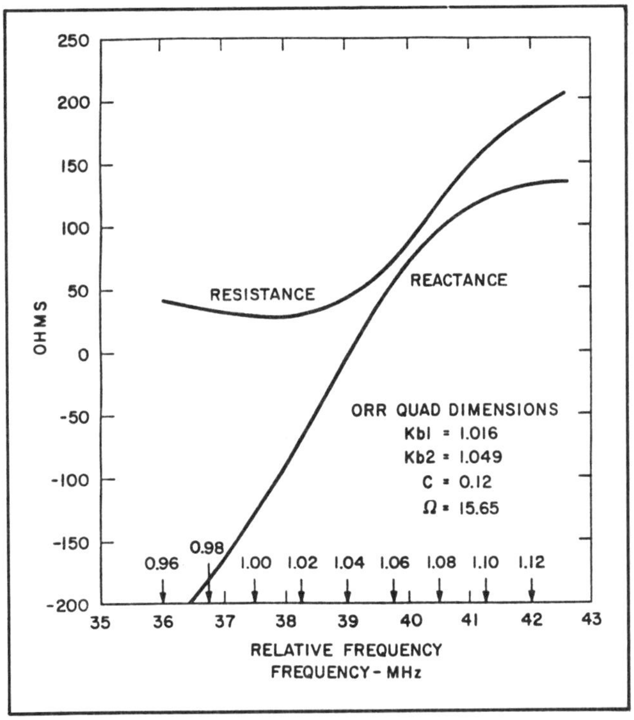

Figure 22. Drive impedance of a quad with dimensions suggested by Orr,[22] as a function of frequency. See text.

would change markedly in size. But the opportunity for operation across an entire band is the most neglected factor in use of quad antennas.

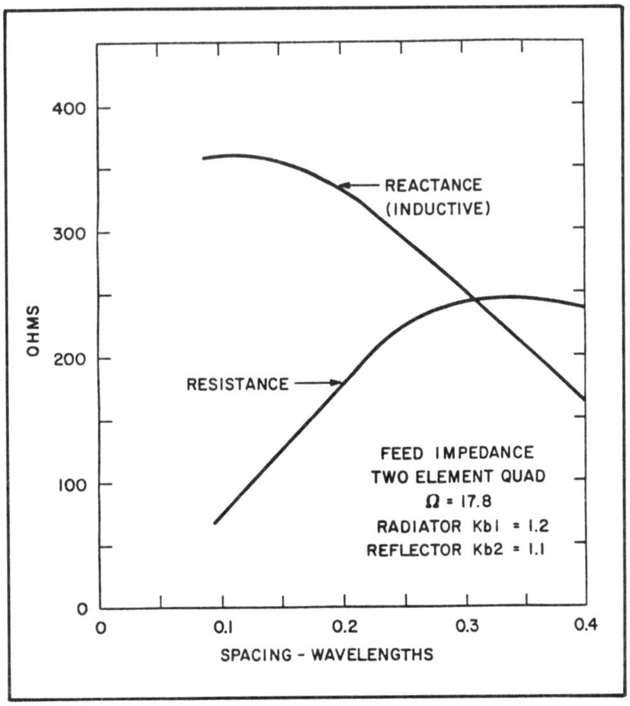

Figure 23. Drive impedance of a large-radiator quad as a function of spacing.

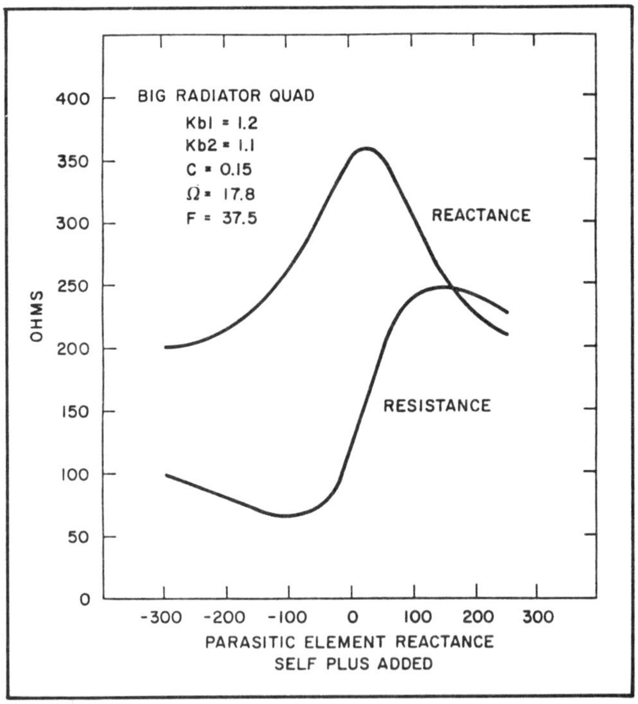

Figure 24. Drive impedance of a large-radiator quad as a function of parasitic tuning. Use this and the preceding figure to estimate drive conditions, and see text for a discussion of expected accuracy.

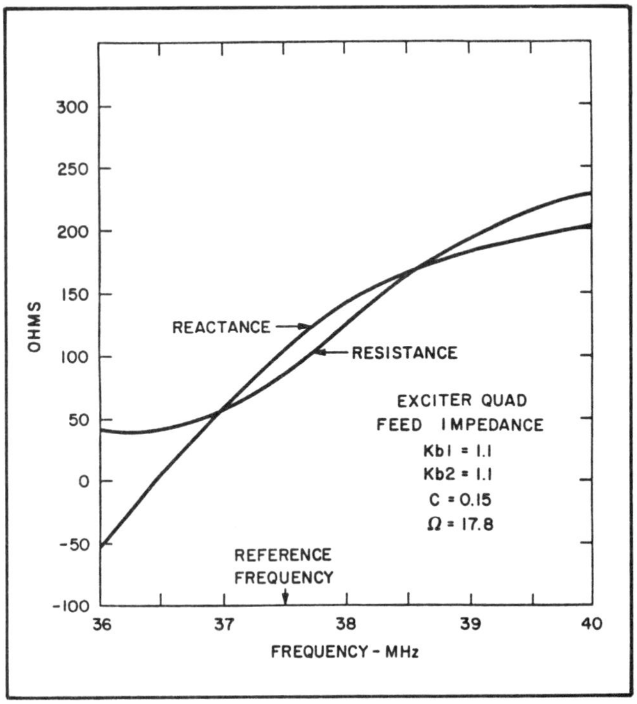

Figure 25. Drive impedance of a semi-large radiator quad as a function of frequency. If the reactance component is canceled by an open stub at the feed point, a 2:1 balun will give a good match to 50-ohm line.

Chapter 6

Multi-element Quads

Performance data for quads with multiple elements

In Chapter 3, Figure 8 showed the gain of a number of circular loop arrays as a function of array or boom length. Part of this figure is reproduced here in Figure 1, but the gain values plotted are for square loops with theoretical values for Yagis developed by Lawson.[19] This figure summarizes the results presented in this chapter. It's also a useful design tool, providing a view of the array size needed for a given design gain.

As you look at Figure 8, note that the gain of quad arrays is consistently above the theoretical gain of Yagis of the same boom length, by about 1 dB. Many have reported this finding. The curve confirms the usual claim that a quad array has about the same gain as a Yagi with one more element. In the smaller arrays, this is a large percentage difference—two versus three elements, or a 2:1 difference in boom length.

At the upper end of the plot, the values shown are for "quagis" rather than for quads. This isn't an indication of lack of usefulness of very large quads, but a reflection of the limits of the available computers. A five-element quad, analyzed to any degree of accuracy using MININEC3, fills a 640-K machine. The linear rods of a quagi director require less memory for the geometry and impedance arrays, allowing analysis of 10 to 12 element quagis to the same accuracy level. Further work on very large quads requires more computer memory and more speed.

When comparing Figure 8 of Chapter 3 and Figure 1 of this chapter, note that the gains of both circular and rectangular arrays are above Yagi gain by essentially the same amount. This occurs despite the theoretical result that the circular loops will have more gain because the radiating area is greater. It appears that this similarity is a result of the analysis course fol-

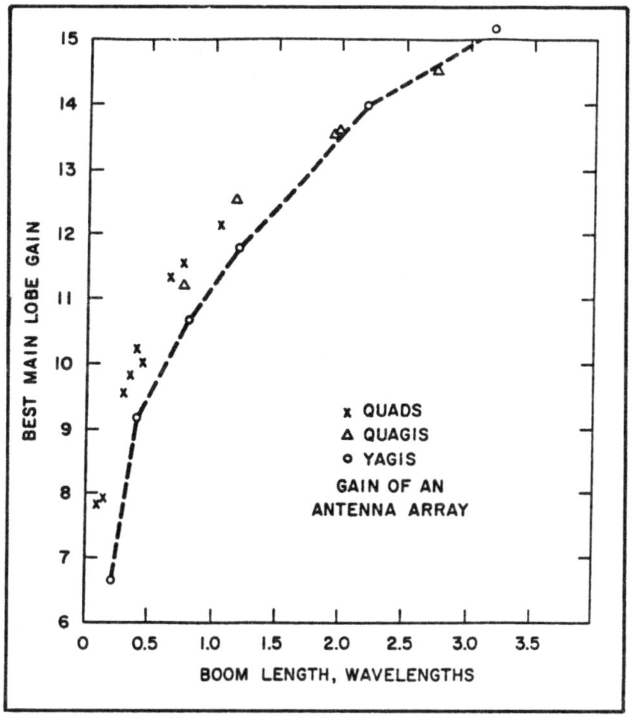

Figure 1. Gain of parasitic arrays versus boom length. Dashed line shows theoretical values derived by Lawson.[19] Points are from MININEC calculation of specific quad and quagi designs. Use this curve for first estimate of size and number of elements. Maximum-gain quad designs will lie about 1 dB above the Yagi line. Compromise designs are usually between this gain and the Yagi line.

lowed. The circular loop data is for a series of loops of specified size and spacing. This analysis of square loops, includes the additional step of determining the condition for maximum gain, which is the point plotted.

Note, also, that the upper or quagi points are essentially on the Yagi curve. This may be correct, as it seems reasonable that a quagi should approach Yagi performance as the number of directors is increased. However, no search for best quagi operation has been made. This book focuses on quads, though aspects of mixed quad-Yagi design are covered.

As a convenience, the square-loop reactance data of Figure 6 from Chapter 4 is replotted here as Figure 2. The origin has been moved to better suit the range of values needed in reflector and director design. Note that the curves for the thicker conductors show curvature on these plots. (See the analysis of circular loops and the curves in Chapter 4.)

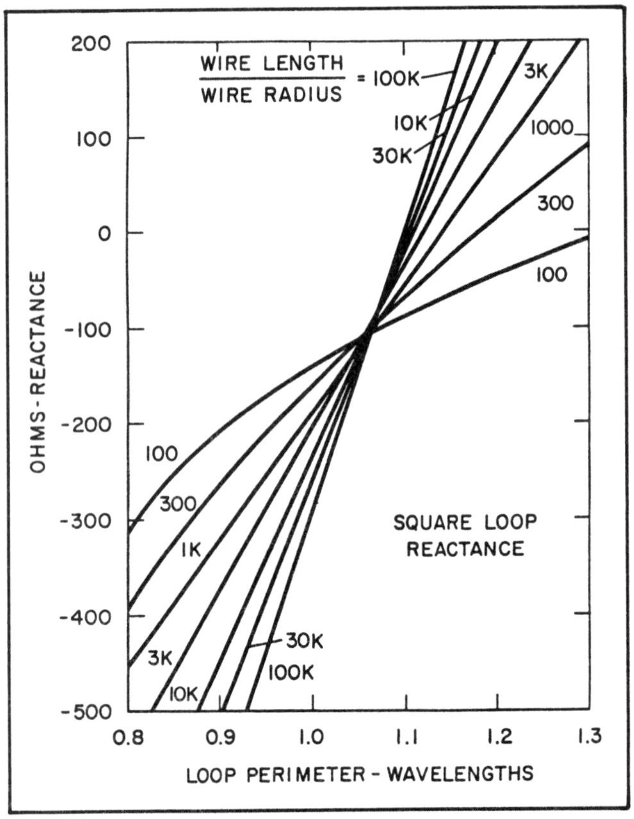

Figure 2. Rearrangement of the data for Figure 6 of Chapter Four to eliminate the step of conversion of loop dimension ratio to the parameter omega. Use to determine the size change needed to compensate for change in conductor size for reflectors and directors, and to estimate the size change of radiators needed to give a pure drive resistance at a given frequency. (See text for scaling rules.)

The Three-element Quad

Figures 3 through 6 give a good picture of quad performance factors, and are also basic design curves. All use the radiator-reflector exciter section whose performance is shown in Figures 18 and 19 of Chapter 5. All are for a director perimeter of 0.9 wavelength.

Comparison of these figures reveals the following:

- The maximum gain increases as the director spacing increases. This is a consequence of the increase in boom length. However, data not reproduced here shows that still wider director spacing will give a *reduction* in gain.
- Maximum gain and maximum F/B ratio involve different director tuning.
- As expected, there is an interaction between spacing and director tuning for both gain and F/B ratio. F/B ratio is much more affected than gain.
- There are marked pattern (and impedance) changes as the director circumference approaches resonance.
- Gains are not highly sensitive to director tuning for large changes in tuning in the capacitive direction.

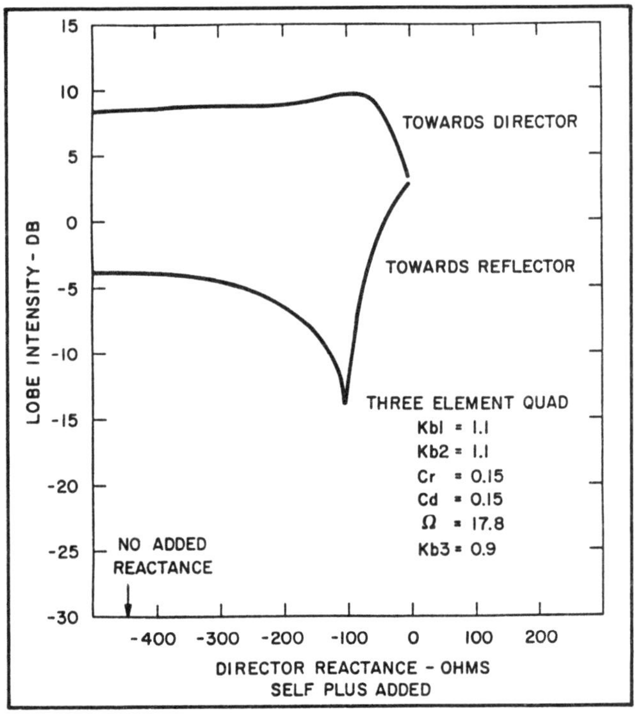

Figure 3. Forward and back-lobe intensities of a director 0.9/0.15, reflector 1.11/0.15 wavelength, three-element quad versus director tuning. The reflector-radiator combination is the "exciter quad" of Chapter 5.

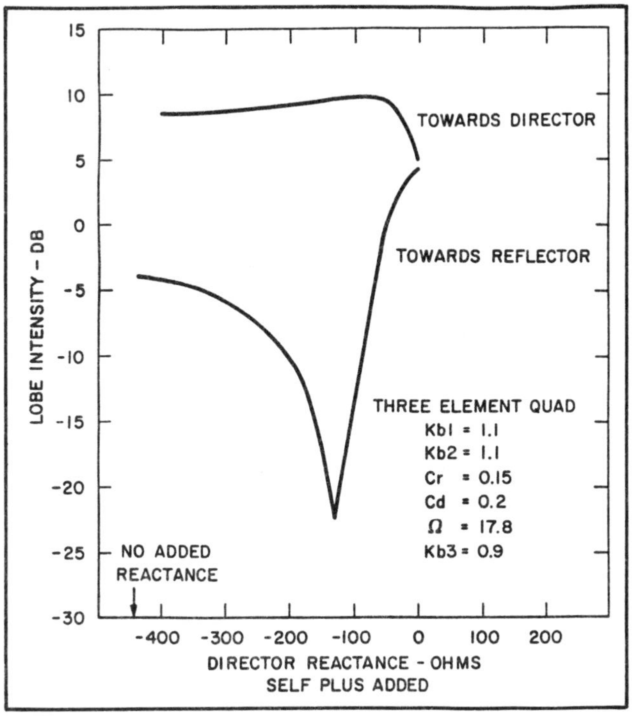

Figure 4. Same as Figure 3 for 0.2-wavelength director spacing. Note the improved F/B ratio and slightly increased gain.

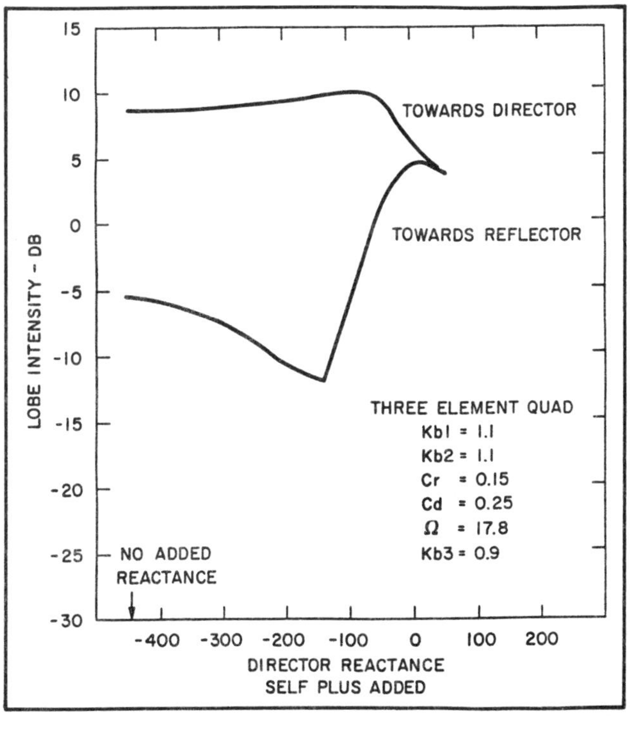

Figure 5. Same as Figure 3 for 0.25-wavelength director spacing. Gain continues to increase, but F/B ratio isn't as good.

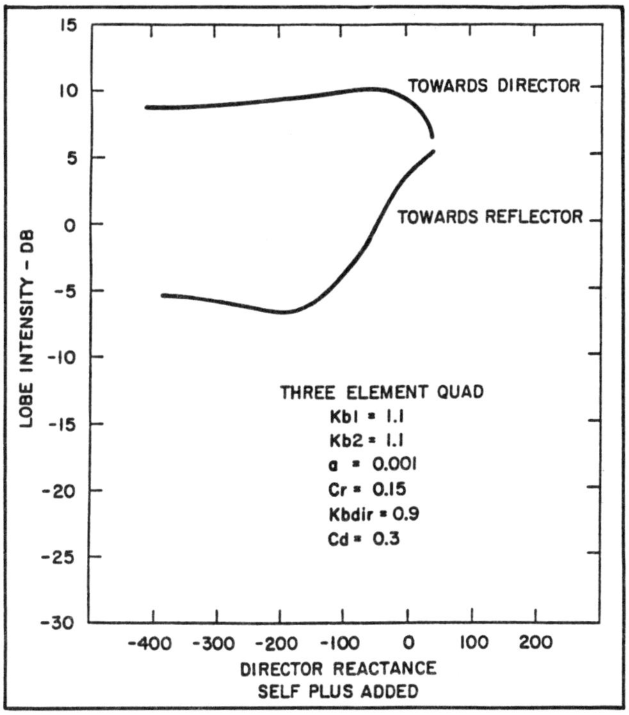

Figure 6. Same as Figure 3 for 0.3-wavelength director spacing. Gain remains the same, and F/B ratio is further degraded. Still wider spacings cause loss of gain and lower F/B ratio.

Figure 7 shows the effect of reflector tuning for a single example of director tuning and spacing. Again, as expected, this shows that gain and F/B ratio are also affected by reflector tuning. Good F/B ratio can be secured for any given spacing by a change in tuning, although the range of conditions giving very high F/B ratio is quite limited.

Figures 8 through 11 show the lobe intensities as operating frequency is varied—the first three for a 0.9-wavelength director perimeter, and the last for 0.95. Good to excellent F/B ratio performance is possible, as well as good gain. These curves also show the bandwidth for a given performance.

These figures use a frequency of 37.5 MHz as the unit frequency. The 37.5-MHz value was selected because of the geometry of quads. For this frequency, a one-wavelength loop has corner coordinates $x = \pm 1$, $y = \pm 1$. For a 0.9-wavelength director, the coordinate pairs are 0.9/0.9. This close correspondence between geometry and dimensions simplifies setting up the problem for analysis, checking for errors, and keeping track of results. Techniques for transferring designs to desired frequencies are shown later.

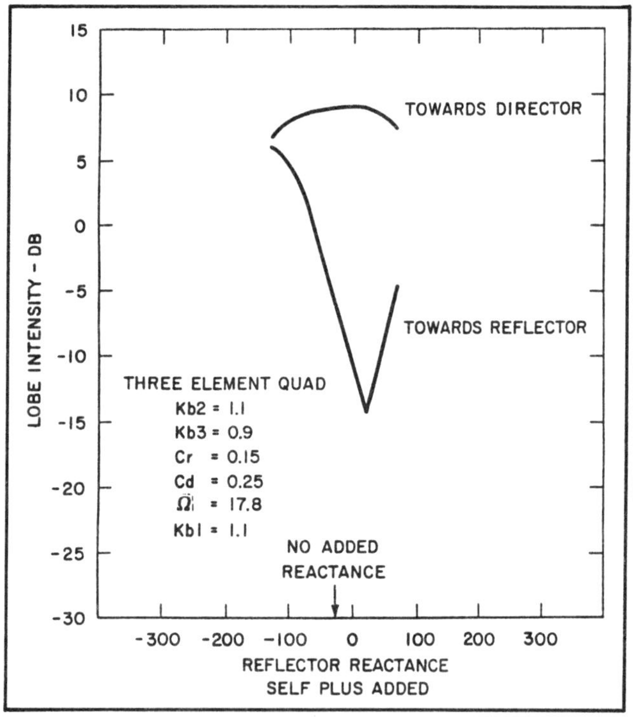

Figure 7. Similar to Figure 3, but with variable tuning of the reflector. Gain is near maximum with no added reactance, but F/B ratio improves somewhat with added inductive reactance.

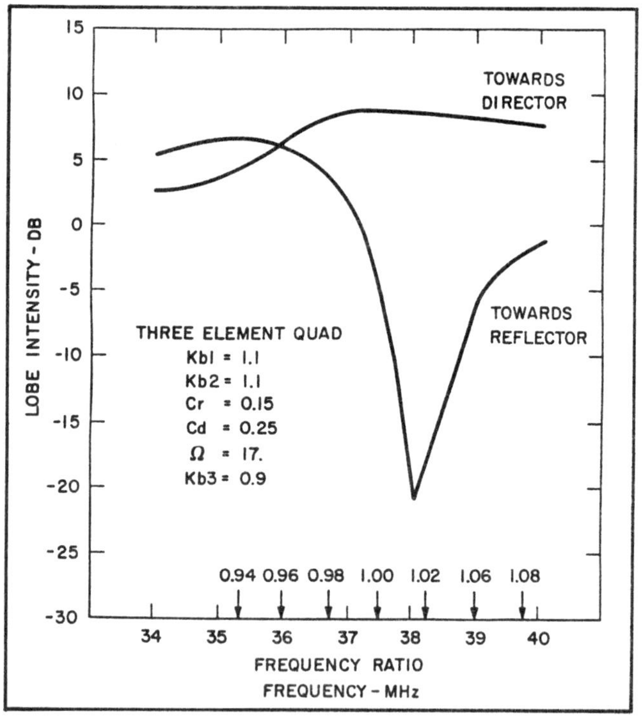

Figure 9. Similar to Figure 5, but for changing frequency. See caption for Figure 8.

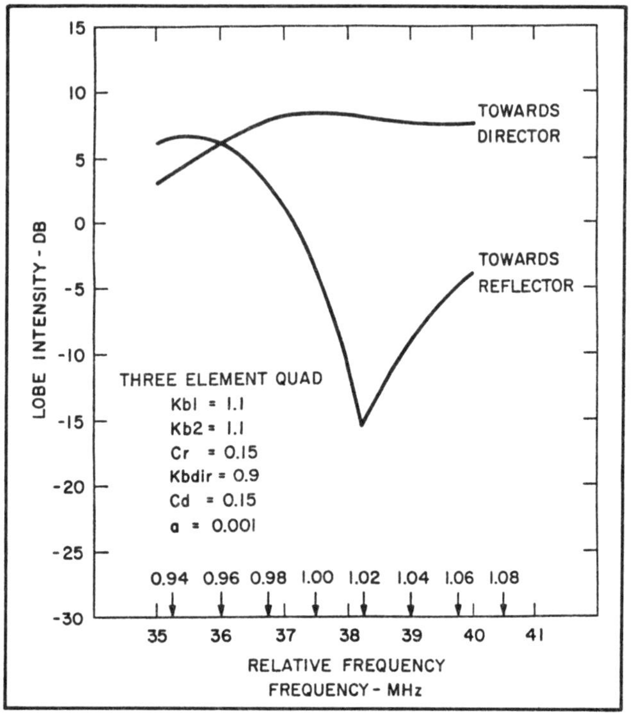

Figure 8. Similar to Figure 3, but for changing frequency. Gain is near maximum at the design frequency, but F/B ratio improves at a slightly higher frequency.

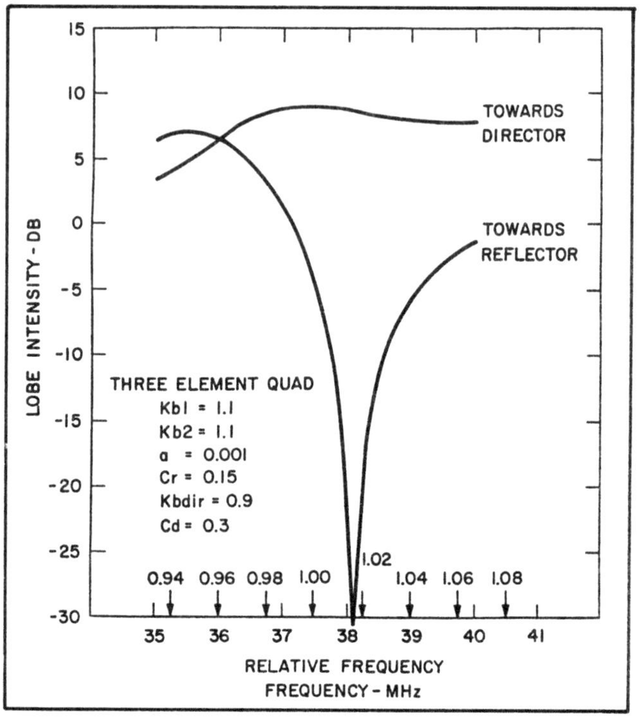

Figure 10. Similar to Figure 6, but for changing frequency. See caption for Figure 8, and later text comments on rear-lobe structure.

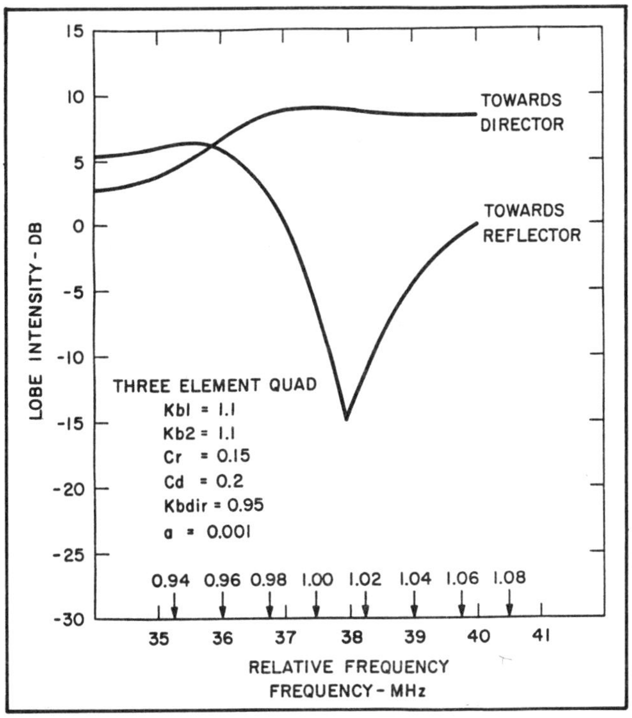

Figure 11. Similar to Figure 4, but for changing frequency and 0.95 reflector perimeter. See caption for Figure 8.

Note that the design with a 0.9-wavelength director at 0.3-wavelength spacing has an extremely high F/B ratio. This seems to be a characteristic of parasitic antennas. For Yagis, Lawson,[19] pages 2-8 and 2-9, shows a markedly better F/B ratio for a spacing of 0.25 to 0.3 wavelength. Though less marked, the circular loop data summarized in Table 1 of Chapter 3 shows the same trend. Since good gain and F/B ratio performance is attainable with the specific values studied, no attempt has been made to expand the range of conditions, or to find the best of all possible operating points.

I should make one point about the F/B ratio and bandwidth relationship. For a modest F/B ratio, say 15 dB, smaller spacings give at least this F/B ratio over a wider band than wide element spacings. This is one of the design choices to be made.

Four or More Element Quads

Except for a few HF "super antenna" installations, it seems that quads with a large number of elements will be used on the VHF, UHF, and even SHF bands. For these, the usual design problem is the gain required. Usually, a high F/B ratio isn't needed, and there is less emphasis on physical size. For these reasons, the investigation of larger quads has concentrated on the wider director spacings.

Figures 12 and 13 show the lobe intensities versus director tuning for a director perimeter/spacing of 0.95/0.25 and 0.9/0.3 wavelength, for four-element quads. Figure 14 is for 0.9/0.3-wavelength directors and a five-element quad. All use the 1.1/1.1 reflector/radiator combination spaced 0.15 wavelength.

These curves show a strong relationship to the corresponding curves for three elements, but there are some differences. For one thing, the range of tuning conditions giving near-maximum gain is smaller. As the maximum-gain tuning point is passed, gain change is much more rapid in the multi-element arrays. And, there is less change in the F/B ratio with tuning. Figures 15 and 16 show the lobe intensities with varying operating frequencies for the four-element designs. Figure 17 gives the same information for the five-element design. These show, again, that the best gain and F/B ratio occur at different frequencies. More important, they indicate that there is very little change in the point of best tuning as elements are added. This

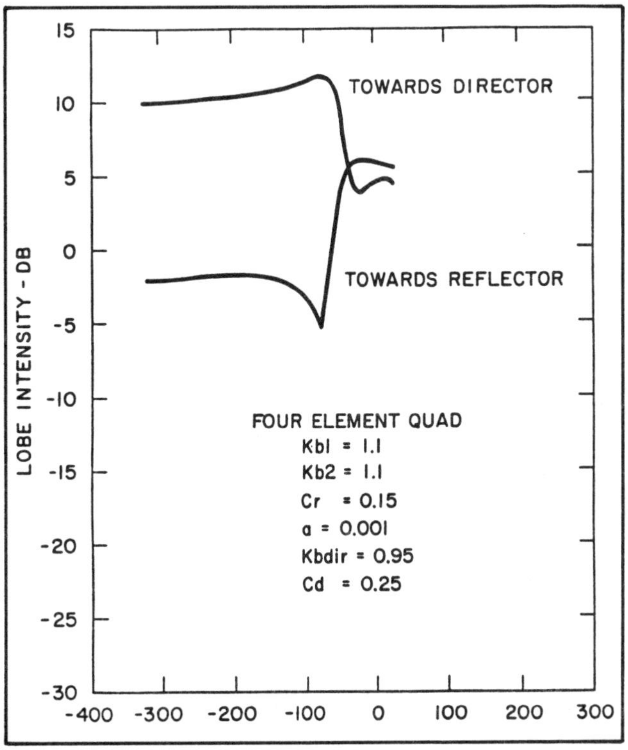

Figure 12. Lobe intensity of a four-element quad as a function of identical tuning of the two directors, with 0.95 directors spaced 0.25 wavelength. As compared to a three-element design, gain has increased, but F/B ratio is less affected.

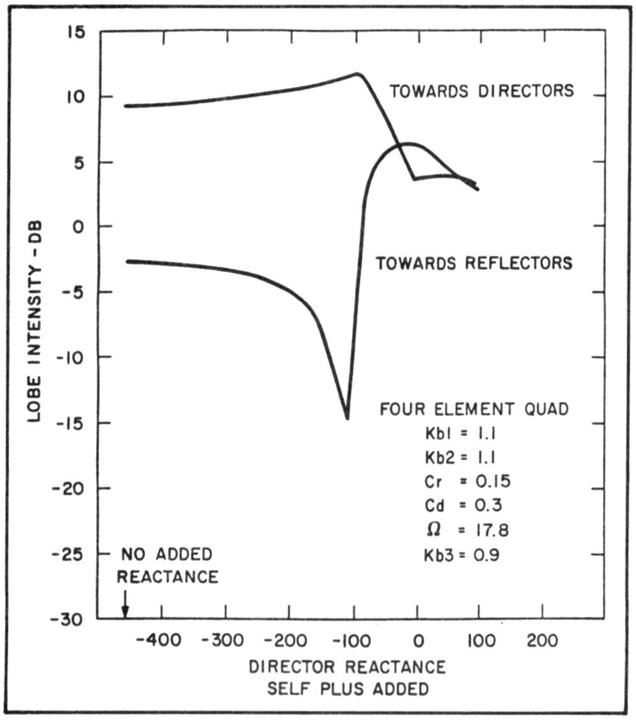

Figure 13. Similar to Figure 12, but for a 0.9/0.3 director perimeter/spacing. Note the sharp changes in gain and F/B ratio as the directors approach self resonance. Directors must remain capacitive to function.

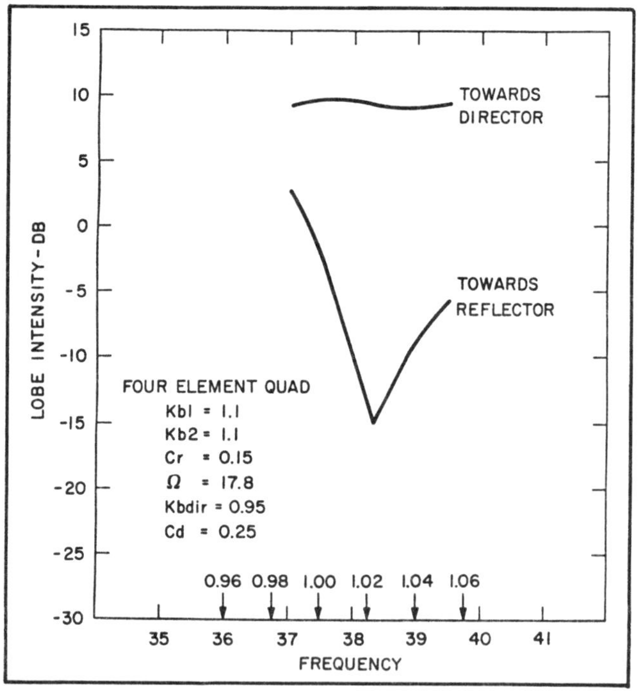

Figure 15. Similar to Figure 11, but for the four-element 0.95/0.25-wavelength director quad. Gain is maximum at the design frequency, but F/B ratio is best at 2 percent higher frequency.

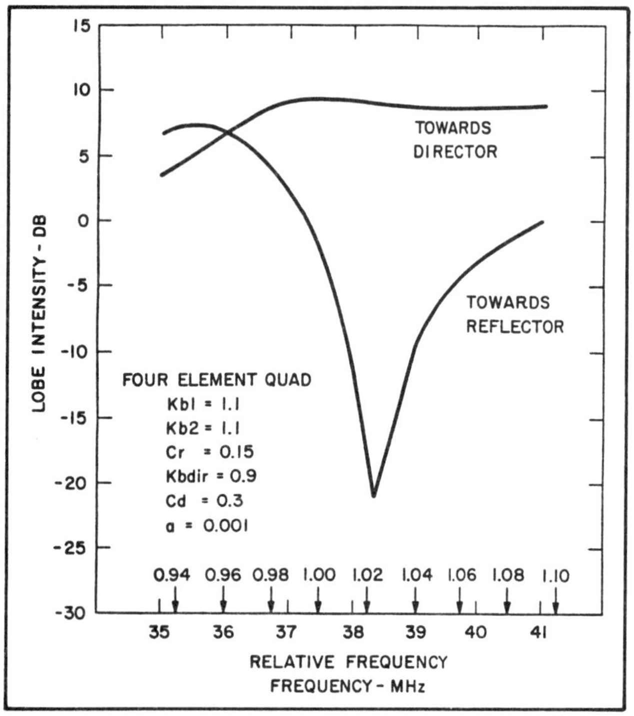

Figure 14. Similar to Figure 13, but for a five-element quad with 0.9/0.3-wavelength directors. Tuning primarily affects gain, and shows the same sensitivity near director self resonance.

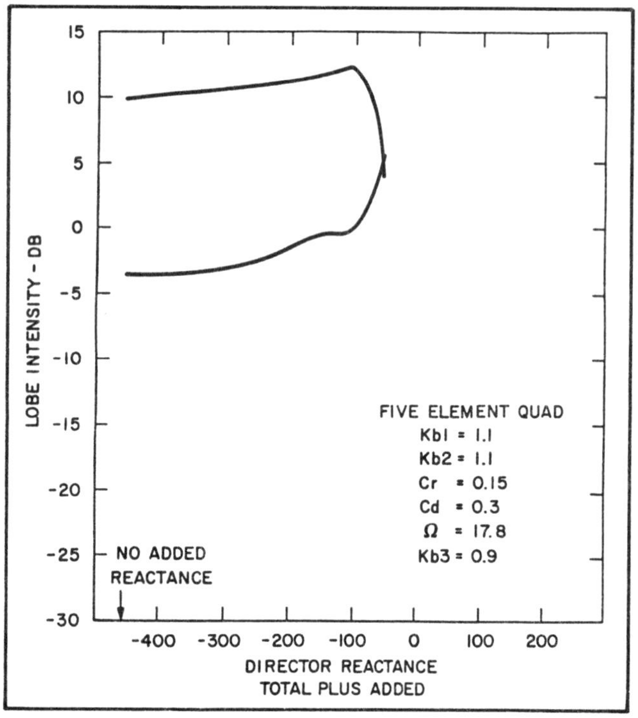

Figure 16. Similar to Figure 15, but for the four-element, 0.9/0.3-wavelength, director quad. Gain is best at 1 percent below and F/B ratio at 2 percent above the design frequency.

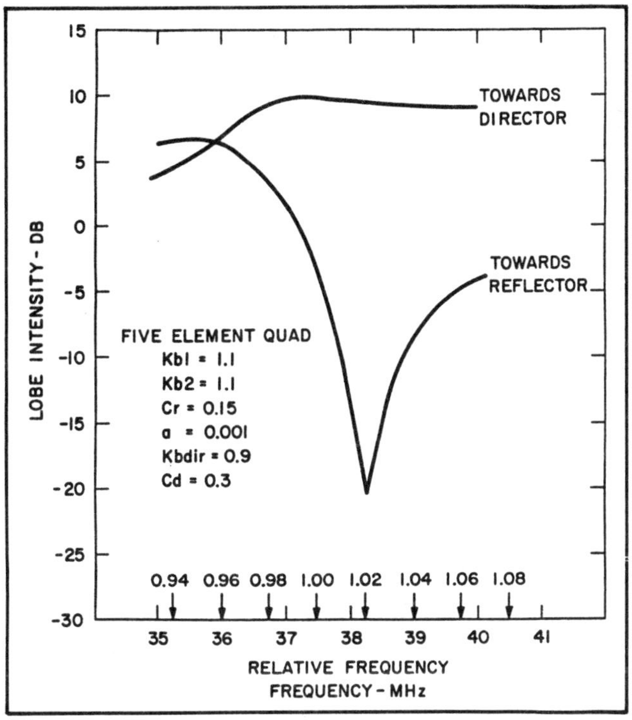

Figure 17. Similar to Figure 16, but for the five-element, 0.9/0.3-wavelength director. Gain and F/B ratio conditions are nearly the same as in Figure 16.

result is to be expected, because the mutual impedance of elements decreases with increased spacing. Consequently, the furthest director has the least interaction with the reflector and radiator.

This factor is important because it allows the gain performance of still larger arrays to be predicted easily and with reasonable accuracy. The process is to add more directors of the same size, calculate the boom length, and read the gain from Figure 1. (The data of Figure 8, Chapter 3, or the results in NBS Technical Report 622 can also be used.) A level 1 dB above the Yagi curve gives the estimate for a maximum gain design; the estimate for the gain of a compromise design is about one-half dB higher.

Drive Impedances

So far, I've ignored the drive conditions of these multi-element designs. This is by intent, because the elements of gain, F/B ratio and bandwidth, and physical dimensions are almost always far more important. It is virtually certain that some form of transmission line matching will be needed for any practical design. Once this premise is accepted, the exact values to be matched are relatively unimportant.

Table 1 shows the calculated feed impedances for the designs shown in Figure 1. Most of the impedances are inductive, a result of using a 1.1-wavelength radiator to increase gain. Both resistance and reactance values vary over a range of about 5:1—usually somewhat less for reactance than for resistance.

As for two-element quads, ladder-type twinlead and a match box make a good low-loss feed with wide-band capabilities. Low-loss, high-voltage 75-ohm line; a stub at the antenna; and a coax-type match box of fair tuning range are nearly as good. If a simple feed is desired, design values can be chosen which allow direct feed to 50-ohm line. A stub at the feed point

PROTOTYPE QUADS DRIVE IMPEDANCE (OHMS)

Frequency (MHz)	3Q.15 Z(ohms)	3Q.25 Z(ohms)	3Q.3 Z(ohms)	a3Q.2 Z(ohms)
35.00	39 − j143	41 − j142	42 − j142	39 − j143
36.00	35 − j57	35 − j56	35 − j54	33 − j58
37.00	56 + j41	51 + j44	49 + j47	50 + j37
38.00	113 + j108	100 + j126	99 + j137	92 + j107
39.00	144 + j137	143 + j175	151 + j194	113 + j156
40.00	146 + j175	165 + j244	185 + j244	116 + j220

Frequency (MHz)	4Q.25 Z(ohms)	4Q.3 Z(ohms)	5Q.3 Z(ohms)
35.00		43 − j141	42 − j141
36.00		37 − j54	36 − j54
37.00	50 + j51	54 + j50	50 + j51
38.00	110 + j136	113 + j135	106 + j143
39.00	160 + j169	164 + j175	167 + j192
40.00	162 + j198	184 + j213	205 + j225

Table 1. Summary of the drive impedances of the quads in Table 1. Use as a guide to expected drive impedances as designs are scaled.

cancels residual reactance. Alternatively, the size of the radiator can be changed to cancel the reactance. The change in size is determined by using Figure 2.

Multiple Reflectors: Bi-directional Beams

A limited investigation of the use of two reflectors to improve the F/B ratio has been made. Both resistance and reactance loading were tried. The maximum improvement found was less than 3 dB. However, in investigating the Quagi, marked changes in lobe structure were found as the Yagi elements were tuned from director to reflector configurations. This suggests that there may be greater improvement if the added reflector(s) is a Yagi element. This is an area which merits more study.

The results of several MININEC runs suggest that a useful bi-directional beam can be developed using a director-radiator-director combination. Each of the resulting lobes is nearly the same as the main lobe of the conventional reflector-radiator combination. No work has been done with more than one director on each side. This is another area where study is needed.

Design Procedures for Quads

The curves and relationships for the quad loop and for multi-element arrays of loops are the tools needed to design quads which meet specified antenna performance requirements. The material which follows describes several different approaches for using these tools. They differ a bit in the accuracy of design prediction; consequently, some are preferred over others. The limitations of each are described.

Always remember that one outcome of a design effort may be the conclusion that the original design objectives weren't realistic. As I've emphasized several times, this happens when one attempts to obtain the maximum amount of gain and the smallest back lobe at the same time. The objectives may need to be modified as the result of design trials, until it's possible to realize the new objectives.

A step common to all approaches is the initial design estimate. To aid in this, I have prepared Table 2. This sets forth the major characteristics of the multi-element arrays studied here. Performance is given in terms of gain and F/B ratio, both at the design point and for the maximum possible. Physical size is easily developed; Figure 1 is useful for this. Bandwidth is estimated from the data curves. Feed conditions are given in Table 1.

With this estimate in hand, the design procedure can be selected from the material in the sections which follow.

Exact-frequency Scaling

Sometimes an initial design estimate will meet the design requirements, but for a frequency other than that

PROTOTYPE QUADS

name	number of elements	boom length	reflector circ.	reflector space	director circ.	director space	f_o (dB) gain	f_o (dB) F/B ratio
3Q.15	3	0.30	1.10	0.15	0.90	0.15	8.40	12.40
3Q2a	3	0.35	1.10	0.15	0.95	0.20	8.70	15.30
3Q25	3	0.40	1.10	0.15	0.90	0.25	8.60	14.10
3Q.3	3	0.45	1.10	0.15	0.90	0.30	8.70	14.00
4Q.25a	4	0.65	1.10	0.15	0.95	0.25	9.50	12.00
4Q.3	4	0.75	1.10	0.15	0.90	0.30	9.10	12.00
5Q.3	5	1.65	1.10	0.15	0.90	0.30	9.70	13.50

name	maximum gain(dB)	at f	maximum F/B ratio	at f
3Q.15	8.40	37.50	23.4	38.20
3Q.2a	8.80	37.40	23.7	38.00
3Q.25	8.70	37.25	29.5	38.50
3Q.3	8.90	37.40	39.5	38.10
4Q.25a	9.60	37.60	24.3	38.40
4Q.3	9.20	37.60	30.0	38.30
5Q.3	9.80	37.40	30.0	38.25

f_o = 37.5 MHz
All designs are compromise type.

Table 2. Summary of the geometry and dimensions of the quads used to prepare the lobe-intensity figures for multi-element quads. See text for usage notes.

which is needed. The design process is then one of scaling the physical size of the original design to the new frequency. The steps are:

- For the original design, determine the frequency at which the desired performance is attained using the appropriate swept frequency curve (such as the one shown in Figure 8). Call this f(old).
- Set the frequency to which this performance is to be transferred. Call this f(new).
- Determine the scaling factor, k = f(old)/f(new).
- Multiply all physical dimensions of the antenna by this factor. Here, "all physical dimensions" include conductor size, as well as location.

If no error is made in the division and multiplication calculations, the performance at the new frequency will be the same as it was at the original frequency.

For example, suppose that you want a "reasonable" 7-MHz antenna, reasonable meaning primarily that the antenna won't be too large. You decide that principal operation will be for CW, but reduction of European broadcast interference on the upper part of the band during phone operation would be nice.

A review of the curves in Chapter 5 shows that the two-element "exciter quad" gives about the most performance for its size. It has good gain at its design frequency. It also has a point of high F/B ratio. A few trials show that the gain can be kept high for CW, and that the high F/B ratio can held in the phone section by choosing the design frequency to be 6.97 MHz. The scaling factor is 37.5/6.97, or 5.38.

Making the multiplications, and changing to feet by multiplying dimensions by 3.28, results in the following (both loops):

length of side	38.8 feet
spacing	21.2 feet

The wire size is 2 × 0.00538 × 39.37 × 1000, or 424 mils, approximately no. 000 copper wire. This wire size is much too large for practical use, but there is a way to obtain an effective diameter. Thin tape can be used instead of wire. Kraus[20] gives the relationship: effective diameter = 0.5 × tape width.

Hard-drawn copper or copper-plated steel tape 7/8-inch wide will work well and the weight will be reasonable. The tape should be spiraled to avoid wind-induced vibration fatigue. (Experimenters might like to try using a 1-inch wide strip of the multi-conductor ribbon cable common in computer installations, or even rotator cable. A steel measuring tape will serve nearly as well.)

Adjusted Frequency Scaling

The large-conductor size in this design example can be compensated for by changing the element size to give the same reactance with a smaller conductor. For the 0.00539-meter conductor radius, omega is equal to 18. Using Figure 2, we find the self-reactance of the loops is about −25 ohms. Suppose the conductor size is to be changed to no. 12—a diameter of 80.8 mils, or a radius of 0.001 meter. This corresponds to omega = 21.5. Again using Figure 2, a self-reactance of −25 ohms is obtained with this thickness factor for loops of 1.09 wavelength perimeter (that is, a size reduction of 0.01 wavelength).

This new conductor size would give:

reflector side	38.5 feet
radiator side	38.8 feet
spacing	1.1 feet

Performance would be within a few percent of that of the 37.5-MHz prototype.

If desired, the size of the radiator can be made the same as that of the reflector. There will be some change in feed impedance, and minor changes in gain and F/B ratio.

Neglected Factor Frequency Scaling

Suppose the conductor size of the preceding example was changed while the loop size remained the same. Previous work indicates that this shouldn't have much effect on gain or drive resistance, but it is likely to change drive reactance and F/B ratio appreciably. If a computer with MININEC is available, this can be checked easily (even on the small 64-K types like the Apple or the C-64).

The results of such an analysis yield:

length of side	38.8 feet
spacing	21.2 feet
conductor size	no. 12 (chosen)

Frequency (MHz)	Gain (dB)	F/B ratio (dB)
7.0	7.6	21.6
7.15	7.3	20
7.3	6.2	9.9

The beamwidth is 64 degrees. The drive resistance varies from 128 to 152 ohms and the reactance from 213 to 211 ohms. A good match for 50-ohm line is obtained with a stub across the antenna terminals and a 3:1 RF transformer.

The result of the change in wire size is typical. Performance remains quite good, but F/B ratio, in particular, and gain are likely to degrade somewhat. They will certainly be at a frequency different than for the exact-scaled antenna. The second technique mentioned, adjusted frequency scaling, is recommended.

Reactance Scaling

Instead of using the swept-frequency plots as the basis for scaling, you may use the reactance plots. The procedure for the reflector and director, in turn, is:

- Establish the element spacing and tuning conditions, using Figures 9 through 12 in Chapter 5, based on design goals (maximum gain, maximum F/B ratio over a given band).
- Using Figure 2, convert the element tuning condition to a per-unit element length for the chosen element diameter.
- Convert the per-unit spacings and lengths to physical dimensions for the desired band by multiplying by 984/f, where f is the frequency for the design goal.

The weakness of this simplified approach is that it neglects the interaction between reflector and director, and between directors, if there are more than one. Partial compensation for these neglected factors can be secured using the curves of Figures 9 through 12 from Chapter 5 for the reflector, and Figures 3 through 6 for the director. This method isn't the best one to use, but it can give acceptable results if you remember that F/B ratio and drive reactance are *most likely* to be the most unfavorable. The method is quite useful in establishing a first design approach for further analysis and modification using MININEC.

As an example of this method, suppose that you need a three-element design with maximum gain at 14.15 MHz.

For the reflector, Figures 9 through 12 in Chapter 5 show maximum gain for a spacing of 0.15 wavelength and −40 ohms reactance. Assuming no. 12 wire is used, omega = 19.8. Figure 2 indicates the reflector size should be 1.09 wavelength.

For the director, using Figures 3 through 6, maximum gain is for a 0.25-wavelength spacing and tuning of −100 ohms. However, the gain is very sensitive to tuning on one side of the curve, so a design reactance of 110 ohms seems the better choice. Taking information from Figure 2, the director size should be 1.06 wavelength. This is a surprisingly large director—the result of rather small conductor size, and of a design for maximum gain.

The design dimensions are:

reflector side	18.95 feet
spacing	10.4 feet
radiator side	18.95 feet
director side	18.4 feet
spacing	17.5 feet

The radiator dimension is set to equal that of the reflector, for nearly maximum gain.

MININEC analysis of this design gives:

Frequency (MHz)	Gain (dB)	F/B (dB)
14.0	9.95	6.7
14.15	10.5	0.4
14.35	9.5	9.5

Beam width is 48 degrees. Drive resistance varies between 19 and 27 ohms, and reactance between 30 and 140 ohms.

This would be a good antenna for quiet band conditions or, if used with enough power, to "drive off the competition." Otherwise, the poor F/B ratio would result in much interference.

Comparative Design

The pattern of a multi-element array of nearly identical elements can be regarded as the product of an *element pattern* and an *array pattern*. For identical spacing and phasing, the array pattern part is the same for all types of arrays. Among other things, this means that Yagi design studies can be used to obtain an approximation of quad performance.

For instance, suppose that full coverage of the 10-meter band (28.85 MHz ±3 percent) is desired, with good gain. Leafing through the Yagi curves in Reference 19 we see that four elements on a 0.75-wavelength boom will give at least 9 dB gain over this band (page 2–14, curve 4 at bottom). Further investigation shows that the design values are reflector = 1.05 times resonant frequency, director = 0.95 times resonant frequency (Table 2.3).

APPLICATION QUADS

band ratio	number of elements	side dir (ft)	side rad (ft)	side ref (ft)	space dir (ft)	space ref (ft)	size	gain lo	gain m	gain hi	F/B lo	F/B m	F/B hi
40M	2	—	38.80	38.80	—	21.20	12	7.6	7.3	6.2	21.6	20.1	9.9
20M	3	15.70	19.20	19.20	17.50	10.50	12	8.9	8.4	7.9	13.2	25.1	18.5
a20M	3	18.40	18.95	18.95	17.50	10.40	12	9.95	10.5	9.5	6.7	10.4	9.5
15M	3	11.20	12.92	12.92	11.70	7.10	18	9.2	9.0	−4.1	19.0	38.0	10.0
10M	3	8.95	9.89	9.89	8.52	5.11	18	9.7	10.6	3.2	23.6	22.4	− 6.0

band	BW (deg)	lo	hi	lo	hi
40M	64	128	213	152	213
20M	62	73	130	89	130
a20M	48	19	27	30	140
15M	48	102	156	134	174
10M	52	158	—	114	—

Table 3. Summary of the geometry and dimensions developed for example quads meeting typical design criteria. Can be scaled to other bands.

Assuming elements of no. 18 wire, omega will be about 20. Figure 17 in Chapter 4 shows the resonant perimeter is 1.105 wavelengths. Multiplying by the 1.05/0.95 factors and converting, the corresponding dimensions are:

reflector side	9.89 feet
spacing	5.11 feet
radiator side	9.89 feet
director side	8.95 feet
spacing	8.52 feet

Here, unlike Lawson's use of a radiator that gives zero drive reactance,[29] a radiator size equal to the reflector size is used for better gain.

The MININEC results for this combination are:

Frequency (MHz)	Gain (dB)	F/B ratio (dB)
28.0	9.7	23.6
28.85	10.6	22.4
29.7	3.2	−6

The beamwidth is 52 degrees. Drive impedance varies quite widely, and is around 158 +j114 at the lower part of the band.

Additional MININEC runs show that the performance of the array remains good down to 26.9 MHz. A recheck of Lawson's curves[19] shows that the range of good gain is from 0.925 to 1.03 times design frequency. Instead of the midband being taken as 1.00, it should have been taken as the mean of 0.925 and 1.03, or 0.9775. This gives a design frequency of 29.5 MHz. To meet the design goals, the dimensions given above should be scaled further by multiplying by 0.9775. For this small alteration, conductor size need not be changed.

This example points out the need for careful examination of prototype performance. It also shows the value of analysis by MININEC or another antenna program as a part of the design. The method allows use of the extensive tables from the NBS and Lawson Yagi studies, and from the designs in the ARRL and RSGB handbooks. The results are especially useful as a departure point for more detailed study using MININEC.

Antenna Dimension Factors

A common method of determining antenna element size is to use a design factor, dividing this by the design frequency to get the physical dimension of the antenna element. Usually this factor is chosen to include the measuring system, accomplishing the change from meters to feet (or inches). The standard quad factors are for a compromise design. Additionally, as discussed earlier, the standard factors don't include consideration of the conductor thickness. Consequently, the points of maximum gain and best F/B ratio are unspecified.

The present study leads to the following factors for the specific case of conductor diameter = 1476/f mils:

Design Goal	Factors For Reflector & Radiator	Director
Maximum BW	1171	862
Maximum gain	1082	886
Maximum F/B	1054	902

director spacing 295
reflector spacing 148

To obtain a fair approximation, the director size should be increased and the reflector size should be reduced by one-tenth of one percent for each unit increase in wire size. For example, the nominal wire size for 14 MHz is approximately no. 10. To change to the more common no. 12, the wire size correction multiplying factors are 1.002 and 0.998 for both director and reflector, respectively.

Prototypes for Scaling

The characteristics of a number of designs suitable for scaling are given in Table 2. Those developed in the preceding examples, plus a few others, are summarized in Table 3. Drive impedances for these are given in Tables 1 and 4.

I hope the designs in these four tables will take care of a large percentage of practical problems, with no more work needed than frequency scaling.

APPLICATION QUADS

frequency (MHz)	Z (ohms)
6.90	77 + j98
7.00	128 + j151
7.10	173 + j177
7.20	199 + j193
7.30	214 + j212
14.00	84 + j108
14.10	100 + j129
14.20	115 + j145
14.30	127 + j160
14.40	137 + j175
14.00	19 + j29
14.10	19 + j62
14.20	21 + j97
14.30	24 + j133
14.40	30 + j173
21.00	102 + j134
21.10	117 + j147
21.20	130 + j157
21.30	142 + j165
21.40	153 + j172
21.50	161 + j178
27.00	102 + j53
27.50	143 + j96
28.00	159 + j115
28.50	134 + j148
29.00	87 + j249
29.50	150 + j446
30.00	363 + j493

Table 4. Summary of the drive impedances of the quads in Table 3. If exact scaling is used, will be exact at the scaled frequencies of the new band.

Patterns

Figures 18A and 18B show MININEC3 calculated H and V-plane patterns for the three-element quad at 37.5 MHz, for H polarization. Figures 19A and 19B show the same patterns for a frequency of 38.1 MHz. Close comparison reveals that the forward lobes are identical over 180 degrees for the H plane and 90 degrees for the V plane. There is appreciable difference over the rest of the circle.

The patterns illustrate, as has always been found, that a simple statement of the gain at 0 and 180 degrees as "the front-to-back ratio" can be misleading. There's almost always change in rear-lobe structure as frequency or tuning is varied. This changes the lobe size and interference susceptibility at other angles. Full evaluation of the antenna requires pattern evaluation. At this time, there is no standard for such evaluation.

For the H plane, the rear-lobe performance at these two frequencies is approximately:

Measurement	Figure 18A F/B ratio (dB)	Figure 19A F/B ratio (dB)
0/180 degrees	13	40
worst case	13	23
average rear	23	28

These figures all indicate that the condition at the higher frequency will result in less interference from the back lobe by 2 to 5 S-units for a single interfering signal, and about 2 S-units for many interference sources.

Figure 20 shows the vertically polarized component in the horizontal plane for this antenna. This component is approximately 28 dB below the H component in this plane. As I've noted, its presence is the

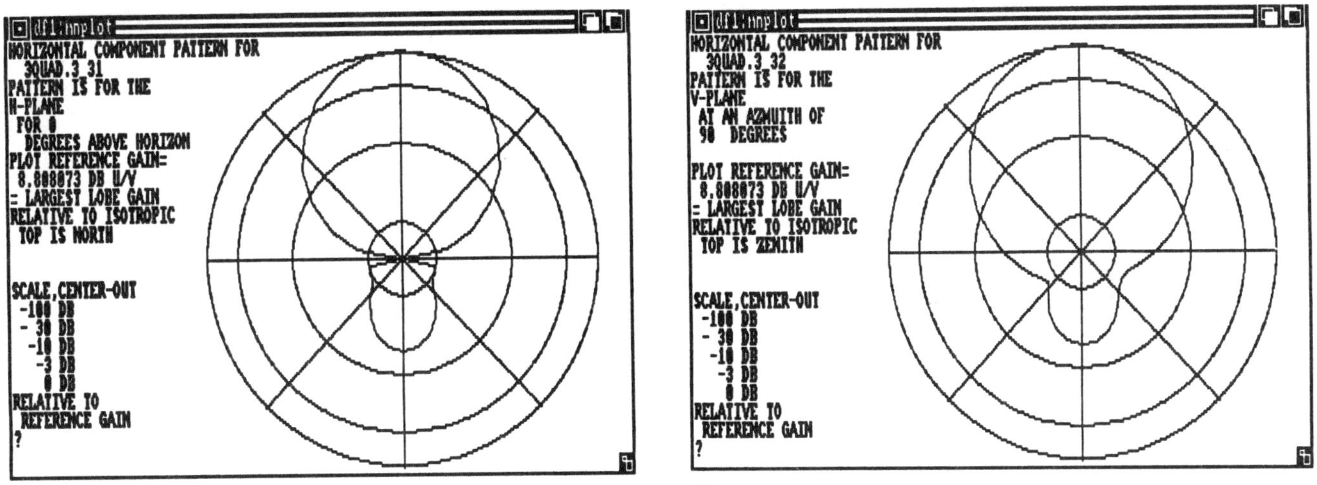

(A) (B)

Figure 18. (A)Horizontal plane pattern for the horizontally polarized component for the 0.9/0.3-wavelength, three-element quad, calculated for a frequency of 37.5 MHz. (B) Vertical-plane pattern. Note that the reference gain is that of the main lobe. This is typical of the patterns away from best F/B ratio frequency.

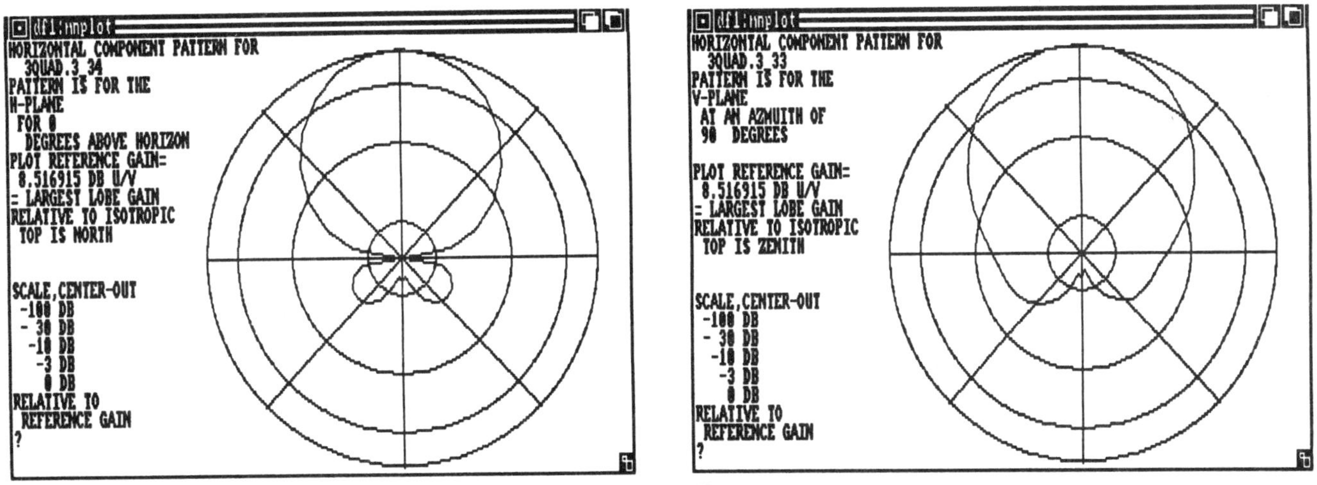

(A) (B)

Figure 19. (A) and (B) same as in Figure 18, but for a frequency of 38.1 MHz. Typical for frequencies close to the best F/B ratio frequency.

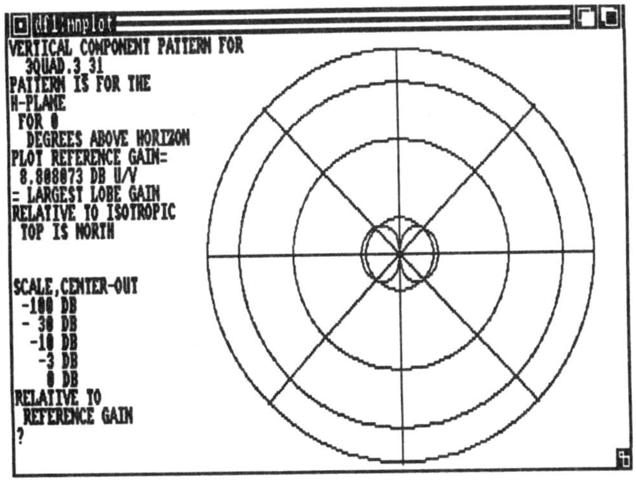

Figure 20. Similar to 18A, but for the vertically polarized component. This component is present in all quads, with some variation in intensity.

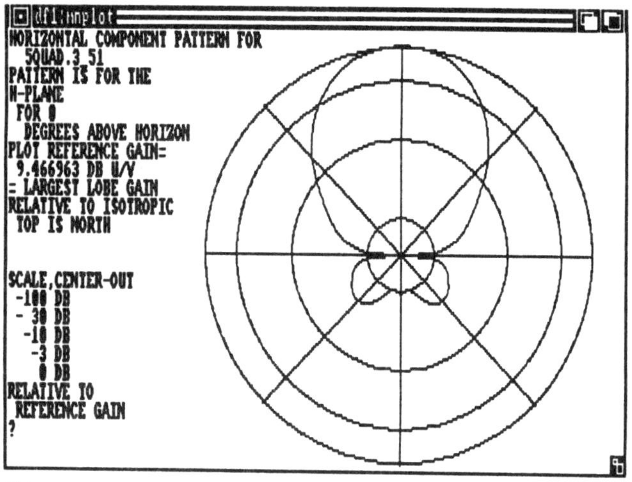

(A)

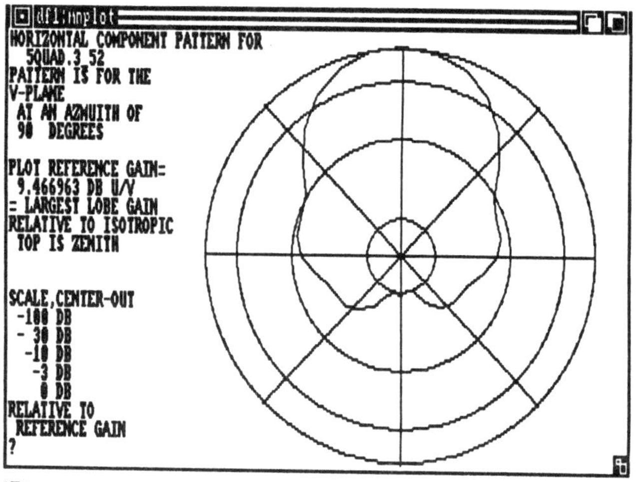

(B)

Figure 21. (A) and (B) similar to Figure 18, but for the five-element quad at 38.25 MHz.

major difference between Yagi and quad performance found so far.

Figures 21A and 21B show the same patterns for the five-element quad. The narrowing of the forward lobe (required for the calculated gain increase) is very noticeable. There are minor differences in the rear lobe structure, but these two antennas clearly belong to the same family.

Chapter 7

Ground Effects on Quad Loops and Arrays

Lobe structure and feed impedance as functions of antenna height

When an antenna is installed at any practical height above ground, a number of differences occur in its operation as compared to ideal or "free space" conditions. The two most important are:

- A major change in antenna pattern, explained by the difference in path length between the direct ray and that reflected from the earth, which is equivalent to the path difference between the ray from the antenna and from an "image" directly below the antenna and at the negative limit of its height. The process is identical to that for any two antennas forming an array.
- Change in feed impedance, largely explained by the proximity of the image antenna, spaced twice the height of the antenna above ground.

In addition, there are detail differences in both feed impedance and pattern because real earth is conductive and is a dielectric. Among other things, this means that the currents in the image antenna differ in magnitude and phase from the true antenna, and that there is a lower signal intensity compared to "ideal earth" conditions.

While there's a reasonably accurate theory for currents on circular loop arrays, no one has yet undertaken an analysis of these above ground. And, as noted earlier in this book, the available theory of square loops isn't suited to practical solution. Thus, there is only a "zero-order" theoretical solution to the problem of quads above ground. This is based on the performance of closely spaced shortened dipoles above ground—the approach used by Lawson in his comparison of quads and Yagis.

A more extensive approach seemed in order for this study. Fortunately this is simple, because the MININEC analysis program includes both free-space and near-earth capability, along with the capability of handling fairly complex earth conditions. (While simple, obtaining results is both tedious and time consuming. Data printouts for the results presented thus far in this book amount to over 2500 pages.)

The analytical approach I followed was to first consider a simple one-wavelength quad loop at varying heights above earth. Because earth effects are most important at HF, these data runs used a frequency of 18.75 MHz, with the loop scaled 2:1 in all dimensions from the earlier work. This scaling means that all earlier free-space results are directly usable. These general runs were followed by a series for typical quad arrays at typical installation heights. Results are presented as performance curves, patterns, and a table giving both performance results and design data.

Drive Impedance

The input resistance of a one-wavelength quad loop is shown in Figure 1 as a function of the height of the loop center above ground. The theoretical drive resistance of a half-wave dipole, taken from the curve in Kraus, is shown for comparison.[20]

The general pattern for the heights plotted is the same for the two antennas. There's a marked reduction in drive resistance for low heights, an oscillation about the free-space value, and a gradual convergence to the free-space resistance at greater heights. As was mentioned in an earlier chapter, the free-space resistance for the quad is larger than for the dipole.

In one respect, this plot is misleading. At a height of 0.125 wavelength, the lower arm of the quad loop would touch the earth. The current in this lower arm becomes large. There is still current in the upper arm of the loop, due to the conduction path around the loop. The combined effect of this conduction path and the currents in the lower arm cause a marked change in the current distribution around the loop. The distribution pattern changes with height, causing large variations in drive resistance. At some height, the loop approaches an open circuit. Because this variation represents an unusual situation, beyond normal quad use, I haven't explored it further.

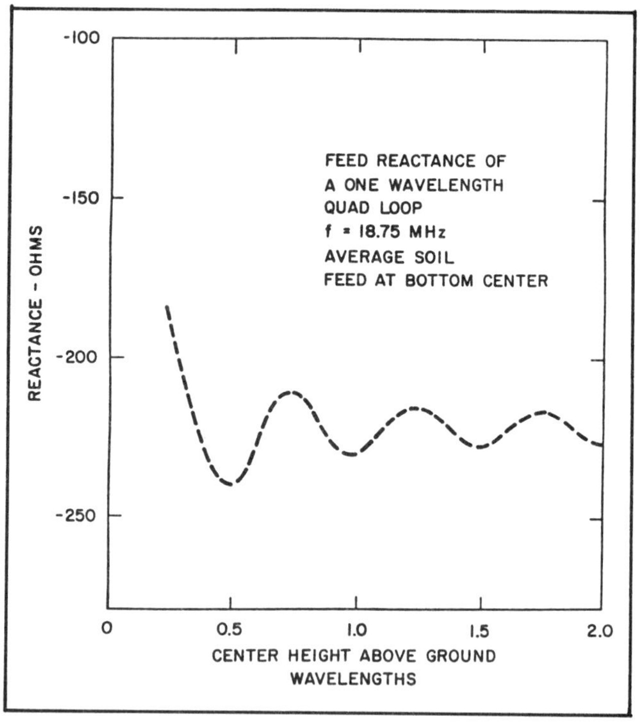

Figure 2. Feed reactance versus height of a quad loop, same conditions as Figure 1. The dipole shows the same pattern of variation.

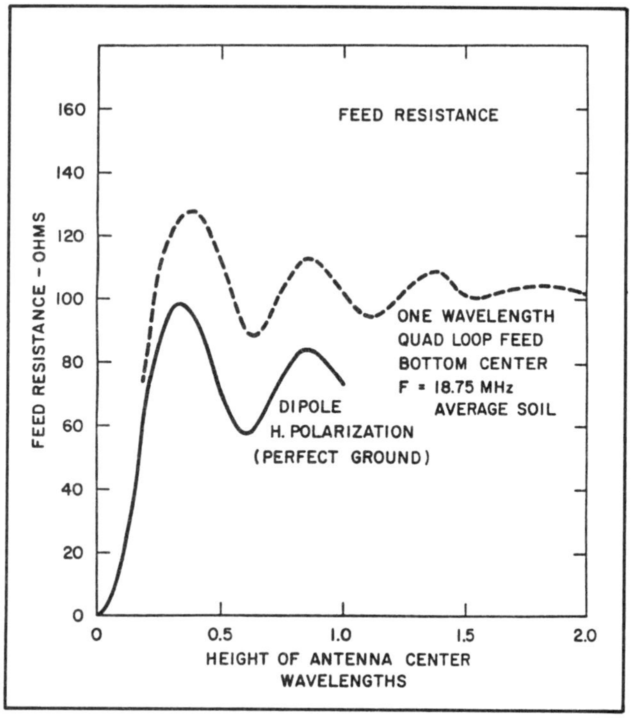

Figure 1. Feed resistance versus height of antenna center. Dotted line: one-wavelength quad loop, fed at bottom center. Solid line: one-half wavelength long horizontal dipole. At 18.75 MHz, average soil.

In a practical sense, the resistance of the quad loop can be approximated by the free-space resistance for all loop-center heights above about one-quarter wavelength. The variation would usually be hidden by the variations due to feed line and support structure proximity.

Antenna texts ignore the variation in drive reactance with height, because it's universally assumed that antennas will be operated at resonance. As shown in earlier chapters, there's good reason to operate quads away from resonance, making the variation in reactance of greater interest.

This drive reactance variation is shown in Figure 2 for a one-wavelength loop above average soil. The same pattern of initial change, oscillation, and convergence to the free-space value is clear. Practically speaking, the variation is of small import because it's less than 10 percent for any height above about 3/8 wavelength. Again, this is less than the change expected from the practical elements of feed and supports.

Feed impedance was also calculated for vertically polarized quad loops fed at the center of one side. Over the range from 0.375 to 2.0 wavelengths height, the bottom and side feed values are within 5 percent of each other. The pattern of variation is the same.

The diamond configuration hasn't been investigated in these studies. Earlier work showed that the difference between the square and diamond was very small—to the point that square-loop results can be taken as applying to both.

Structure of the Radiation Pattern

With a dipole close to earth, maximum radiation is directed toward the zenith. The elevation angle of the maximum component moves toward the horizon as the antenna elevation is increased. Figure 3 shows the variation of elevation angle for a one-wavelength quad loop, with the dipole radiation angle for comparison. Height is from the center of the antenna. Both values are from MININEC calculations for average soil.

As expected from the presence of the upper part of the quad, its radiation angle is always lower than that of the dipole. The difference is quite marked at small elevations, and this appears to be the origin of the quad reputation for good performance at low height. The difference decreases as height increases.

Figure 4 shows the main-lobe signal intensity (gain)

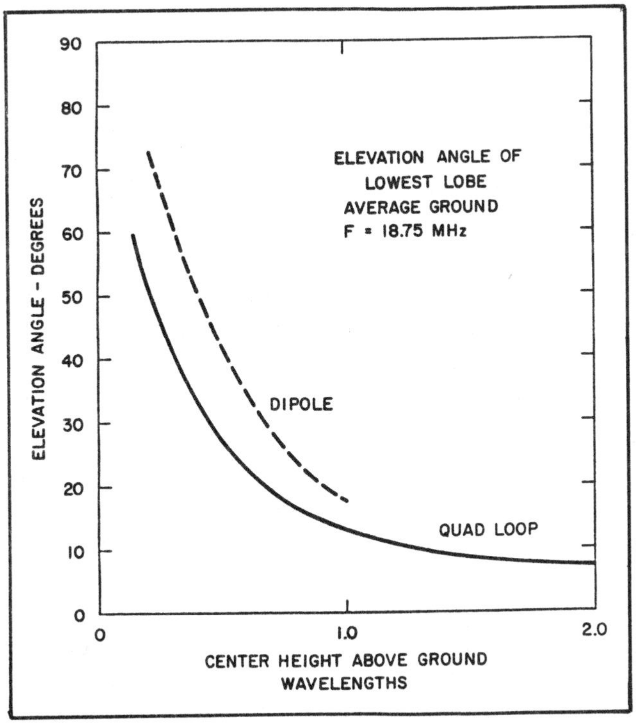

Figure 3. Elevation angle of the lowest lobe versus antenna height. Solidline, quad; dotted line, dipole. Same conditions as Figure 1.

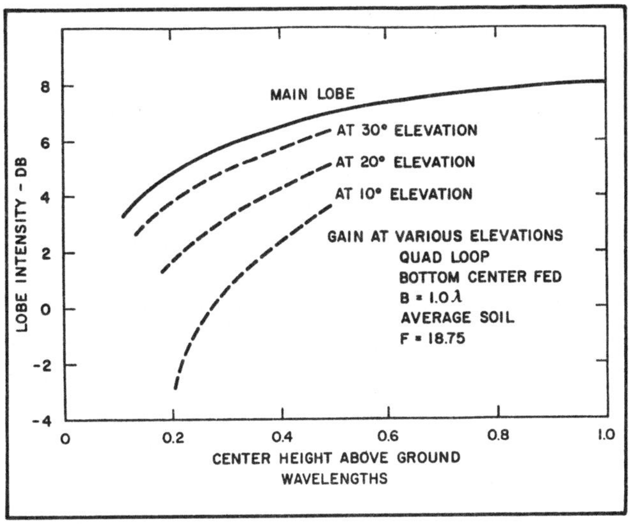

Figure 4. Main lobe intensity (gain) of a quad loop versus height (solid line). Conditions same as Figure 1. The dotted curves show gain at specific elevation angles.

as a function of height, and the intensity at three specific elevation angles, typical of those important for low-angle DX propagation. The lower heights show a loss as compared to high antennas. This is the result of higher ground currents which lead to higher loss. However, the lobe can still be stronger than free-space intensity—down to as low as 0.15 wavelength.

Figure 5 illustrates the same type of plot for a dipole. Close comparison of this and Figure 4 shows that they are nearly identical. The lobe intensity of a dipole is higher than that of the quad at very low heights. However, as is evident in Figure 3, this greater intensity

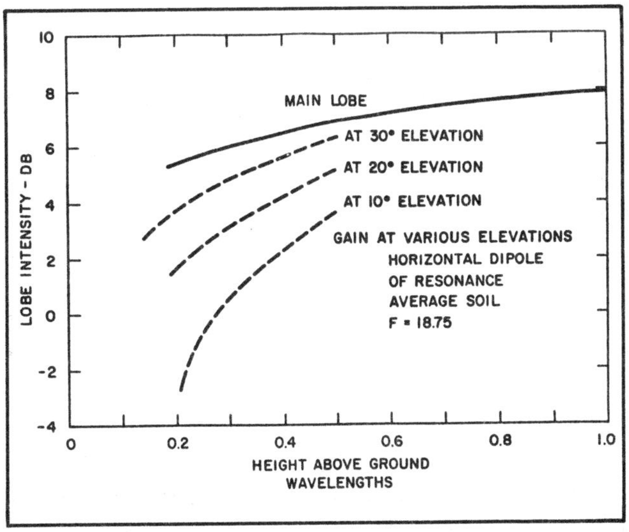

Figure 5. Same as Figure 4, but for a dipole. Note the close similarity of the two curves.

is at a higher elevation angle. At specific low-elevation angles, the gain is identical for the two antennas, to within a small fraction of a dB.

To explore this pattern change further, a number of additional conditions were investigated. Figure 6 shows the intensity of the main lobe at specific elevation angles for one-wavelength loops over the HF band. The loop height is adjusted to keep the lower part of the loop at a constant 3.0 meters above average ground. This represents a scale-height range of 0.16 to 0.43 wavelength.

Two major factors are combined here. For higher frequencies, the relatively greater height increases the gain at all elevations from main lobe center down. The other factor has to do with the increase in gain at lower frequencies; that is, low heights are less adverse at low frequencies—at least for average soil. This finding is in line with the fact that the dielectric constant increases, and the loss factor decreases at lower frequencies. In effect, poor soil becomes less poor as the frequency decreases (see, for example, Reference 26).

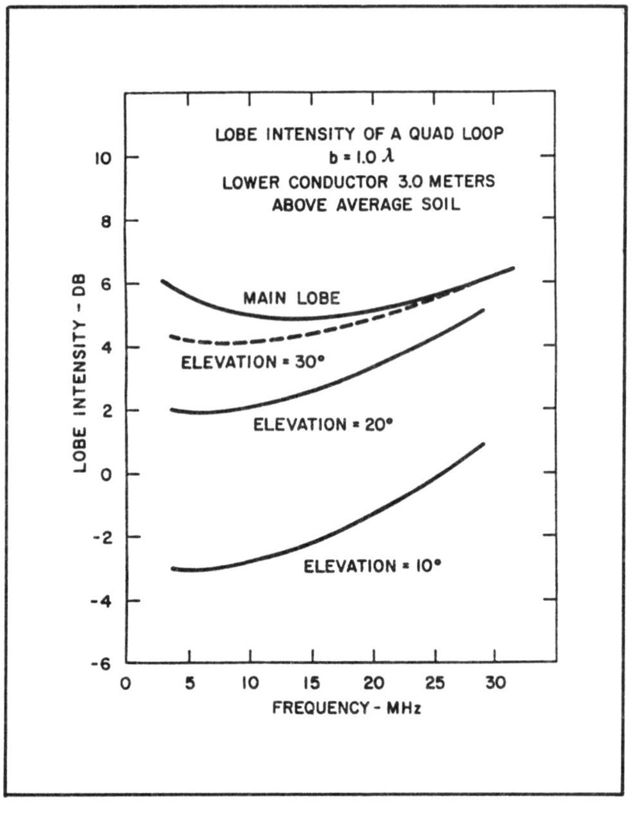

Figure 6. Lobe intensity (gain) versus frequency for one-wavelength quad loops with the lower arm at a constant 3.0 meters above ground. Top curve: main lobe gain, others at specified elevation angles. See text for a discussion of curve shape.

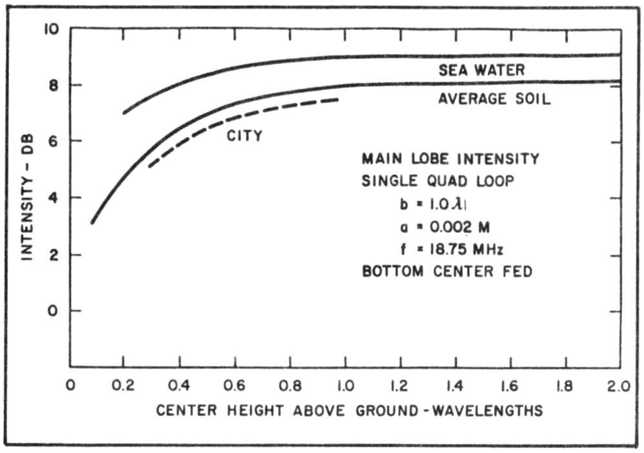

Figure 7. Main lobe intensity versus center height of a one-wavelength quad loop, showing effect of soil type. Top curve, sea water; second curve, average soil; lower curve, typical city earth. Note that MININEC models conductivity change, but not dielectric constant.

In this connection, you must always remember that analysis programs like MININEC are models, designed to give acceptable accuracy over a reasonable range of conditions. MININEC3 uses perfect ground conditions to calculate element currents, but does model conductivity change for far-field pattern calculations. The change in dielectric constant is ignored. Calculation results are guides, and must be used with special caution *under extreme conditions*. In this particular case, it appears that the model understates the amount of reflection gain increase at low frequency.

The calculated effect of soil type is shown in Figure 7. Gain reduction is appreciable at all heights for poorer soil, and much less so for high-conductivity sea water.

The effect of soil type is also shown in Figure 8. The horizontal axis is not truly a scale, but an indication of relative earth quality for propagation. The small change in signal at great antenna height as compared to the greater change at low heights is very evident. The curves for dipoles lie below the corresponding ones for quads. The 2.0-wavelength curve was so close to the quad curve that it isn't shown separately.

The final curve in this series, Figure 9, shows the lobe intensity for vertical polarization, with the loop fed at side center. The lobe intensity is well below that for horizontal polarization at all heights. However, the elevation angle of the main lobe is markedly lower for vertical polarization, as indicated by the angle of maximum radiation shown at the arrows on the curve. At 10 degrees elevation, vertical polarization is stronger for all heights less than 1/3 wavelength.

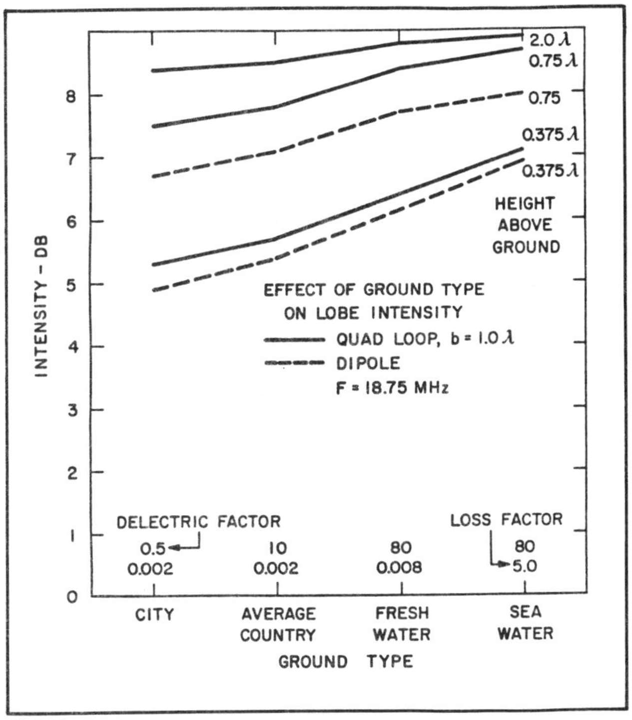

Figure 8. Effect of ground conditions on gain. The horizontal scale is arbitrary, and is roughly proportional to communication value of the soil. Soil values are shown above the type. Curves are labeled with antenna center height. Values would change somewhat at other frequencies.

A limited number of runs were made with vertical polarization and sea water. The loss in gain is much less than that for average soil. For example, at 0.5-wavelength height, the gain at 5 degrees above the horizon is 7.8 dB. Even at 0.15-wavelength, the gain is 3.3 dB at 10 degrees elevation. Island dwellers wanting good DX with modest antenna height should try vertical polarization. An early popular loop of this type, the Bi-square Beam, gained its good reputation from Pacific island operation.

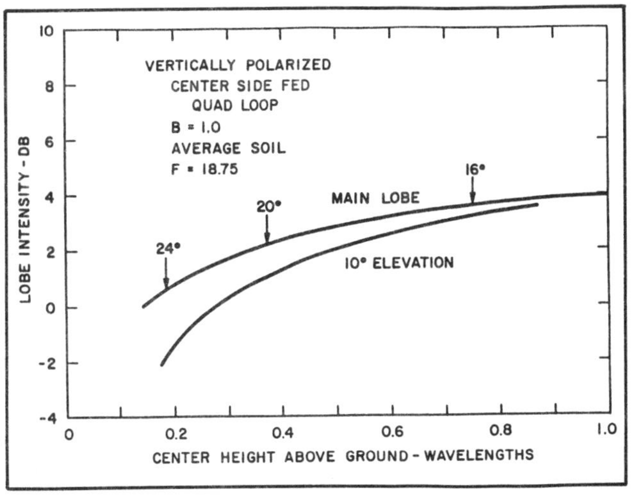

Figure 9. Gain of a quad loop versus center height with feed at side center. Upper curve, for the main lobe, carries arrows to show the elevation angle of the lobe center. Lower curve is for 10 degrees elevation. Compare with Figures 7 and 3 to see the effect of feed point change.

Pattern Effects of Ground

Figure 10 shows the vertical plane patterns for the series of loops with the lower conductor at three meters above average earth—the type used for Figure 6. At the lowest frequency (Figure 10A), with a scale height of 0.16 wavelength, the pattern is close to that of a dipole. The radiation to the zenith decreases, and the angle of maximum radiation moves toward the horizon as the frequency (and scale height) increase (Figures 10B through 10E). At the same time, the lobe intensity changes, as shown in Figure 6.

These antennas were fed at the center of the bottom arm, giving dominantly horizontal polarization. There's a substantial vertically polarized component in some directions. However in this plane, the vertical component is over 100 dB below the horizontal component.

Figure 11A shows the vertically polarized component for the 3.5-MHz loop fed at the center of one side to obtain vertical polarization. The elevation angle of maximum radiation has been reduced markedly from the bottom-fed case, and there's no component toward the zenith. However, there is also a marked loss in lobe intensity, from 5.9 dB for the bottom feed to 0.5 dB for the vertical feed. As previously stated, the gain loss for salt water is much less severe. The cross-polarized or horizontal component is also present, and doesn't disappear in this plane. Its intensity is shown in Figure 11B. It is some 20 dB below the main lobe.

As illustrated in Figure 12, the pattern for the 7-MHz antenna at three meters lower arm height is much the same. As a result of the greater scale height, the elevation angle of the main lobe has decreased and the cross-polarized component is showing some directivity. However, its intensity is lower. Higher-frequency patterns show this general progression, but the differences are small, so they aren't given here.

Typical pattern cuts across the main lobe are shown in Figure 13. These are for the 3.5-MHz antenna at three meters lower arm elevation, at an elevation angle of 20 degrees. At this elevation, the lobe is less than

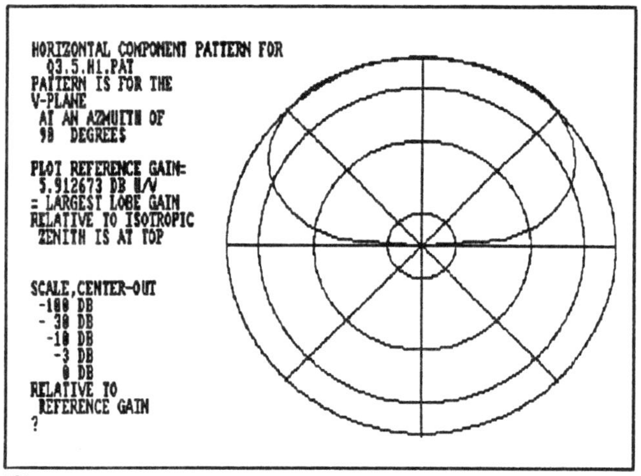

(A)

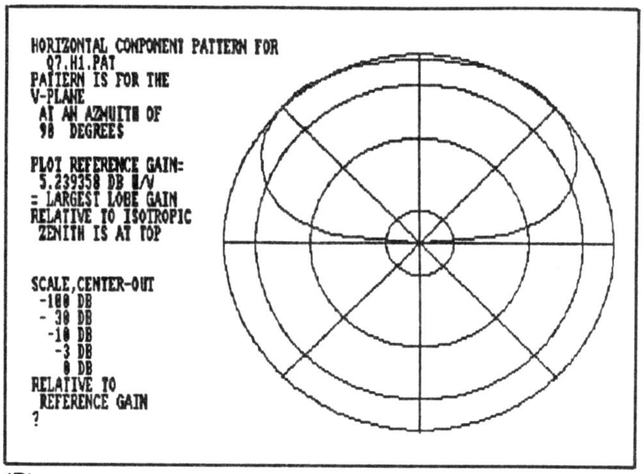

(B)

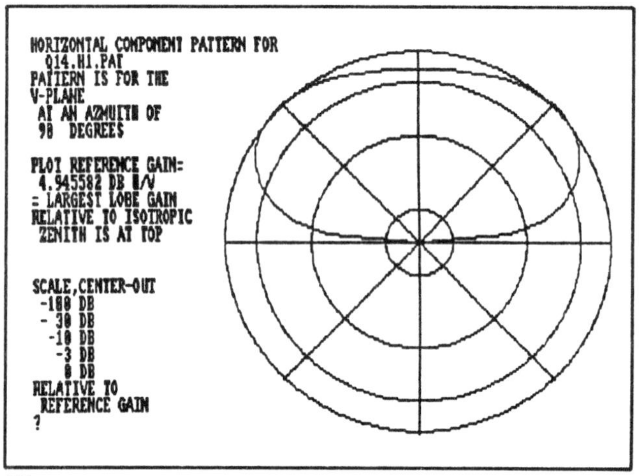

(C)

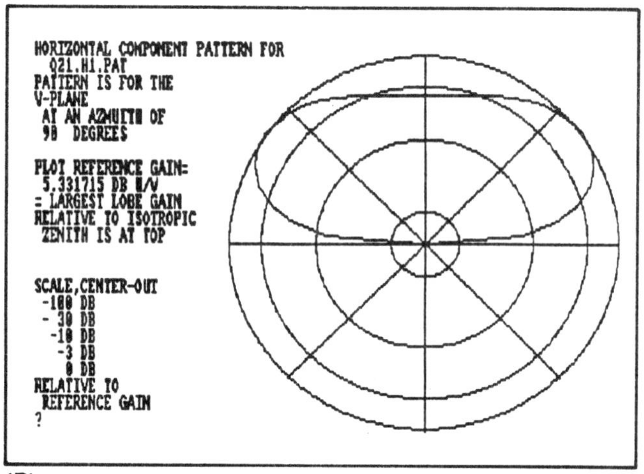

(D)

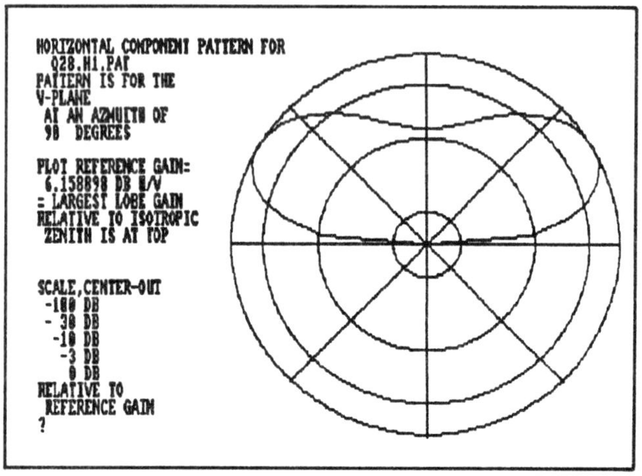

(E)

Figure 10. Family of patterns showing the vertical pattern changes for the bottom center-fed antenna series at constant 3.0 meters lower arm height. All are for one-wavelength loops: (A) at 3.6 MHz, (B) at 7.1 MHz, (C) at 114.15 MHz, (D) at 21.2 MHz, (E) at 28.5 MHz. Patterns are for the horizontally polarized component at right angles to the loop plane. The vertically polarized component is more than 100 dB less intense in this plane.

3 dB below the lobe maximum of 4.4 dB. The effect of the vertically polarized component in filling in the radiation in the plane of the loop is evident by comparing Figures 13A and 13B.

The cross-polarization lobe filling is further evident in the plots of Figure 14. The V component of 10-degrees elevation is shown in Figure 14A. Figures 14B, 14C, and 14D show the total component for 10, 20, and 30 degrees above the horizon. Both the H and V components increase with an increase in elevation, es-

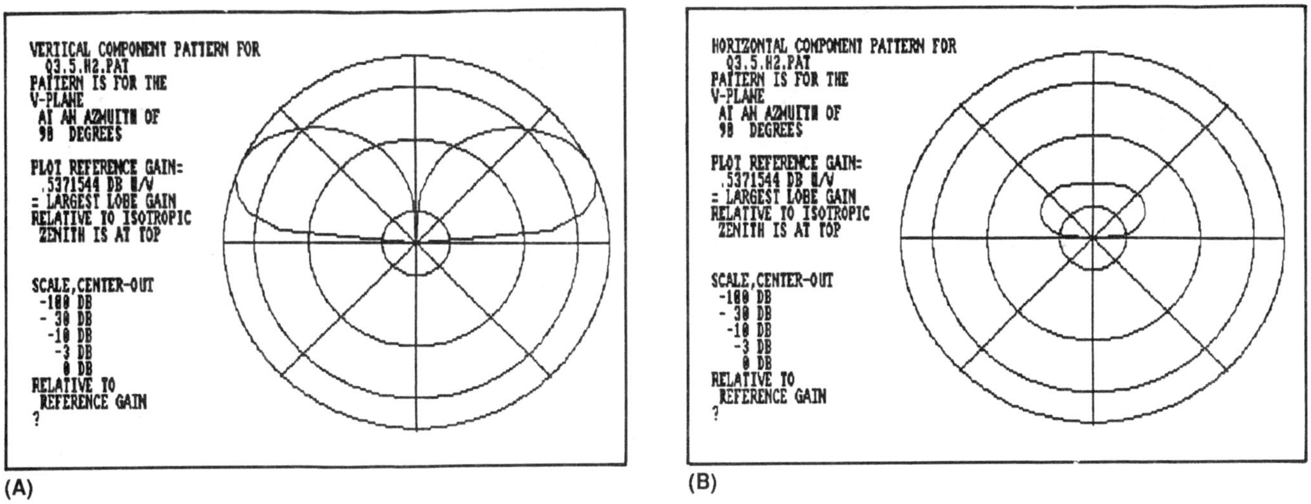

Figure 11. Patterns for the 3.6-MHz loop with feed at side center. (A) Vertically polarized component in the vertical plane, right angles to the loop. (B) Horizontally polarized component, same plane.

sentially by the same ratio. A result is that the front-to-side ratio is nearly constant.

These typical plots show relatively small differences in patterns and gains, with the quad loop showing somewhat better performance at low elevation angles.

Patterns for Specific Band Designs

Figure 15 shows calculated patterns for a 7.1-MHz two-element quad, bottom center fed, at a height of 0.375 wavelength, or 52.5 feet to center. The vertical plane pattern is shown in 15A, a 30-degree elevation cut for horizontal polarization in 15B, and the same cut for vertical polarization in 15C. Antenna gain is respectable (9.8 dB); F/B ratio is a very good 24 dB.

The V component in 15C does show some directivity, but this is along lines at 45 degrees to the main lobe. This component is some 20 dB below the main lobe. At some angles the V component does reach within 3 dB of the main lobe, but this intensity is always at high elevation angles.

Figure 16 shows the pattern of this antenna with feed at the side center giving primarily vertical polarization. The main lobe, shown for the vertical plane at 16A, has dropped in elevation to 15 degrees, but the maximum lobe intensity has dropped by 4.3 dB. There's some difference in vertical pattern shape, with a deep null at the zenith, and the beginning of a secondary lobe at approximately 70 degrees elevation. The

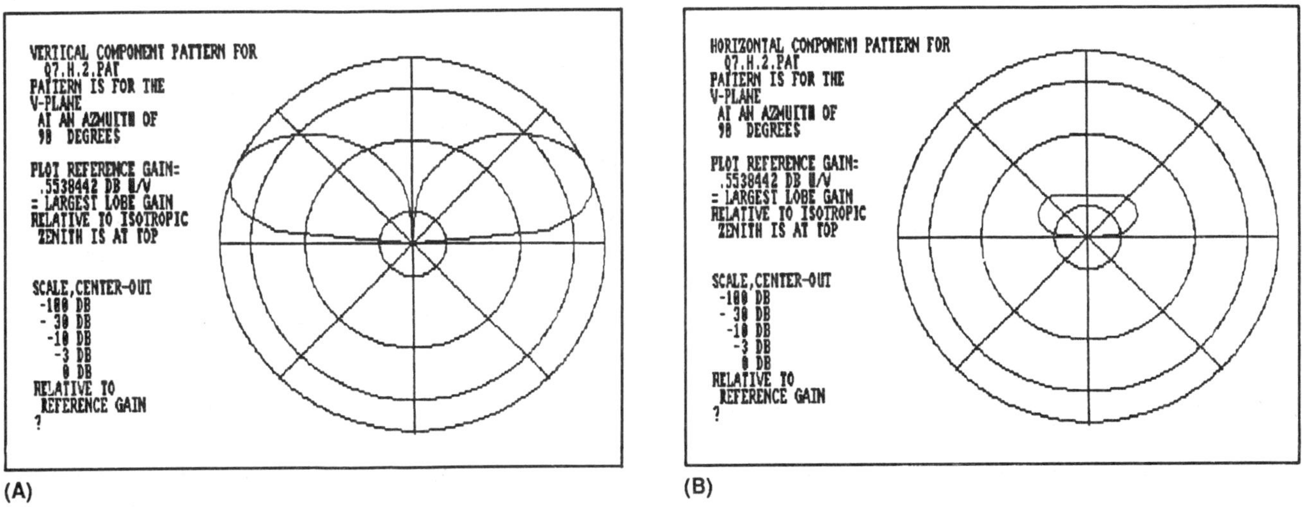

Figure 12. Patterns for the 7.1-MHz loop with feed at side center. Conditions same as Figure 11.

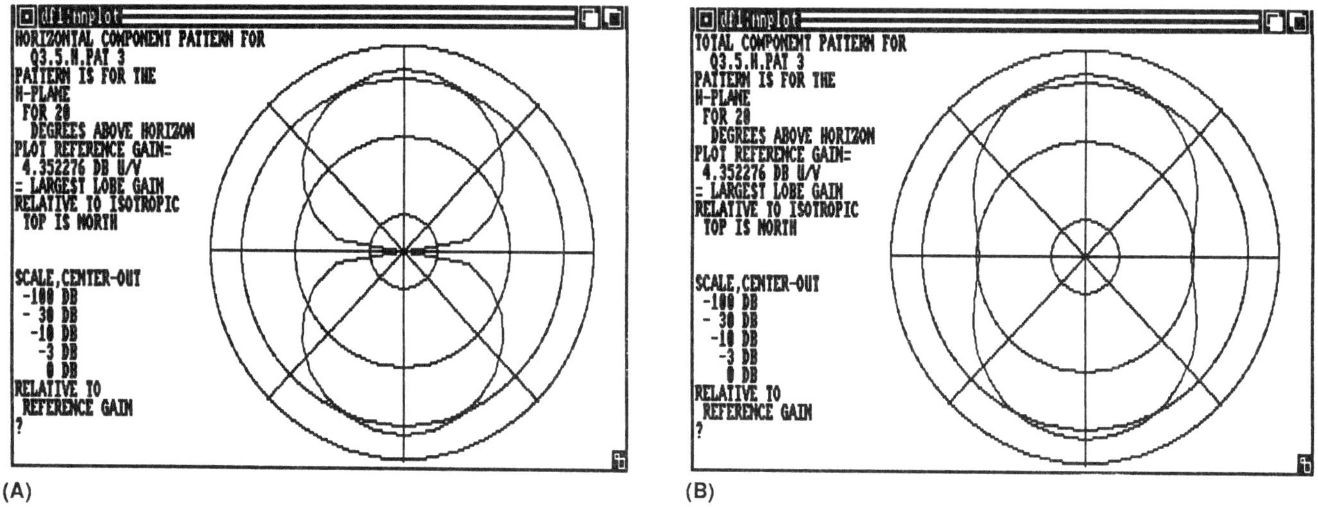

Figure 13. Patterns for the 3.6-MHz loop fed at bottom center. (A) Horizontally polarized component at 20 degrees above the horizontal plane. (B) Total radiation component, same condition. Note the fill-in of the radiation in the plane of the loop by the vertically polarized component.

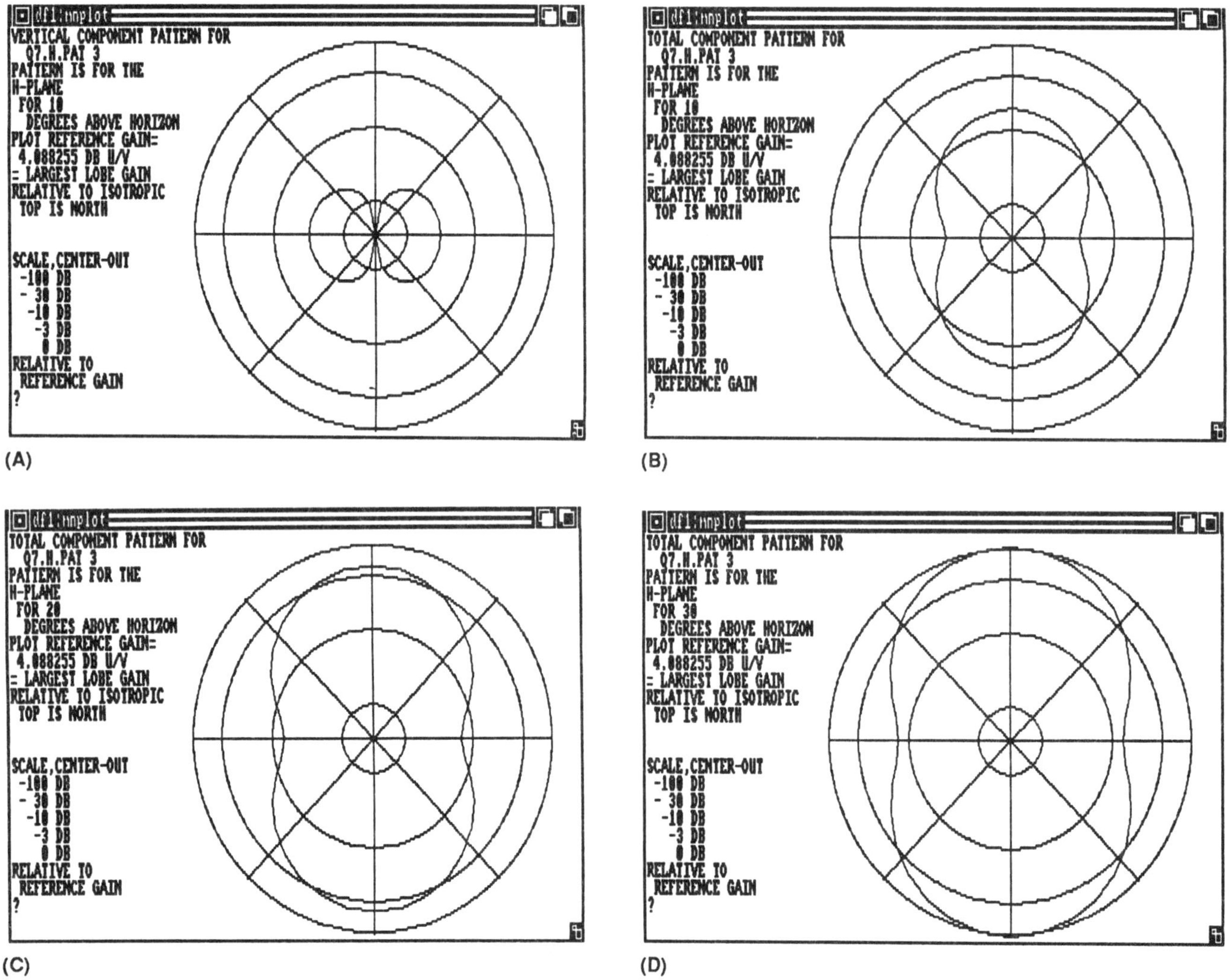

Figure 14. Patterns for the 7.1-MHz loop fed at bottom center. (A) Vertically polarized component at 10 degrees above the horizontal plane. (B) Total component, same elevation. Note the 90-degree difference from A. (C) Same as B, but 20-degree elevation. (D) Same as B, but 30-degrees elevation, the angle of the main lobe.

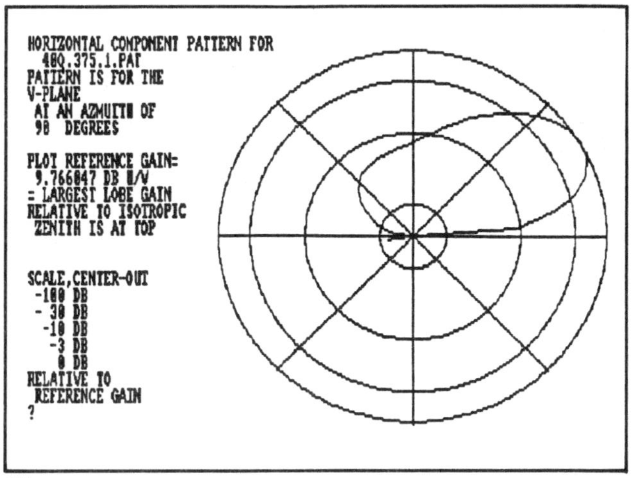

(A)

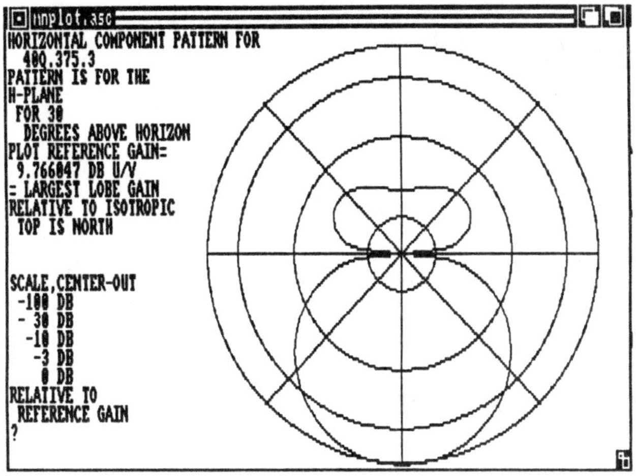

(B)

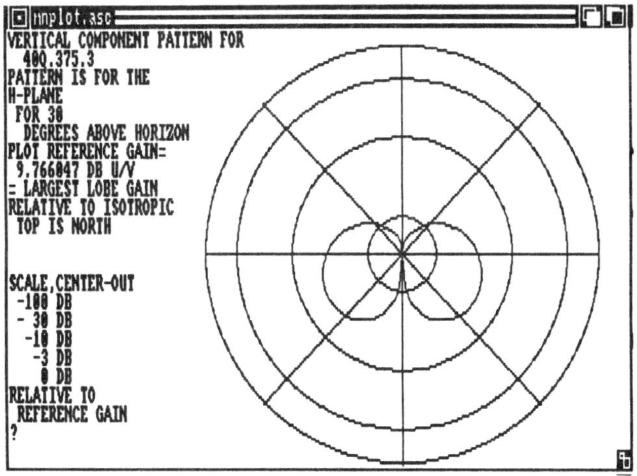

(C)

Figure 15. Patterns for a 40-meter two-element quad at 52.5 feet center elevation, bottom center fed. (A) Horizontally polarized component in the vertical plane, at right angles to the elements; the total component is identical. (B) Same component at 30 degrees above the horizontal plane. (C) Vertically polarized component 30 degrees above the horizontal plane.

horizontal 15-degree cut at 16B shows a slightly wider forward beam, and a considerably different rear lobe structure with no side null. However, the F/B ratio is about the same.

Large 40-meter quads aren't common. Also, the average height of amateur antennas worldwide seems to be about 40 feet. Figure 17 gives data for a more common situation. It shows the vertical plane pattern for a 20-meter, two-element quad, 40 feet high, with bottom arm feed (height is 0.573 wavelength). Gain is 11.3 dB, with the forward lobe maximum at 25 degrees elevation for average soil.

Lawson[19] gives a gain of 13.2 dB at an elevation angle of 27 degrees for the main lobe of a three-element Yagi at 0.5 wavelength above ideal ground. MININEC-calculated gain change in going from ideal conditions to average soil is − 1.3 dB for a dipole 0.5 wavelength high. Thus, the three-element Yagi and a two-element quad have essentially the same performance. The Yagi has a slight edge in gain, and the quad a slight edge in low-lobe angle. Practically, the two should be identical—a finding reported many times in antenna-construction accounts.

Turning now to more extensive antenna installations, Figure 18 shows patterns for a three-element quad 1.29 wavelengths (60 feet) high. Figure 19 shows those for the same antenna at 1.44 wavelengths (100 feet).

The vertical patterns at A in each figure, show the normal growth in number of lobes. As compared to the Yagi patterns given by Lawson,[19] there are some differences in relative strength of the various lobes, but none that seem to be important.

The two horizontal cut patterns at 18B and 19B are

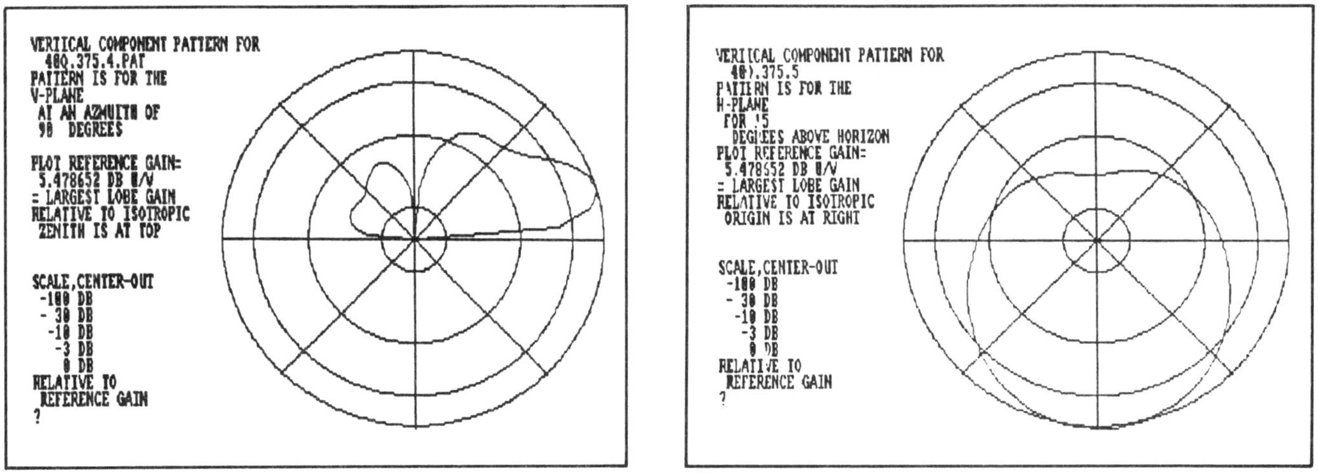

(A) (B)

Figure 16. Patterns for the 40-meter quad of Figure 15, but fed at side center. (A) Vertically polarized component in the vertical plane; (B) same component at 15 degrees above the horizontal plane. Compare gain and lobe angle of the two feed conditions.

identical, although the cut angle is 15 degrees above the horizon for the 60-foot height and 10 degrees for the 100-foot height.

This antenna is one of the compromise designs. Gain is respectable, almost exactly 14 dB (including the reflected component from average ground). F/B ratio is reasonable, the largest rear lobe is essentially at isotropic antenna level. Either gain or F/B ratio could be increased by tuning or rescaling dimensions. Overall, this antenna at these heights gives very respectable performance.

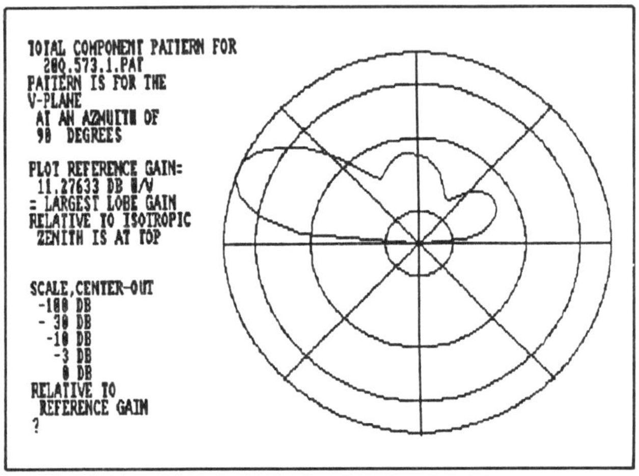

Figure 17. Vertical plane pattern for the total radiation from a two-element 20-meter quad at 40 feet height. Horizontal pattern is essentially identical to that of Figure 15. Note that this and following diagrams are oriented 180 degrees to those of previous figures.

These figures include added information. Figures 18C and 19C show the pattern of the second lobe for horizontal plane cuts off 55 degrees elevation for the lower antenna and 30 degrees for the higher. The reasons for the differences have not been explored completely. They are partly due to (a) the presence of the vertically polarized component and (b) the effect of the image antenna on pattern.

The vertically polarized component at an elevation angle of 30 degrees is shown in Figure 19D. As for the earlier case, this component shows some directivity, again with the lobes lying at 45 degrees to the main lobe. Specifically, the lobes are nearly 16 dB below the H-lobe component, and show a F/B ratio of 40 dB to the 180-degree line—some 20 degrees to the rear of the 45-degree line. The lobes of this vertically polarized component are down some 16 dB below the H-component lobe, at 45 degrees to it, and are some 100 dB down directly on axis. It's difficult to see how this component can be important to performance in the practical sense. A possible explanation for some reported performance under poor conditions is that this vertical component provides a built-in diversity system, which becomes important when the fade polarization is instantaneously vertical. For this to be effective, there would have to be simultaneous changes in the path horizontal arrival angle, or the antenna would have to be operated off axis with respect to the major lobe.

A reasonably complete literature survey hasn't yielded any data useful in testing this possibility.

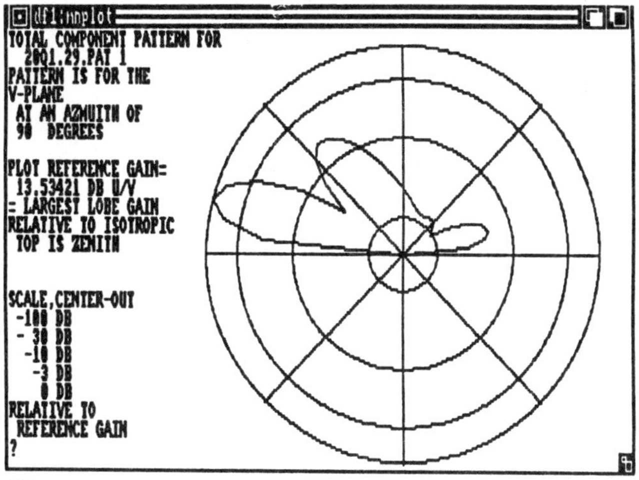

(A)

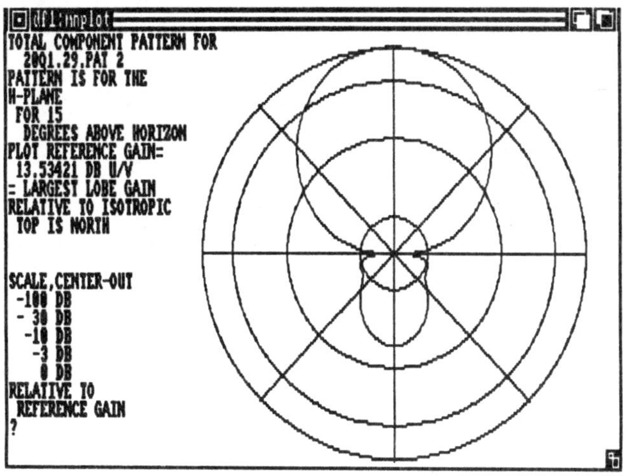

(B)

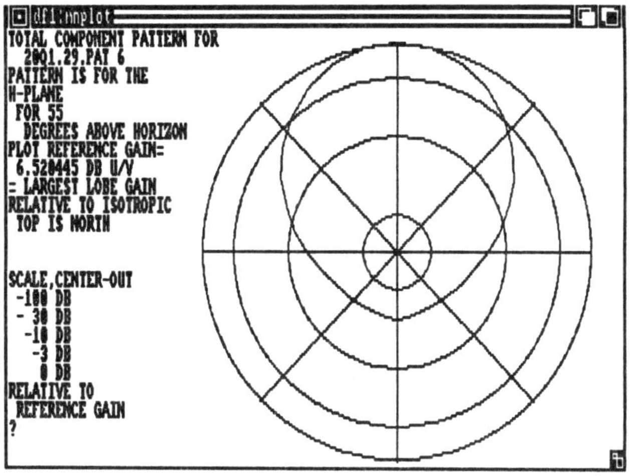

(C)

Figure 18. Patterns for a 15-meter three-element quad at 60 feet height, fed at bottom center. (A) Total radiation in the vertical plane, at right angles to the elements. (B) Total component for the main lobe, 15 degrees above the horizontal plane. (C) Total component for the second lobe, at 55 degrees above the horizontal plane.

Cross-polarization diversity is reported to reduce fade severity, and simultaneous H and V-polarization transmission is used on the FM broadcast band to improve fringe-area coverage. However, I have found no data on single-antenna diversity.

Experimentally inclined amateurs might want to explore this matter further. One might want to make a direct comparison of both quad and Yagi signals with antennas offset from the direct path. Another area to pursue is that of true polarization diversity tests with crossed Yagis or the dual-feed quads to be described later.

Returning to the pattern measurements, Figure 20 shows the vertical plane pattern of a 15-meter three-element quad designed for good F/B ratio, at 1.29 wavelengths (60 feet). Coincidentally, the gain is the same as that of the 20-meter quad at 1.44 wavelengths; the F/B ratio is excellent at 30.8 dB. However, this value is for the design frequency only, and won't be as good at other frequencies in the band.

Figure 21 shows patterns for a four-element 10-meter quad at relatively low physical height—0.87 wavelength, or 30 feet. This is a reasonably wideband design, giving 13.8 dB gain at the design frequency, and a F/B ratio of 12 dB. Gain could be increased at the expense of bandwidth by retuning. On this band, high F/B ratio is rarely needed.

Note the presence of additional small lobes in all of the patterns shown for this antenna. These appear as the number of elements increases, and can be difficult to control in large multi-element arrays. (See the discussion on antenna scaling in an earlier chapter.)

The last patterns (Figure 22) are for a 2-meter three-element quad with 1/8-inch conductor size. It is assumed mounted at 5.88 wavelengths (40 feet). This antenna was designed for maximum F/B ratio and

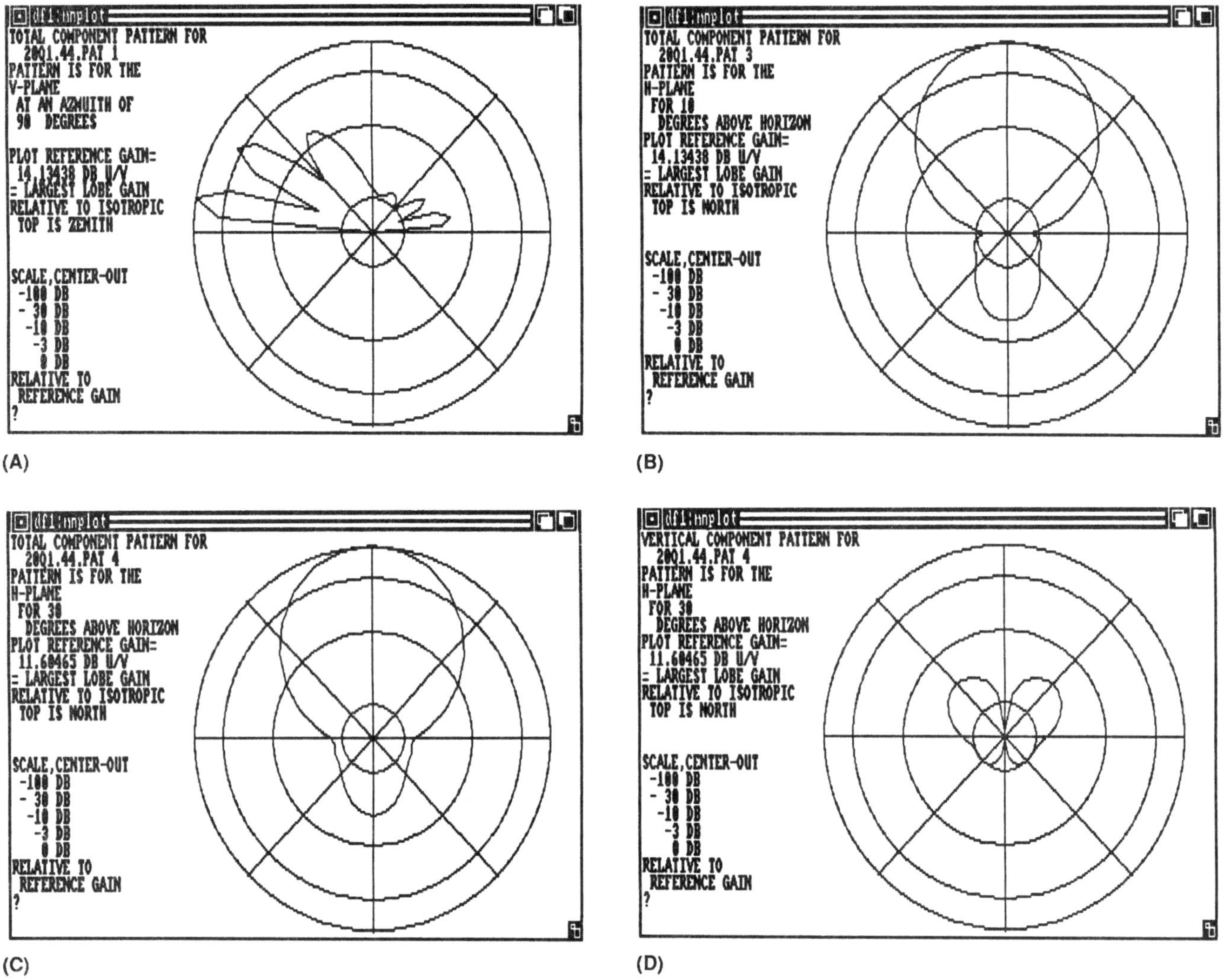

Figure 19. Patterns for the antenna of Figure 18, but at a height of 100 feet. (A) Total component in the vertical plane. (B) Total component for the main lobe, at 10 degrees above the horizontal. (C) Total component for the second lobe, at 30 degrees above the main lobe. (D) Vertically polarized component for the second lobe, at 30 degrees above the horizon.

is an attempt to exceed the characteristics of a three-element quad designed and built by KB4GW for fox hunting.

Figure 22A shows the vertical plane pattern for the total radiation, with bottom feed. Figure 22B shows the vertically polarized component, with feed at the side. Note the differences in lobe depth and in zenith radiation.

Figure 22C shows the vertical component radiation pattern for a horizontal cut 2 degrees above the horizon, with side feed. This type of pattern is desirable for DX work, because the single minimum is well defined and at 180 degrees to the main lobe. Higher gain designs will have a narrower main lobe, but will show two minima, about ± 135 degrees from the main lobe, making bearing definition more difficult.

Designs geared to such a specific performance goal will usually require several successive trials. These patterns are for the second such attempt. My first attempt had good F/B ratio at a frequency 1.2-percent too high. Even this design, with a paper F/B ratio of nearly 28 dB, can be improved; a F/B ratio of over 35 dB is attainable. However, it's probably better to use construction and physical tuning for the next step, because the effects of feed and support aren't included in the model.

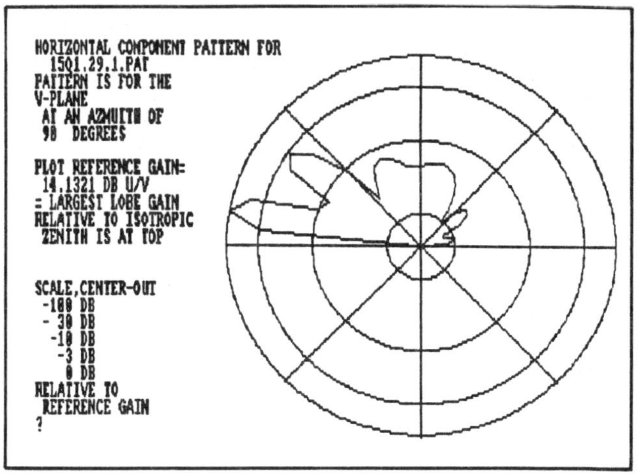

Figure 20. Horizontal component pattern for a four-element 15-meter quad at 60 feet height. Feed at bottom center. This is a high F/B ratio design.

Data for Design Purposes

Table 1 summarizes the design and performance characteristics of this series of example quads. These dimensions should be followed closely if the stated performance is needed. Scaling of size and spacing can be used to move the design points to other parts of the band. If specified performance on different bands is needed, scaling should be complete, including conductor size. Loop size only can be scaled to compensate for conductor size change, as described earlier in the book.

The Bi-square Beam Revisited

The bi-square beam is composed of a pair of top-open diamond loops at right angles to each other, which are fed at the bottom, one at a time. Usually the beam is close to the ground, but the design has occasionally been used at considerable height.

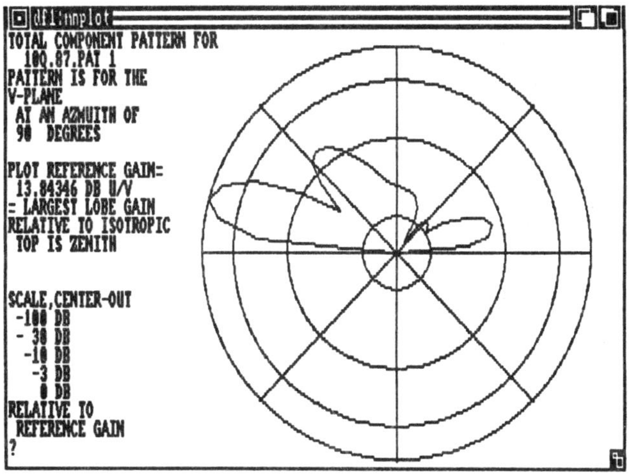

(A)

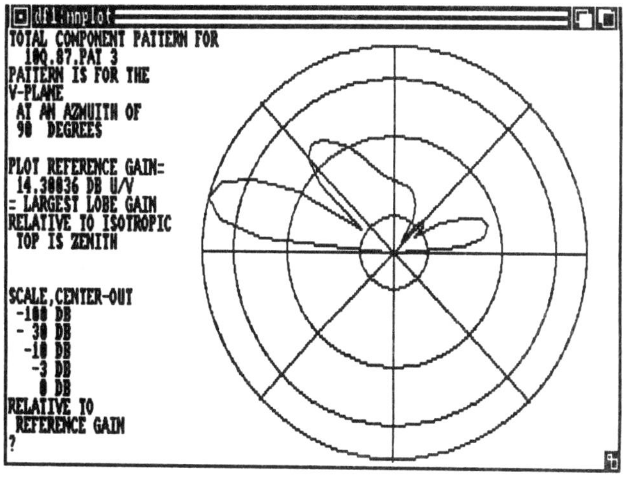

(B)

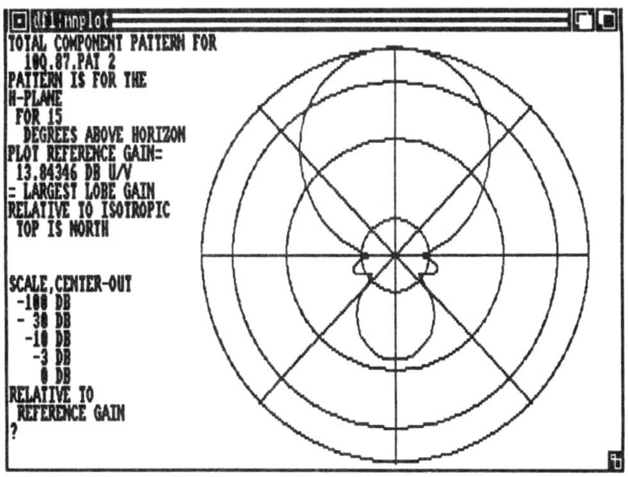

(C)

Figure 21. Patterns for a good bandwidth four-element 10-meter quad at 30 feet height, feed at bottom center. (A) Total component pattern in the vertical plane at right angles to the elements, for average soil. (B) Same as A, but for antenna above seawater. Note the change in gain and pattern detail. (C) Total component at 15 degrees above the horizon. Note the presence of minor lobes, typical of multi-element arrays.

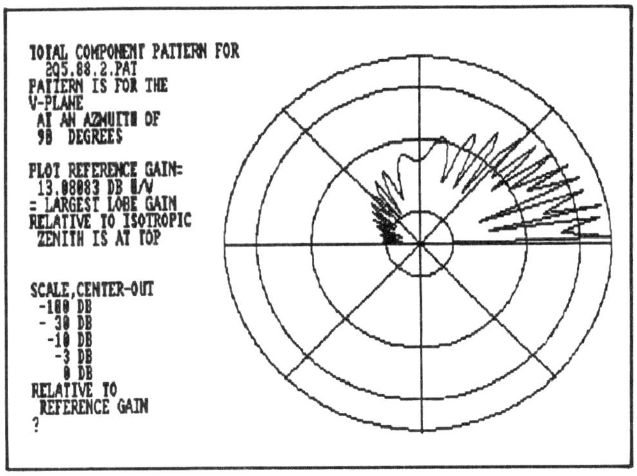

(A)

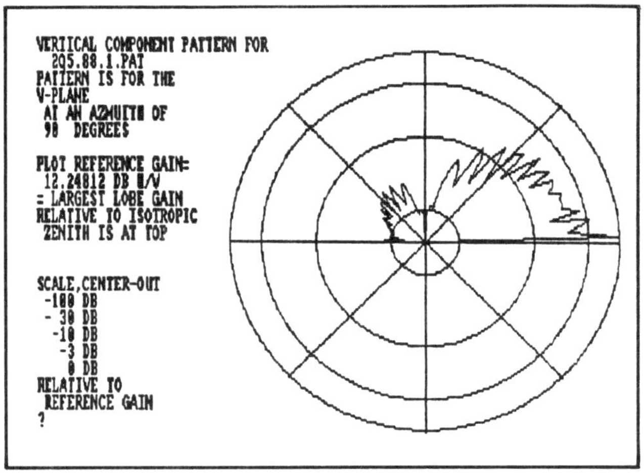

(B)

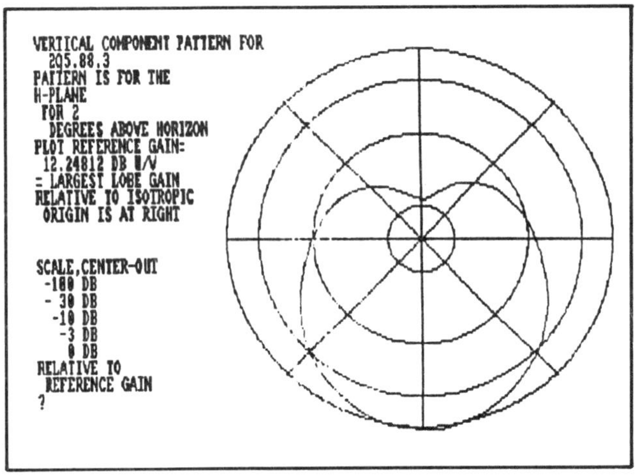

(C)

Figure 22. Patterns for a three-element 2-meter quad designed for DF work, and mounted at 40-feet elevation. (A) Total component in the plane at right angles to the elements, with feed at bottom center. (B) Vertical component pattern, antenna fed at side center. Note changes in lobe structure and gain. (C) Vertical component pattern 2 degrees above the horizon, side center feed. When used for fox-hunting at a height of 6 feet or so, the vertical patterns will show fewer lobes, with the horizontal pattern nearly the same.

Table 2 summarizes typical performance. This unit is 17.44 feet (5.317 meters) on a side, corresponding to 1-wavelength circumference on 20 meters. The bottom is 9.84 feet or 3 meters above average soil.

Because the top is open, the beam is voltage (high impedance) fed on the fundamental and all even harmonics. The drive reactance is high, and changes rapidly with small changes in frequency. Quarter-wave stub matching is usually used with these beams, but open wire line with a Hi-Z balanced to unbalanced match unit is more flexible, allowing multi-band operation.

On the fundamental the gain is about the same as a dipole, with the lobe at 30 degrees elevation, very good for a low antenna. Gain on the second and fourth harmonics is about that of an isolated two element quad, again with a lobe around 30 degrees. There is a considerable vertical component on the harmonics. pattern plots are shown in Figures 23A, B, and C and 24A, B, and C.

One possibility is to use this antenna with the dimensions I have given here as a wide band antenna covering the bands from 20 down to 6 meters. Performance would also be usable on 40 meters, but the feed resistance would be low. Another possibility is to scale the design by 2, giving low angle dipole gain on 40 meters, good gain on the bands from 20 to 10 meters, and usable results on 80 meters.

These antennas are vertically polarized on the design frequency and harmonics. As shown earlier, their performance is better over good soil, and especially over sea water. They got their reputation for good performance from post World War II operation from various Pacific islands.

The Horizontal Quad

As noted in Chapter 1, a quad has components of radiation in the plane of the loop. While this is an adverse factor in normal quad use, from the view of side

Frequency (MHz)	Elements	Sides (ft.) Director	Radiator	Spacing (ft.) Director	Reflector	Wire (inch)	Height (feet)	Fig. no.
7.1	2	none	38.82		21.18	0.080	52.50	15
14.1	2	none	19.57		10.67	0.040	40.00	17
a14.1	3	15.70	19.19	20.94	10.47	0.105	60.00	18
b14.1	3	15.70	19.17	20.94	10.47	0.105	100.00	19
21.2	4*	11.18	12.92	11.74	7.05	0.040	60.00	20
28.5	4*	7.77	9.50	10.36	5.18	0.105	30.00	21
147.0	3	1.48	1.91	1.73	1.04	0.125	40.00	22

	Z_{in} (ohms) R X	First Lobe Gain	Elevation (degrees)	F/B (dB)	Beam Width (degrees)	Second Lobe Gain (dB)	Elevation (degrees)
7.1	167 + j208	9.77	30	24	75	− 2.20	90
14.1	194 + j163	11.26	20	14	80	− 1.80	90
a14.1	70 + j98	13.53	15	14	70	6.69	50
b14.1	71 + j99	14.13	10	14	60	9.94	35
21.2	126 + j154	14.13	10	31	60	10.73	35
28.5	78 + j101	13.84	15	12	70	6.20	50
147.0	155 + j142	12.24	2	28	85	9.50	8

Two-meter antenna side-center fed. Others bottom center.
*Directors of equal size and spacing.

Table 1. Design, assumed installation, and performance data for the series of quads of Figures 15 through 22. If used at other heights, there will be a corresponding change in the vertical lobe structure, plus small changes in gain and feed impedance. Dimensions may be scaled for other bands (see text).

QRM pickup, it opens possibilities for low elevation horizontal antennas.

Table 3 summarizes the performance of a square loop 20 meters or 66 feet on a side, mounted 3 meters above average soil. Even on 80 and 40 meters, the gain is about that of a dipole in free space. On 20 and above, the gain is about that of a beam in space. The antenna is horizontally polarized, and is voltage fed on 20 meters and higher, and current fed on the lower bands.

Horizontal pattern cuts at several frequencies are shown in Figure 25A, B, C, and D. Radiation approaches being omnidirectional, with the number of lobes changing with frequency, as explained in Chapter 1. Nulls are present, of course, but typically average about 10 dB below the main lobes.

BI-SQUARE BEAM

	14.15	28.30	42.85	MHz
Main lobe				
Gain	2.64	10.35	8.71	dB
Elevation	30 deg	30 deg	40 deg	deg
Az	0 deg	90 deg	90 deg	deg
2nd lobe				
Gain		0.35	5.00	dB
Elevation		80 deg	80 deg	deg
Drive				
Resistance	810.00	1310.00	1090.00	ohms
Reactance	− 3260.00	− 920.00	− 1060.00	ohms
Beam side	17.44 ft	or	5.32 M	
Bottom	Height	9.84 ft	or 3 M	
	Average	soil		

Table 2. Major performance characteristics of a bi-square beam. Design is typically for 20 meters. Performance is shown for fundamental, second, and third harmonic. Performance will be useful for intermediate bands. Designs for other frequencies will be different unless all dimensions are scaled by the ratio old frequency/new frequency. Over good soil or sea water, these beams have good performance.

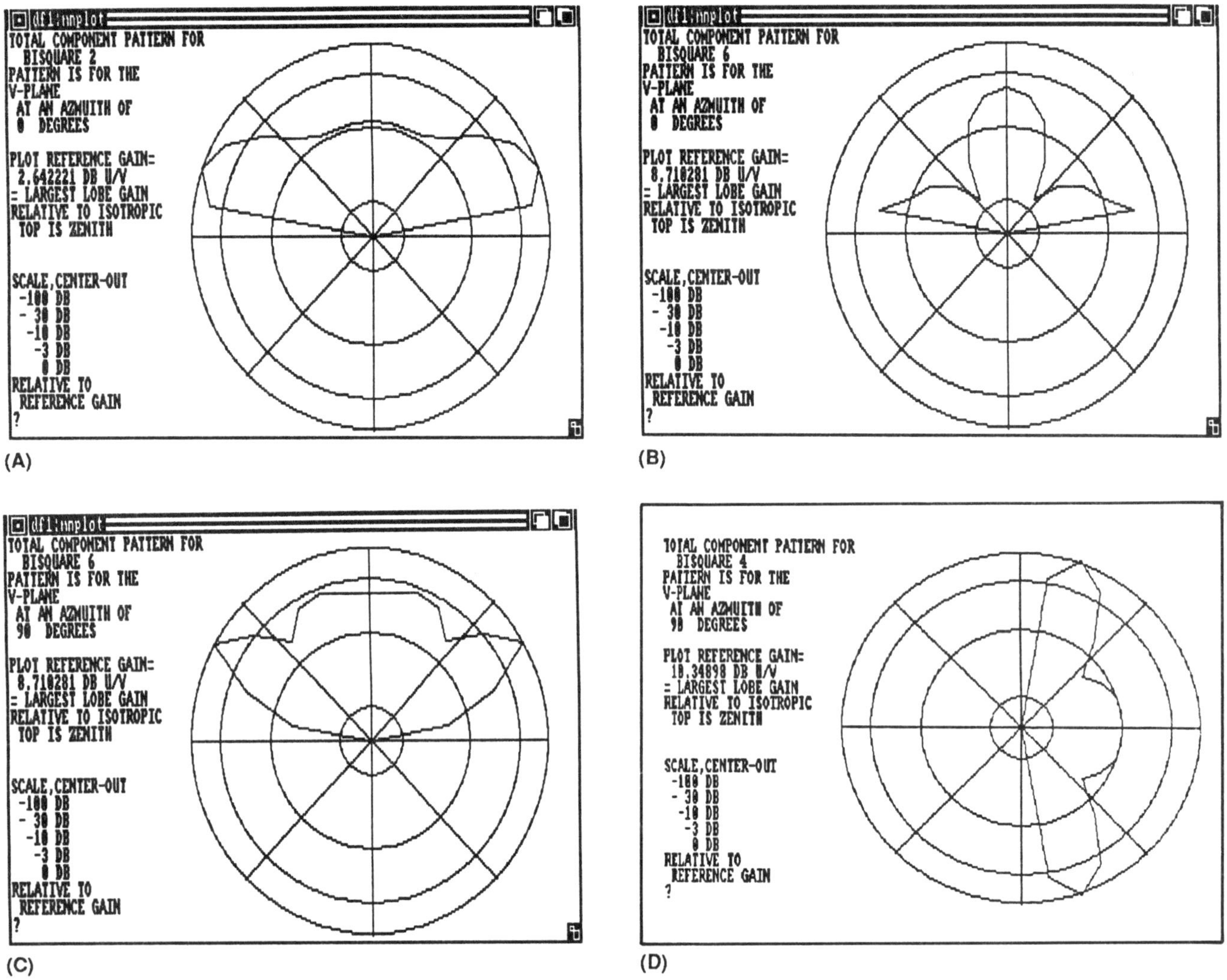

Figure 23. Vertical plane patterns for a 14-MHz bi-square beam in the plane of the loop, and at right angles to it. (A), (B), at fundamental frequency. (C), (D), at second harmonic. Note the relatively low major lobe.

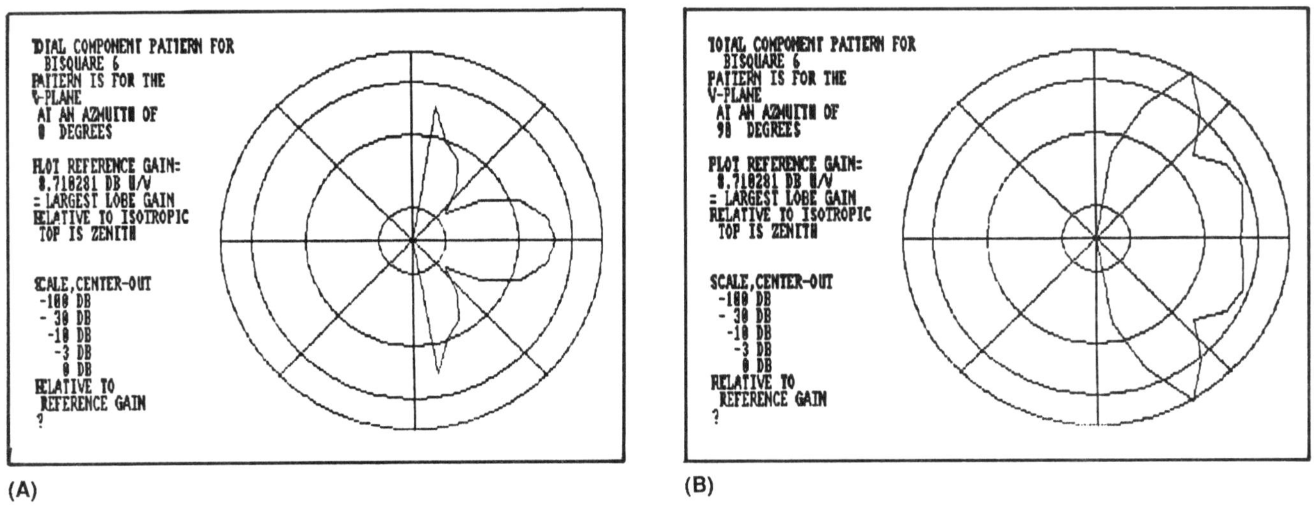

Figure 24. Vertical plane patterns for the bi-square beam at the fourth harmonic. (A), in the plane of the loop; (B), at right angles to it.

HORIZONTAL QUAD

MHz Frequency	Ohms Drive R	Ohms Drive X	30 deg Gain	deg at Az	Number Lobes	DB Max gain	At deg Elevator
1.8			2.75	90	1	7.60	90
3.5	4	– 248	1.49	90	1	19.33	90
7.0	25	– 389	5.71	0	2	8.02	
10.5	75	– 752	5.55	40	3	6.46	90
14.0	222	– 1036	3.86	90	3	9.20	90
17.5	464	– 858	6.37	50	3	9.65	90
21.0	659	– 683	8.56	60	3		
24.5	942	– 884	8.68	55	4		
28.0	799	– 697	8.22	70	4		
31.5	737	– 353	8.42	20	4		
35.0	614	– 216	9.29	30	5		

Sides 65.6 ft or 20 meters
Height 9.85 ft or 3 meters
Wire radius 0.196 inches

Table 3. Major performance characteristics for a horizontal quad loop nominally resonant at 40 meters, mounted close to ground. While primarily useful for intermediate range (E-layer) propagation, these are useful for F-layer DX on the higher bands. Performance will be different if scaled to other bands unless all dimensions are scaled by the ratio old frequency/new frequency.

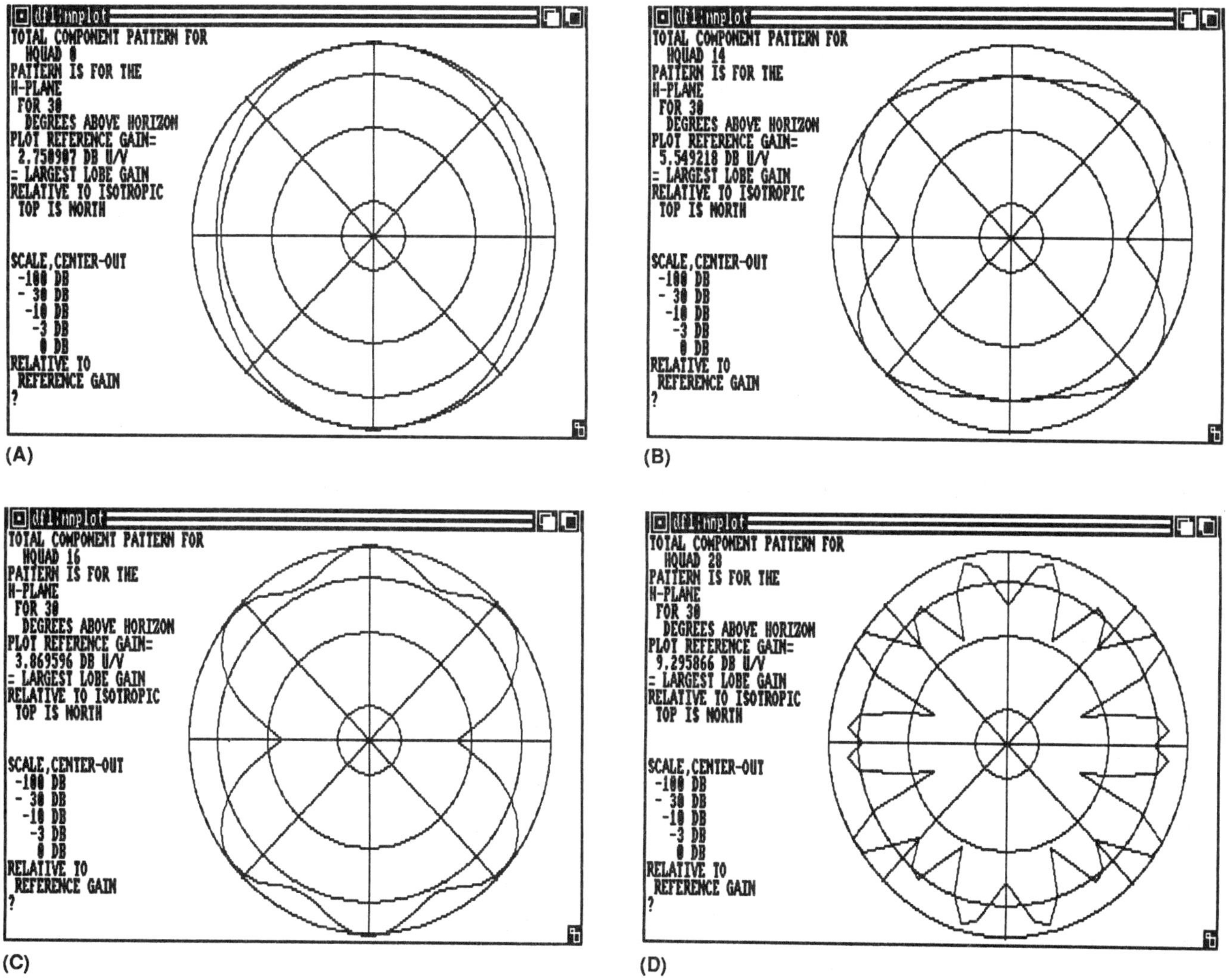

Figure 25. H-plane pattern for the plane 30 degrees above the horizon for a horizontal quad loop nominally resonant at 80 meters, and 3 meters above average earth. (A) at 1.8 MHz. (B) at 10.5 MHz. (C) at 14 MHz. (D) at 35 MHz. Note that the pattern remains approximately omnidirectional, with the number of lobes increasing as frequency increases, as discussed in Chapter 1.

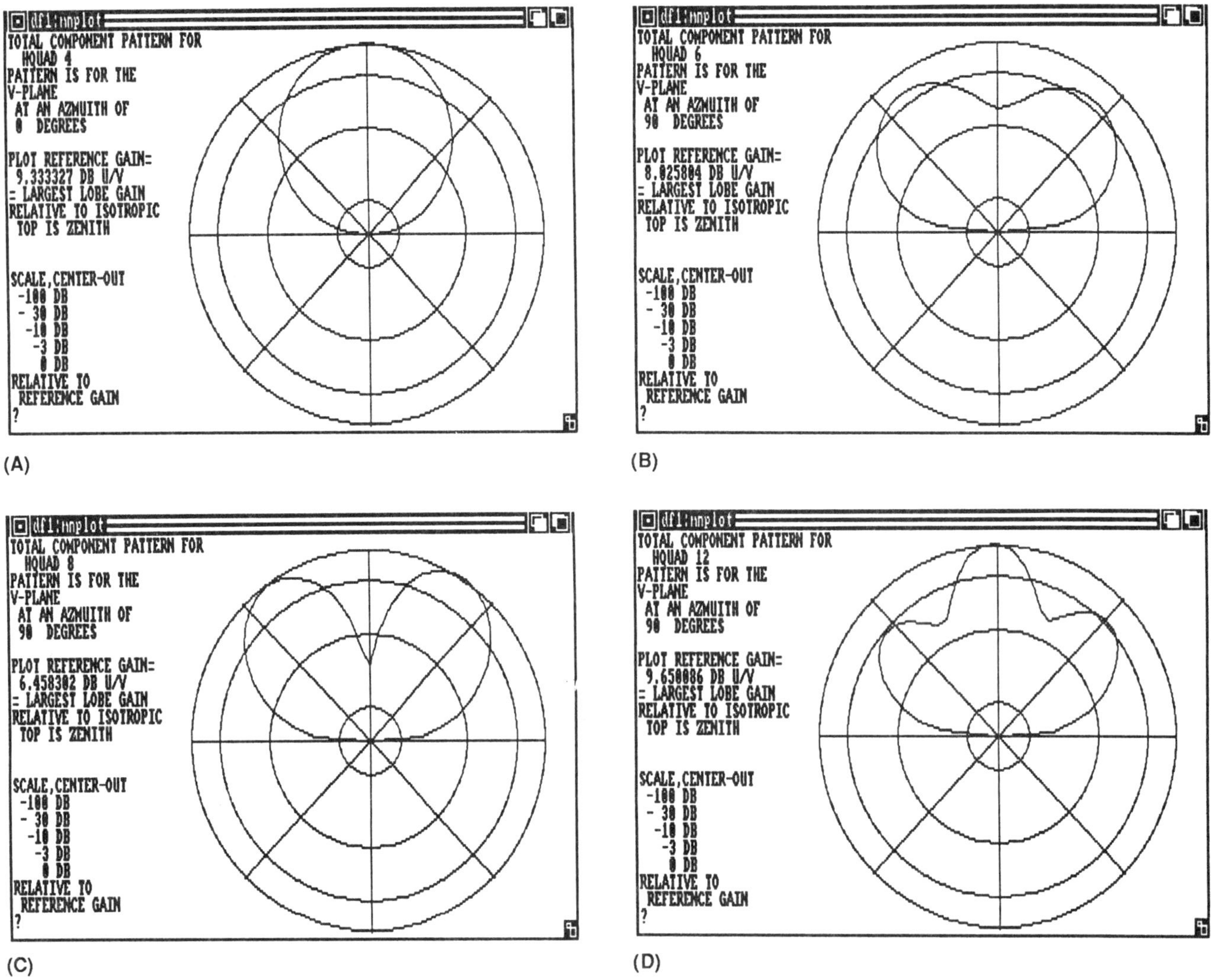

Figure 26. Vertical plane patterns for a horizontal quad loop nominally resonant at 80 meters and 3 meters above average earth. (A), at 80 meters, in the plane containing the feed point of the loop. (B), at 40 meters, at right angles to the feed point. (D), at 17.5 MHz, at right angles to the feed point. Patterns for other frequencies are variations of these examples. Because of the relatively high lobe angles, this type of antenna is primarily useful for intermediate range E-layer propagation.

Vertical plane patterns are shown in Figure 26A, B, C, and D. At low frequencies the main lobe is to the vertical, so the antenna primarily good for short distance E-layer communications. At higher frequencies, the lobe angle moves toward the horizontal. Performance would be good out to maximum E-layer hop distance, and acceptable, although not great, for F-layer DX.

The performance isn't greatly sensitive to exact shape. A diamond or triangle with the same wire length will approach the same performance (see the discussion an data on squashed diamonds and rectangles in Chapters 1 and 4). Antennas of this family can solve troublesome site problems, even with the fine wire needed for ''invisible antennas.''

Chapter 8

Delta Loops and Arrays

Design data for delta and triangular antennas

The delta loop, or inverted triangle, wasn't developed primarily to obtain a better radiator, but to try to remove some of the mechanical problems of the quad. Two sides of a triangle are used as support tubes with the bottom ends held firmly on a boom. Wire is used to connect the top ends as the third side of the triangle. Thus, the Greek letter delta is formed.

Delta Loop

An idealized delta loop is shown in Figure 1 with the nomenclature used here. In the practical loop, the sides are slightly bowed to keep the top (usually wire) under tension. This difference is ignored for this analysis. Also, this examination of the delta loop isn't as extensive as that for the quad. Partly to compensate for this, a number of the curves show both quad and delta-loop data. This helps clarify the family resemblance of the two.

Self Impedance

The self- or driving-point resistance of an isolated delta loop is shown in Figure 2. Figure 3 shows the self reactance. The curves are very close to those for the corresponding quad loop. The major difference is a shift of the parallel resonance points to a larger perimeter. These are for the same conductor size for the three arms.

When the top arm is smaller, there is a transformer effect related to that of the gamma match using different radiator and match-rod sizes. The influence of this is partly illustrated by comparison with Figure

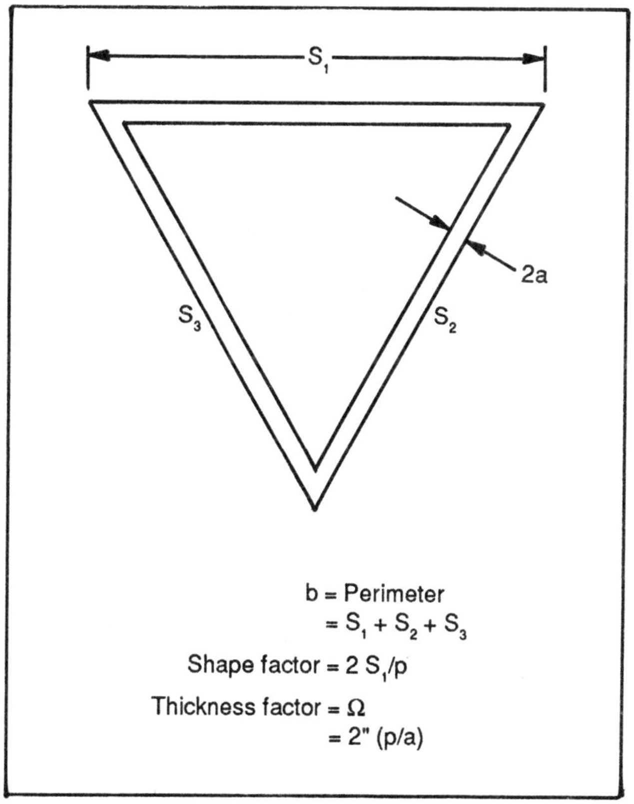

Figure 1. Schematic and nomenclature for a delta loop. In most designs the top is a wire of different diameter than the sides. See text for a discussion of the effect of this and other features on the results of the analysis.

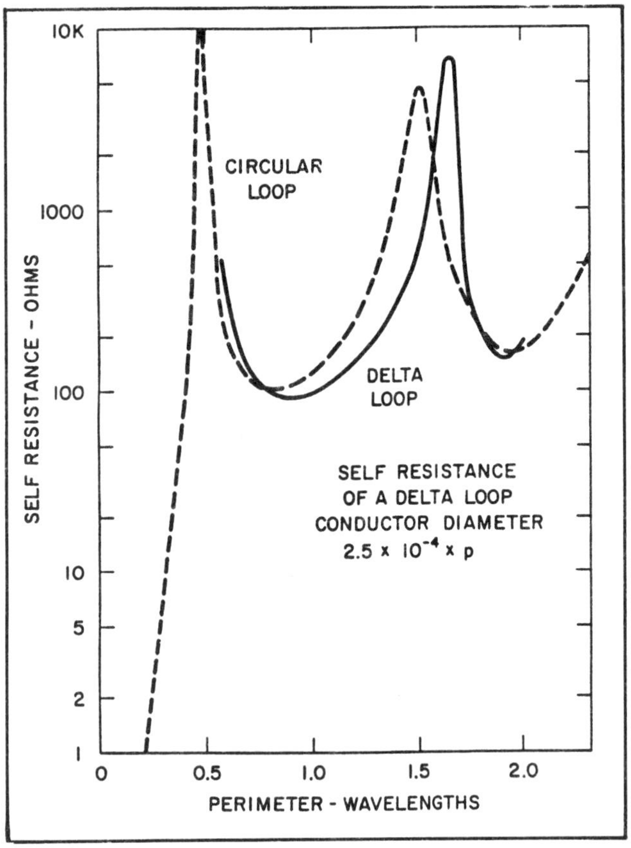

Figure 2. Self resistance of an isolated delta loop with that of a quad loop for comparison.

4, which shows loop drive resistance with two different conductor combinations. One effect of this different conductor size is that the drive impedance changes as the feed changes from the large conductor to the small one.

There is also an analysis problem: MININEC assumes that current is constant over a segment. This isn't a problem with center feed, but problems do occur with end feed and with feed at a corner. Because of the last factor, nearly all of the results presented here are for top-center feed. Be aware that the feed impedance for the normal delta loop design with bottom-corner feed will be different. (This shouldn't be a problem, because such designs are normally fed with an adjustable gamma match.)

Figure 5 shows the effect of thickness factor on self resistance, with quad data for comparison. As expected from the smaller loop area, the delta self resistance is lower. Figure 6 shows the self reactance as a function of perimeter for several conductor sizes. The curves are generally similar to those for quads (Figure 6, Chapter 4). The zero reactance or resonant points are plotted in Figure 7, with quad data for comparison.

Gain

The gain (perpendicular to the delta loop) versus loop perimeter is shown in Figure 8 with quad and dipole gain for comparison. As expected from area considerations, the maximum gain is lower. However, the variation of gain is different from that for a quad, and resembles that of a dipole. This suggests that the delta is very useful as a multiband radiator. For the usual design, operation on, say, three adjacent bands could be obtained with three gamma matches fed from a common point. Open-wire feed to the top would give a good wideband design.

All of the above data are for idealized equilateral delta loops. The effect of changing shape is shown in Figures 9 and 10. As for quads, the shapes which separate the high-current points increase the gain and

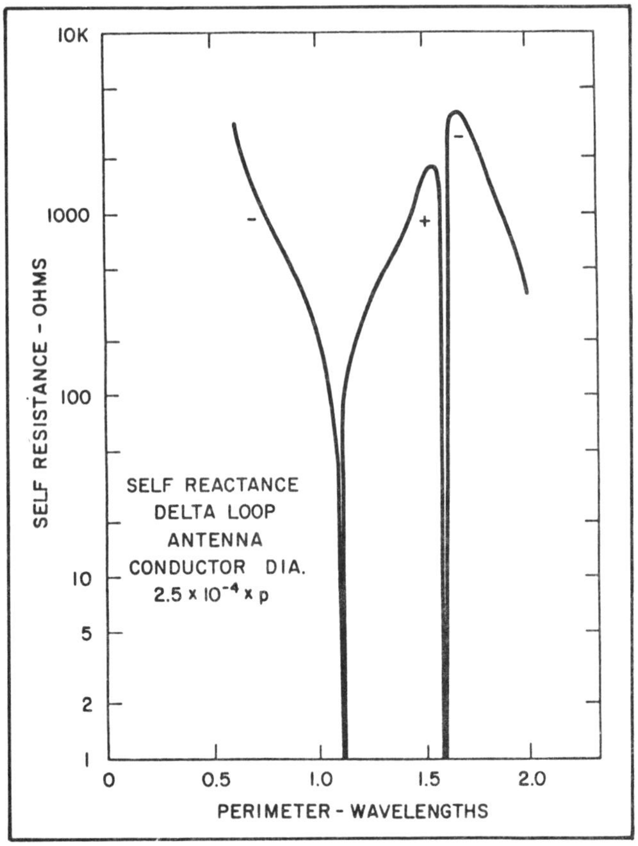

Figure 3. Self reactance of an isolated delta loop. Compare with Figure 6 of Chapter 4. The points of zero reactance and the magnitudes of the peaks are sensitive to conductor size.

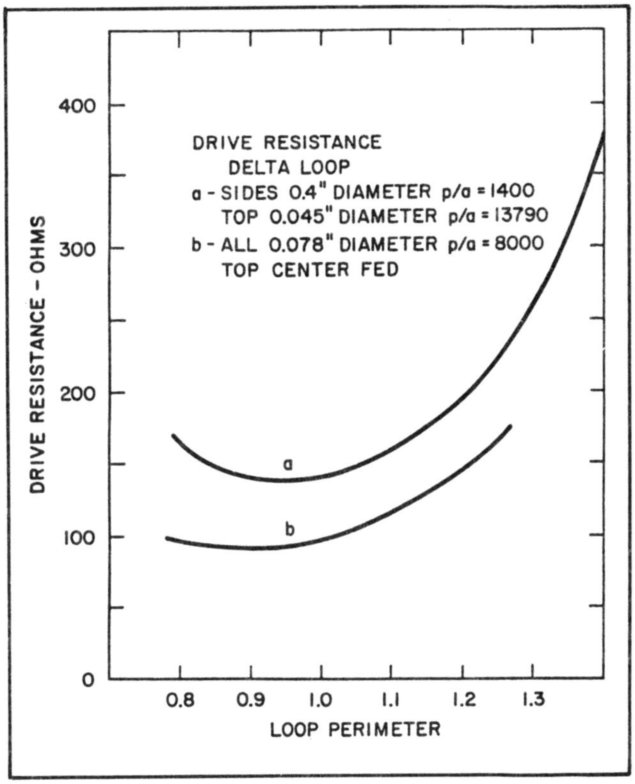

Figure 4. Variation of self (drive) resistance with loop perimeter. The drive resistance of the normal type depends on the feed point (see text).

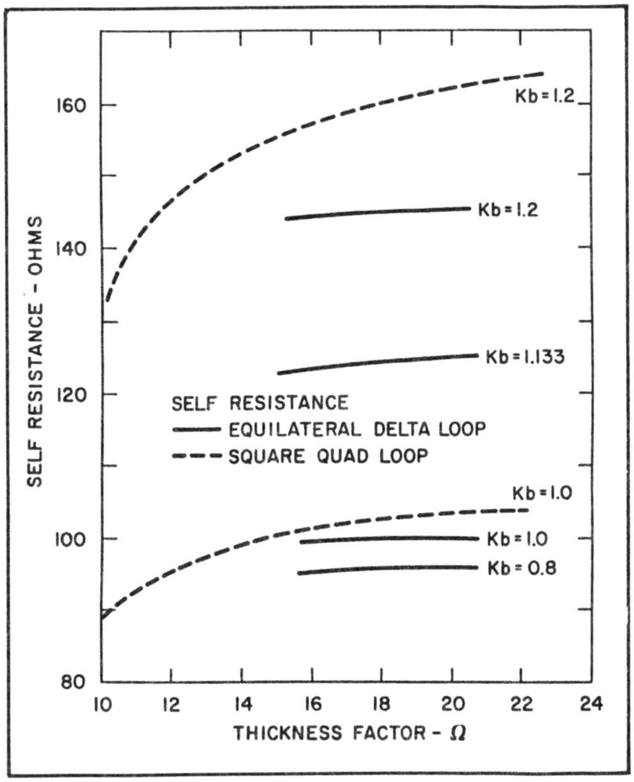

Figure 5. Variation of self (drive) resistance with conductor size with quad data for comparison.

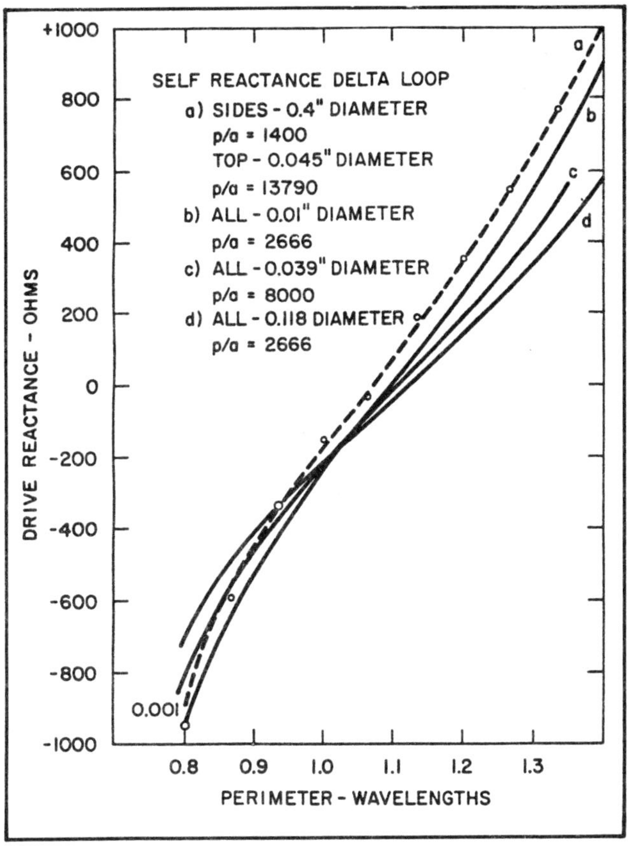

Figure 6. Variation of self (drive) reactance of an isolated delta loop as a function of perimeter. Use this data to determine the loop size giving the parasitic reactance needed to obtain desired performance.

reduce the drive resistance. (Drive reactance is not greatly affected.)

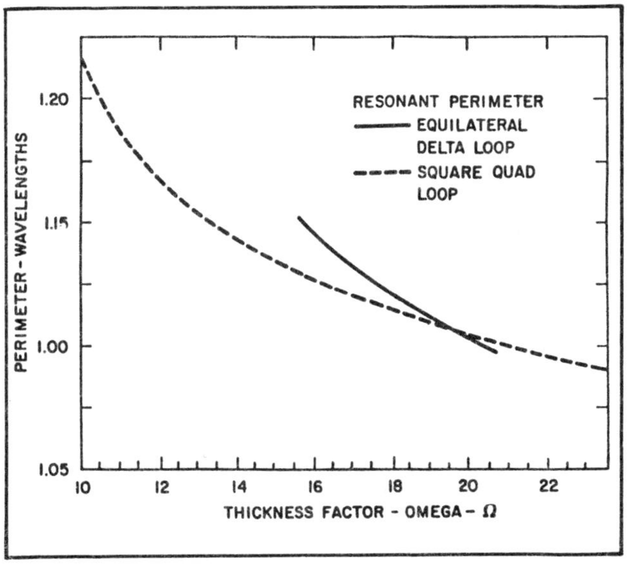

Figure 7. Perimeter size for resonance (zero self reactance) versus conductor size. Quad data is for comparison.

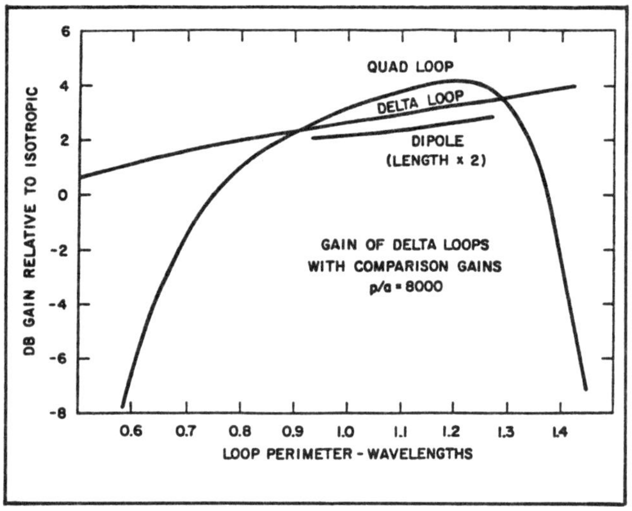

Figure 8. Variation of gain of quad and delta loops versus perimeter with dipole data for comparison. This is for the lobe at right angles to the loop plane or dipole. All curves go to numerical zero (infinity dB) gain at a perimeter of 2.0 wavelengths.

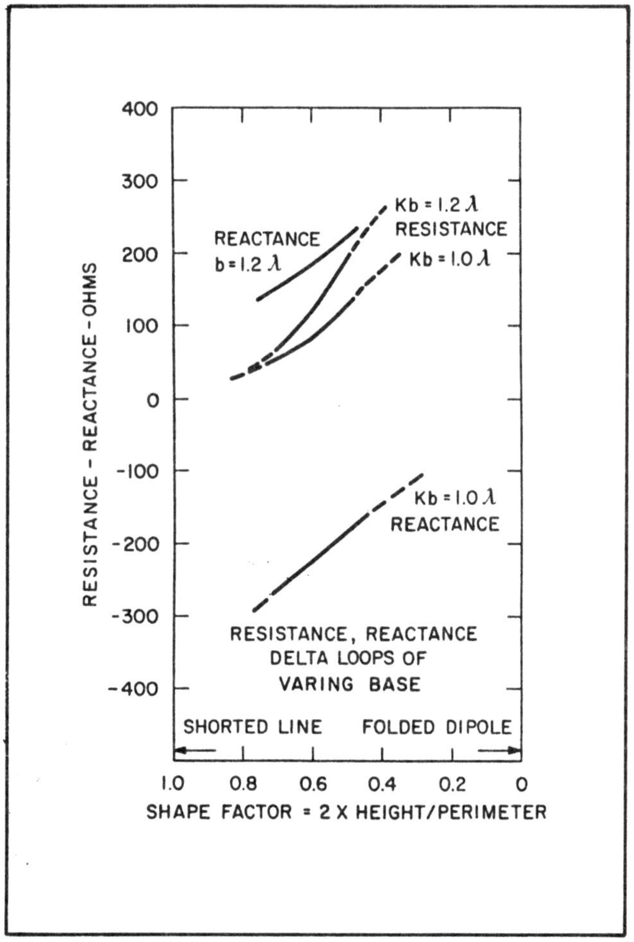

Figure 9. Effect of change in shape factor on drive resistance and reactance of delta loops. Moving the two high-current points further apart reduces the drive resistance.

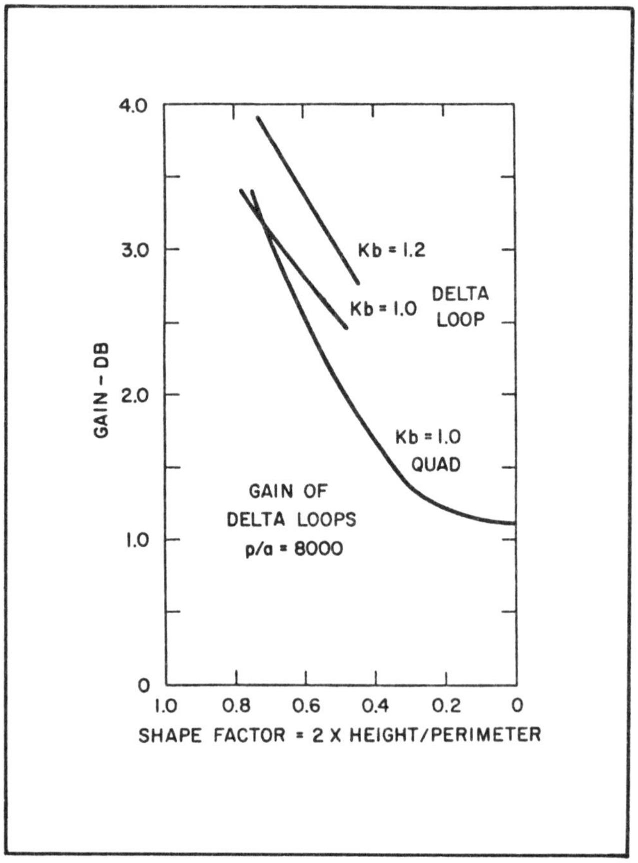

Figure 10. Effect of change of shape factor on delta-loop gain. For factors greater than about 0.9, the gain rapidly falls to numerical zero (infinity dB).

Patterns

In the following, all patterns are for top-center feed. Figure 11A shows the H-plane pattern for a one-wavelength delta. This pattern is essentially identical to that of a quad, but the gain is about 0.3 dB lower. The vertically polarized component is shown in Figure 11B and is maximum in the plane of the loop, as for the quad.

Figure 12A shows the upper half of the vertical component in the vertical plane of the loop. The lower half is a mirror image of this. The H-component is zero. Figure 12B shows the total component at right angles to the loop; the vertical component of this is very small, about 80 dB lower.

Figure 13A shows the H-component pattern in the H-plane for a 1.5-wavelength loop. The V-component is at least 10 dB lower. The half vertical plane pattern is shown in Figure 13B. As for the quad, symmetry evaluation shows that the perpendicular lobe disappears on-axis at perimeter = 2.0, proving the sides have the same diameter. With the usual wire-and-tub-

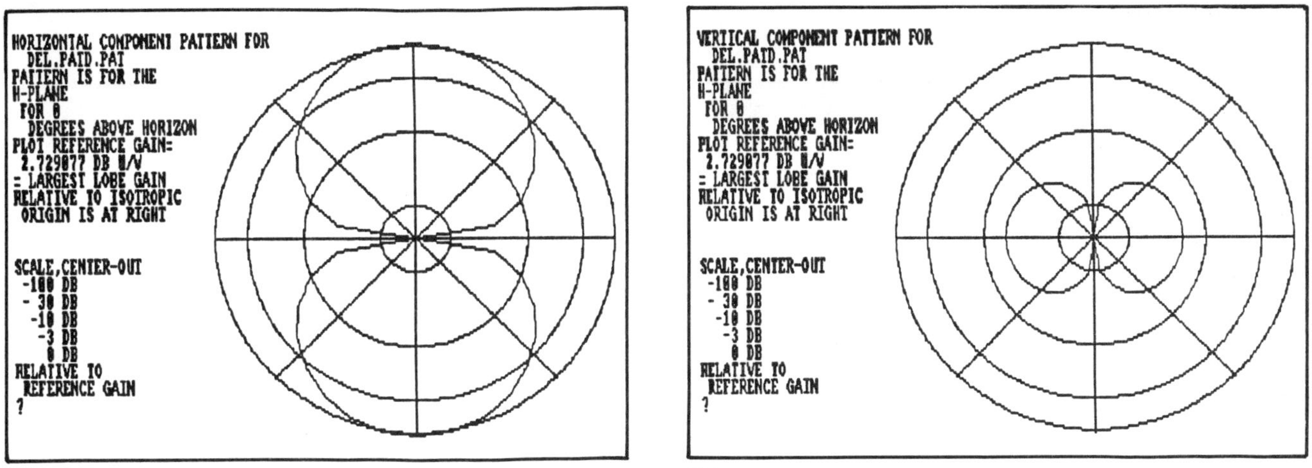

(A) (B)

Figure 11. Typical delta loop patterns. The horizontally polarized component in the horizontal plane is at (A), with (B) showing the half pattern of the vertically polarized component. The other half of the pattern is the mirror image.

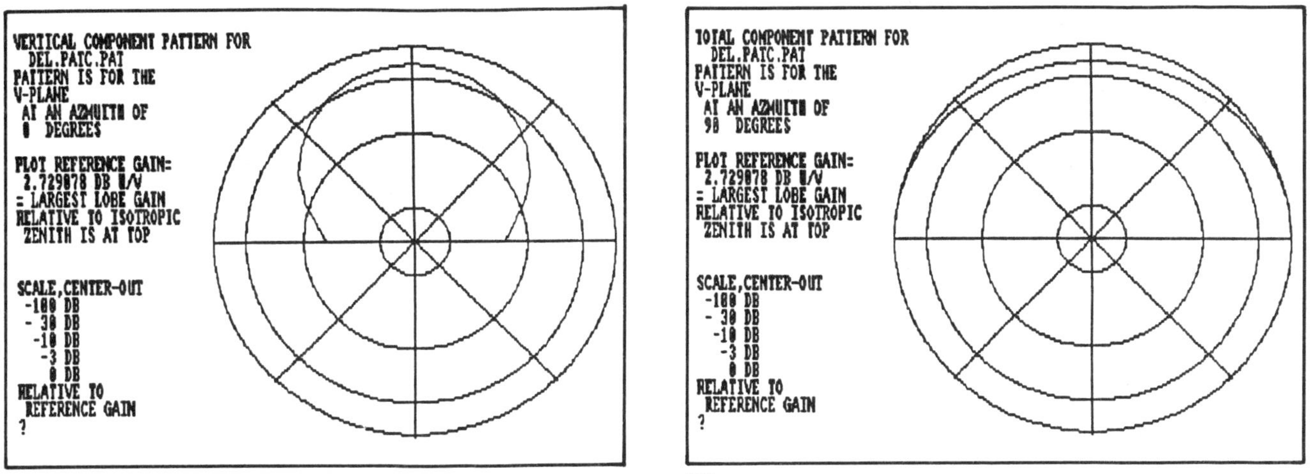

(A) (B)

Figure 12. Typical vertical plane half patterns. (A) shows the vertical component in the plane of the loop, (B) the total component at right angles to the loop.

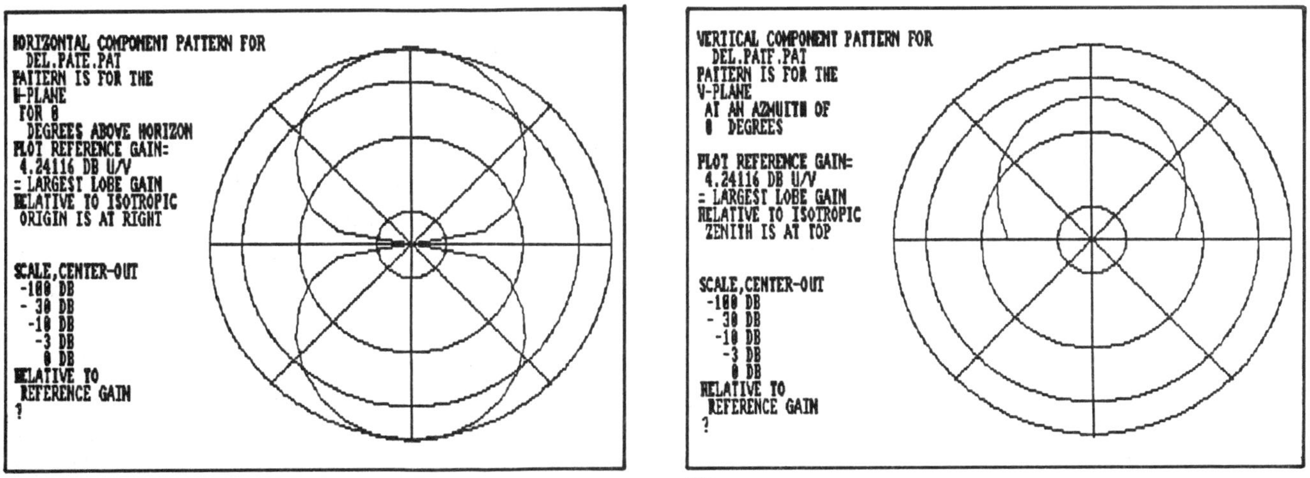

(A) (B)

Figure 13. Patterns for a 11.5-wavelength perimeter delta loop. (A) is the H-component in the horizontal plane; (B) the V-component half pattern in the plane of the loop. Note the increase in gain as compared to Figure 11.

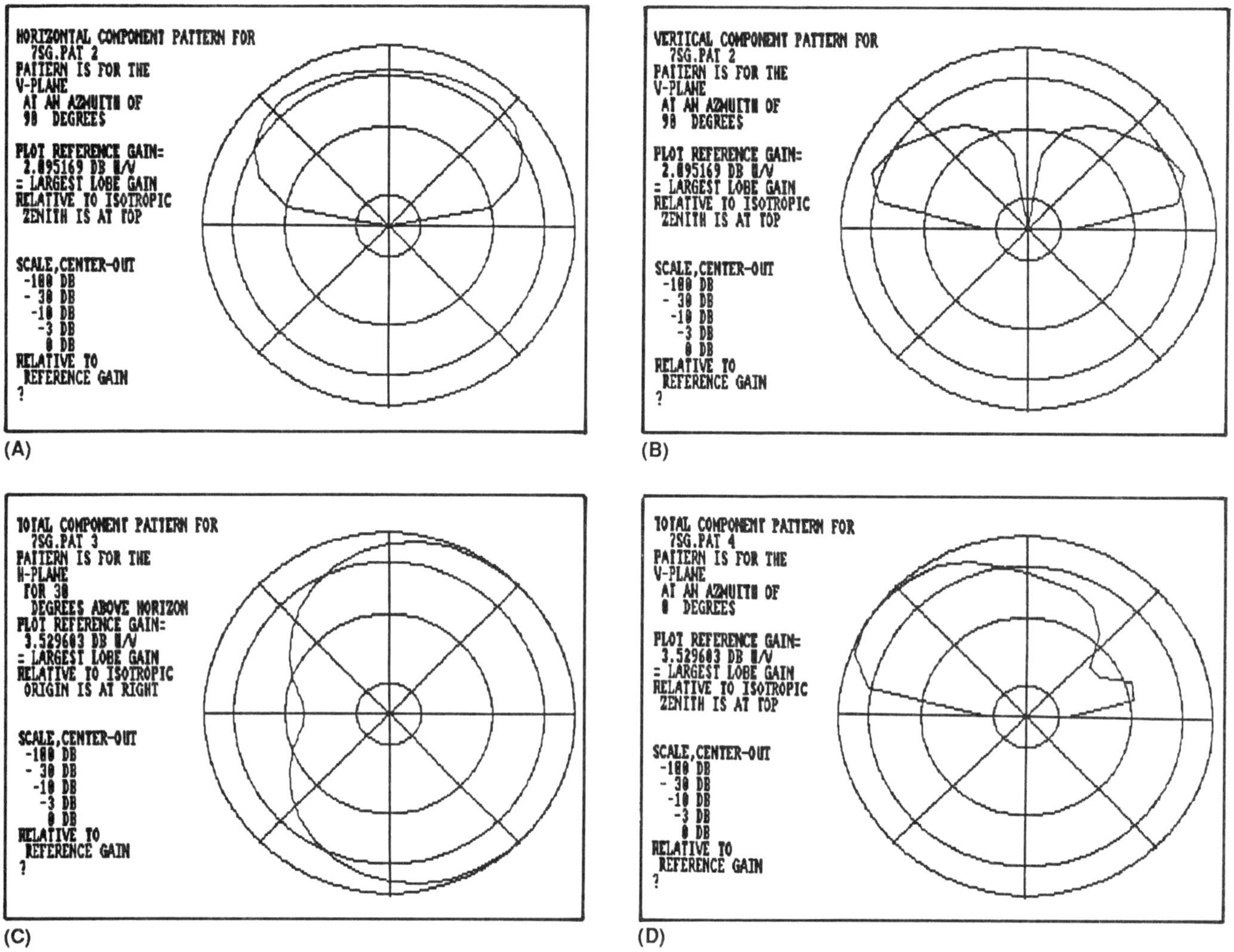

Figure 14. Patterns for a right-angle triangle loop of one wavelength perimeter. The two at (A) and (B) are bottom center fed; (C) and (D) are long side center fed. The horizontal plane pattern corresponding to (A) and (B) is essentially identical to Figure 11A, but with lower gain. The radiation with side feed is primarily vertically polarized.

ing design, currents are not symmetrical, and a residual lobe occurs at perimeters with integer numbers greater than unity.

Shape Variations

It isn't necessary to keep the triangle apex down, nor is it necessary to use only triangles having three (or two) equal angles. Of the many orientations and shapes possible, those which are suitable for suspension from a single tower seem to be the most practical. Three of this family have been examined in a preliminary way, but this has been enough to suggest that they may be very useful on the lower amateur frequencies. If the possibilities are of interest, I recommend that MININEC be used to examine specific designs.

Figures 14A through D show patterns for a right-angle triangle. Figures 14A and B are bottom center fed; C and D are long-side center fed. The vertical component is composed of a typical 60-foot tower, the slant side of a section of 3/16-inch guy insulated from the ground, and the horizontal bottom a piece of no. 12 wire 8 feet above ground. These dimensions give a one-wavelength perimeter at 7 MHz. MININEC analysis shows interesting possibilities.

The patterns in Figures 14A and 14B are for feeding at bottom center, and are for the vertical plane at right angles to the loop. The radiation angle is low, about 20 degrees above the horizon for the perpendicular to the triangle, and about 45 degrees in its plane. The interesting feature of this arrangement shows in D, a horizontal cut 30 degrees above the horizon. The

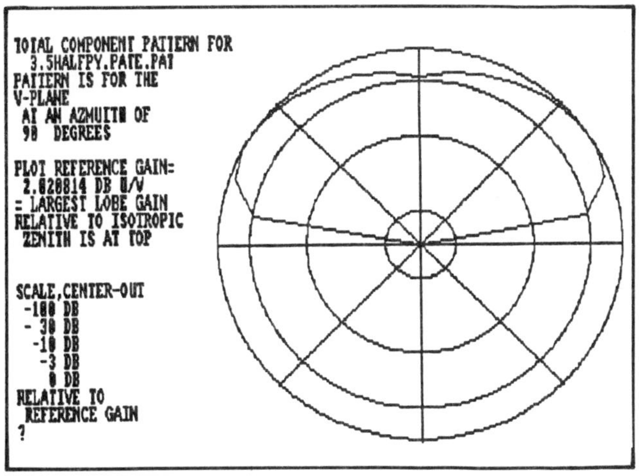

(A)

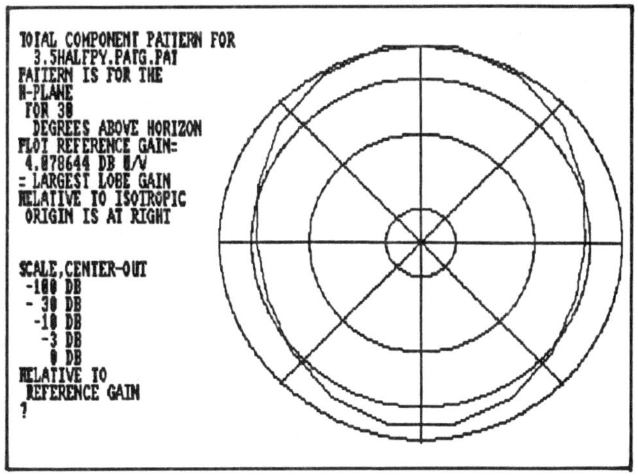

(B)

Figure 15. Patterns for an equilateral triangular loop, with the plane of the loop slanted from vertical to allow a 40-meter loop to be supported from a 40-foot tower, with the bottom 6.56 feet above ground. Feed is at bottom center. (A) shows the total component in the vertical plane at right angles to the bottom wire; (B) the pattern at 30 degrees above the horizon.

total component pattern becomes directive, maximum being in the direction from the tower to the guy. The front-side ratio is less than 1 dB, but the F/B ratio is 13.5 dB. The design should be good in rejecting back-lobe QRM, as can be seen in 14D—the total radiation in the vertical plane of the triangle.

These patterns include the effect of a grounded tower. The effect of tower size, grounding of the guy or tower, or type of soil hasn't been studied.

Several pairs of ferrite half-beads of the type used for computer cable interference elimination can be placed on the guy, just below the horizontal wire attachment. These would introduce a fairly large reactance, giving a RF-choke effect, reducing the current on the grounded part of the guy to a negligible amount. While this has not been tested in detail, it is the principle of the W2DU balun.

A few runs were made with other feed locations, but these need to be studied further. Also, a few tests have been run for a triangle composed of an insulated guy, a wire further from the tower, and a horizontal wire. The need for further study of such types arises because computational problems develop. A sharply acute angle places two conductors close together: for part of their length they can act more like a transmission line section than a radiator. A few results show promising directivity, but more work is needed to determine the limits of acceptable error before drawing conclusions as to usefulness of these acute-angle antennas.

If available supports aren't high enough to keep a triangle antenna in a vertical plane, it can be sloped from the vertical. For example, a 40-foot tower will allow a 40-meter equilateral loop, and an 80-foot tower an 80-meter one. The possibilities are interesting. Figures 15A and B show two patterns. Figure 15A is for the vertical plane at right angles to the loop, with feed at one side center. Figure 15B is for the horizontal cut 30 degrees above the horizon, with bottom-center feed. There is a slight increase in gain toward the loop bottom, the point further from the tower support.

All of these special shapes show less gain than is possible with a high horizontal dipole. However, the gain difference is not great, and the possibility of pattern shaping and attendant interference reduction indicates that more experimentation and analysis are needed.

Delta-loop Array

The classic delta-loop beam uses two delta loops, usually cut to the relationships:

radiator length = 1005/f
reflector length = 1030/f

which are the same as usually used for quads.

Mutual Impedance

The mutual admittance between two 1.0-wavelength equilateral delta loops is indicated by Figures 16 (mag-

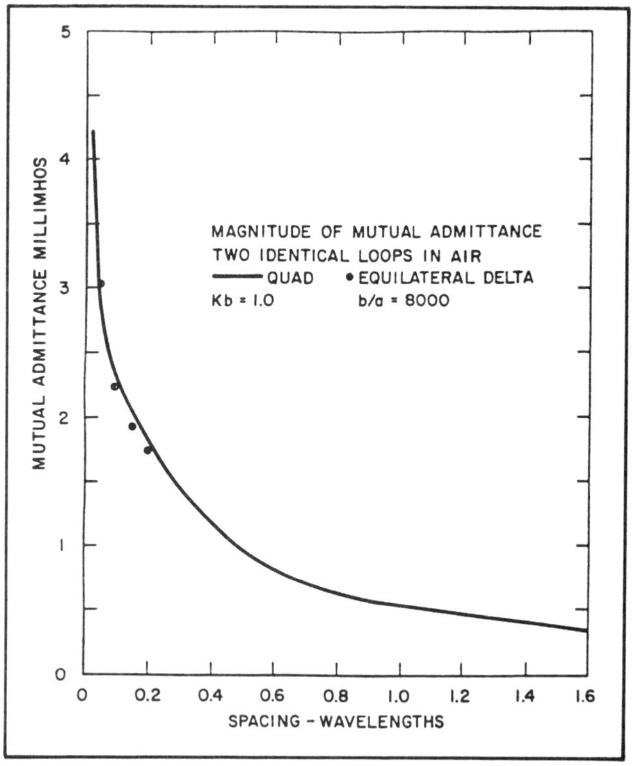

Figure 16. Magnitude of the mutual impedance of two identical loops. Data for an equilateral delta are shown as points with quad data for comparison.

nitude) and 17 (angle). For comparison, and since the complete range was not investigated, the quad curve is shown, with the delta values as points.

Delta-loop Patterns

The family of polar plots Figures 18 1A through 5G shows the effect of element size and spacing on the delta-array pattern. As for the quad, best overall two-element performance is secured with the radiator-reflector combination. Spacing for this should be rather close, with as little as 0.05 wavelength being usable. Best director results are secured with wider spacings.

The overall performance is summarized in Figure 19, showing the intensity of the front and back lobes as a function of spacing. Again, the curves are quads, the delta values are shown as points. Maximum gain occurs at a spacing of 0.15 wavelength, and maximum F/B ratio at 0.1. The lobe intensities are shown again in Figure 20 for the usual range of use. These are for the delta array only.

Tuning the Delta Loop

Figures 21 through 24 show the effect of tuning the parasitic element on the front and back lobes of the delta-loop array. As for the quad, tuning is different for maximum gain and for maximum F/B ratio. For most applications, a spacing between 0.1 and 0.15 wavelength, with the parasitic element just longer than resonant frequency at midband, would appear best.

To review, the procedure for design for specific performance is:

- •Select the performance, either maximum gain, maximum F/B ratio, or a compromise such as maximum bandwidth.
- •Read the parasitic reactance from the curve for the spacing planned.
- •Determine the element size from Figure 6, interpolating for the conductor size planned.

For a typical tube/wire array at 0.15-wavelength spacing, the reflector size will be 1033/f for maximum gain, and 1063/f for maximum F/B ratio.

If a director is used, a factor in the range 980/f to 1005/f appears suitable, the smaller factor being for best F/B ratio. It is recommended that MININEC be used to examine a proposed design; this study hasn't included multi-element arrays. As for the quad, a

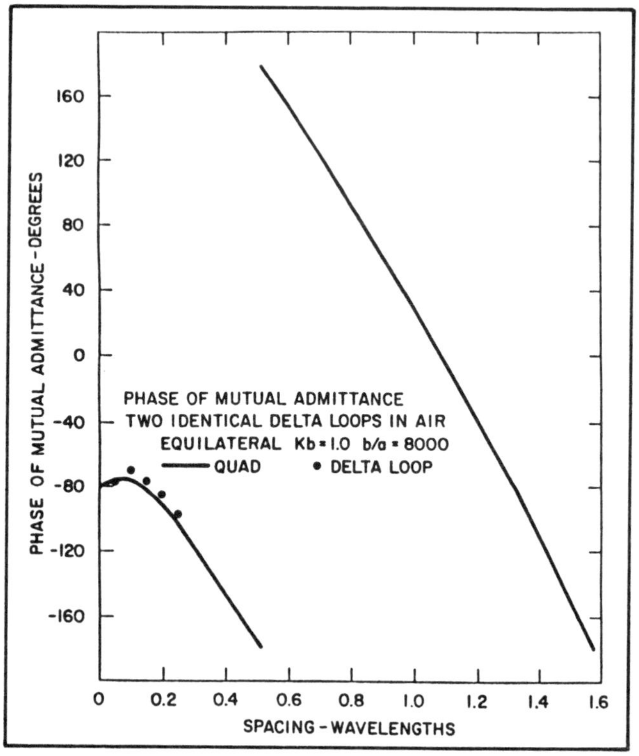

Figure 17. Angle of mutual impedance of two identical loops, again with data for a equilateral delta shown as points with quad data for comparison.

small improvement in gain will result with the radiator the same size as the reflector.

Feed Impedance

The calculated feed impedance for two identical delta loops is shown in Figure 25. Part A shows the effect of separation: as this increases, the values will converge to the isolated-loop values indicated by the arrows. The general pattern will resemble that of Figure 20 in Chapter 5.

The variation of drive impedance with element tuning is shown in Figure 25B. Both drive resistance and reactance change by a factor of over two. The same pattern has been seen in quads, (Figure 24, Chapter 5).

These curves, plus the equivalent ones for the quad, suggest the following guide for tuning loop arrays:

- Measure the drive impedance, using an R/X noise bridge (or a better type), recording tuning condition and impedance.
- Plot the resistance and reactance against tuning conditions.
- If there is a minimum in the resistance curve, the corresponding tuning is at or close to maximum gain tuning.
- If there is a maximum or minimum in the reactance curve, that tuning is at or close to maximum F/B ratio tuning.
- If there is no minimum or maximum, tuning is too far from optimum.

Note that the maximum F/B ratio is about half way between a resistance maximum and minimum.

This method of tuning has not yet been tested on a practical antenna. The MININEC data analyses so far suggest that it is quite general for parasitic arrays.

A 15-meter Delta Array

Suppose that a single-band, two-element delta loop is desired for installation on a 40-foot tower. Maximum F/B ratio at 21.1 MHz is chosen as the design criteria. From the curves, this condition requires a parasitic reactance of 20 ohms, obtained at 1.08-wavelength circumference. Scaling the tubing-wire delta loop gives the following dimensions:

length of side	8.35 feet
tube diameter (average)	0.7 inch
wire size	no. 12

The spacing selected is 0.15 wavelength, or 6.88 feet. The radiator is chosen to be same size as the reflector.

MININEC analysis shows that the gain of this array is good, 11.65 dB above isotropic. The F/B ratio is 22.1 dB. Feed resistance is 170 ohms; feed reactance 194 ohms inductive for top-center feed; bottom feed would be less. The reactance would be essentially zero if a radiator of 0.99-wavelength circumference had been used.

The patterns for this array at the design frequency are shown in Figures 26A through C. The horizontal plane pattern at A is typical of parasitic antennas at or near maximum F/B ratio. At this height, the vertical-plane pattern at B shows three lobes: the top near 90-degrees elevation, the second just below 44 degrees, and the lowest at 12 degrees (C shows the forward lobe with better plotting resolution). There is some radiation in the plane of the loops, as shown at D. This is about 10 dB down for the center lobe.

This antenna would be a respectable DX performer. For emphasis on shorter-range contacts, the height should be reduced to about 1/2 wavelength, or 20 to 25 feet. And, of course, greater height would give still lower bottom-lobe radiation angle, increasing DX capability.

Performance will be the same on other bands if the element size and spacing are scaled by the ratio 21.2/f(new). This includes tubing and wire-size scaling. (These sizes were chosen to give strength suitable for most parts of the United States; for hurricane and heavy icing areas, larger tubing is needed.)

A 3.6-MHz Triangle Array

Directive antennas on 80 meters are still scarce. The sloping delta gives promise of changing this. While only one of the many possibilities has been explored in this study, the results indicate that immediate practical use is worthwhile, and there are possibilities for improvement through experimentation and analysis.

The array chosen for this study is a pyramid composed of two equal-sized triangular loops, slanted for single-point support and to reduce height. The top of the loop was set at 80 feet, just below the top of an 80-foot tower. It was assumed that short booms were used to support the loops, the booms being 6.56 feet from the tower center. The bottoms of the loops were assumed to be 46 feet from the tower center and 9.84 feet above ground. Bottom length was set at 91 feet. The total length of wire is 280.55 feet. The wire size

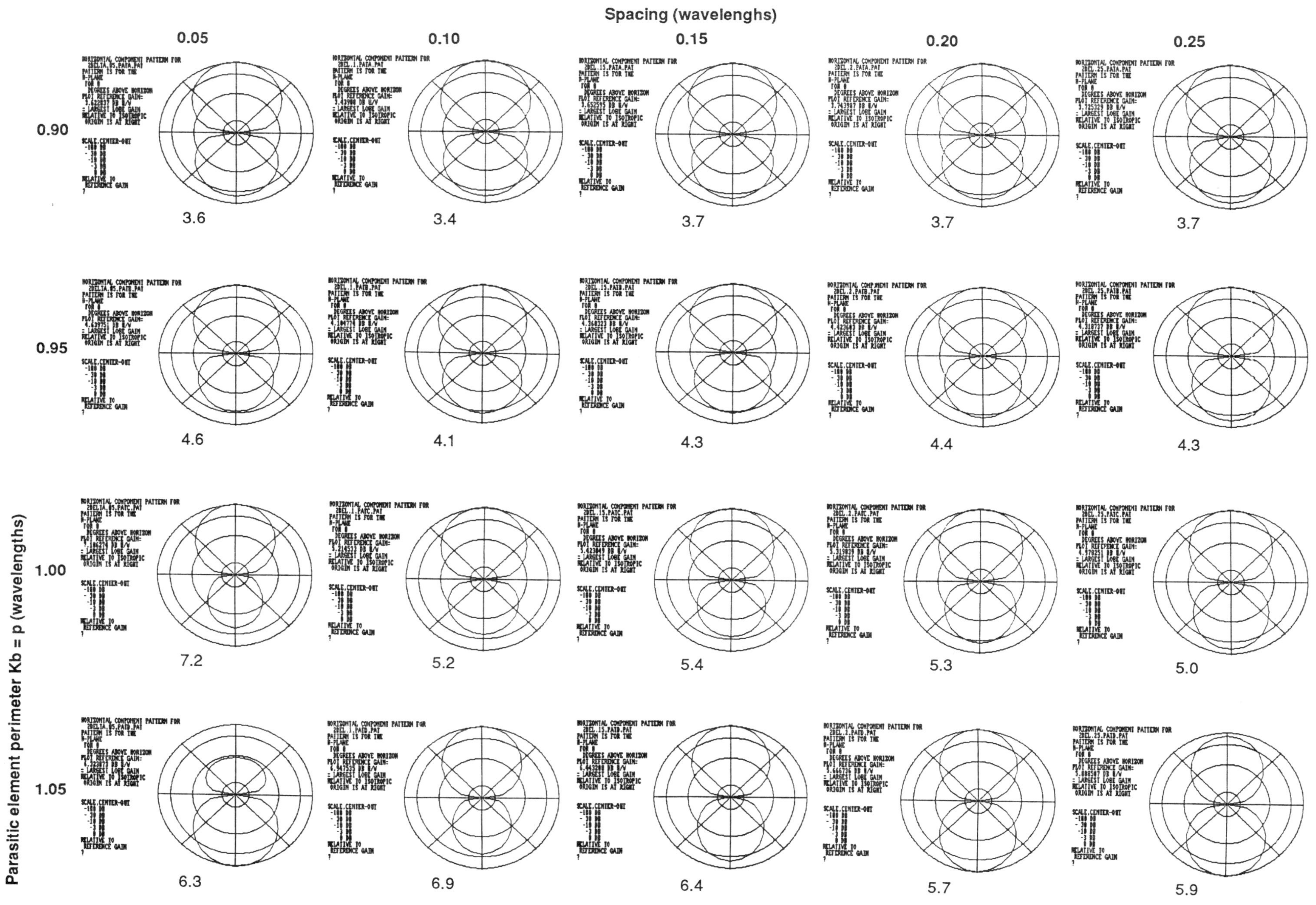
Spacing (wavelenghs)
0.05
0.10
0.15
0.20
0.25
Parasitic element perimeter Kb = p (wavelengths)
0.90
3.6
3.4
3.7
3.7
3.7
0.95
4.6
4.1
4.3
4.4
4.3
1.00
7.2
5.2
5.4
5.3
5.0
1.05
6.3
6.9
6.4
5.7
5.9

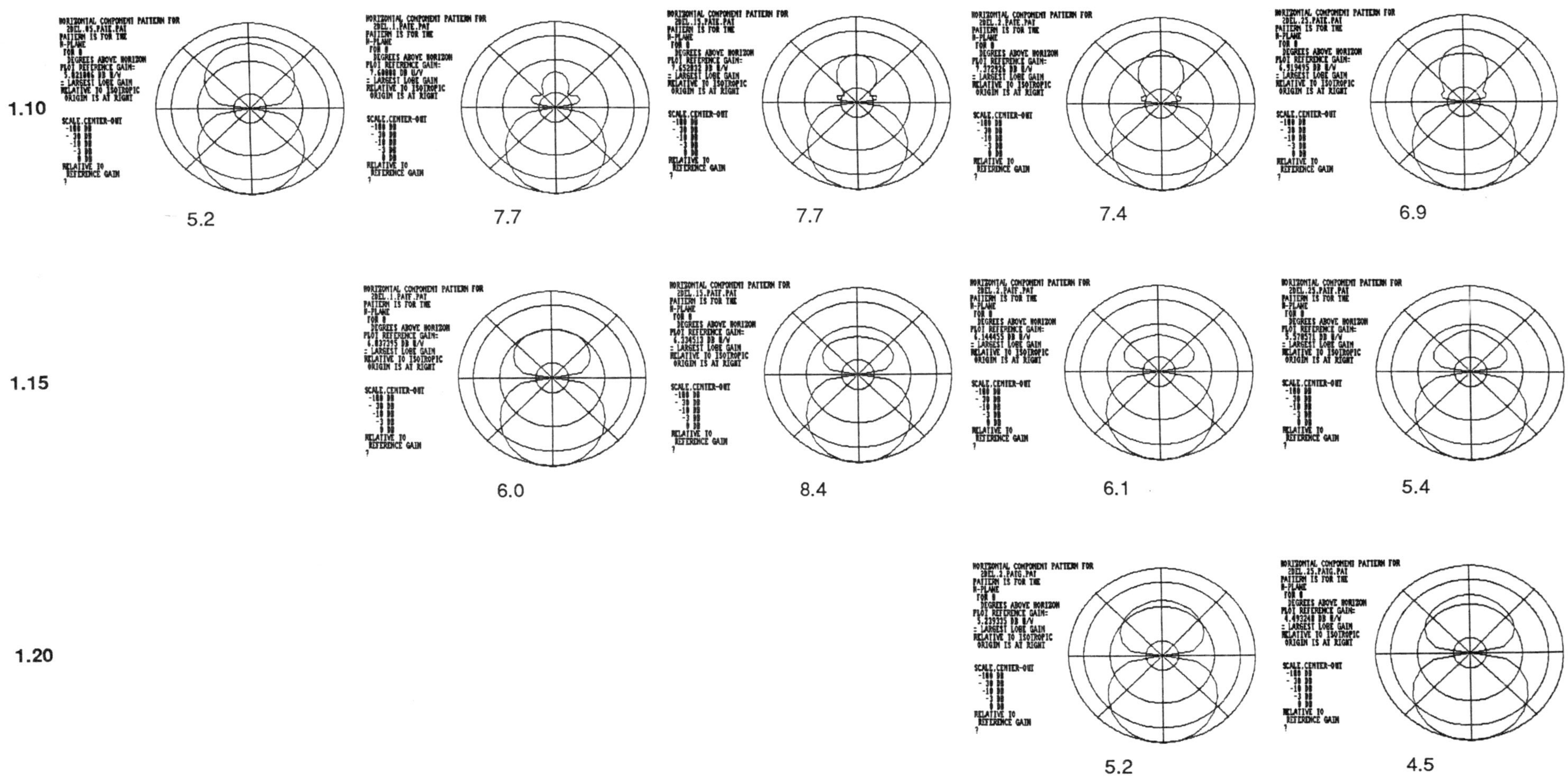

NOTE:

1. Pattern family is 2 element delta loop arrays p/a = 8000.
2. Number below figure is largest lobe gain. Plot is normalized to this value.

Figure 18. Family of patterns of equilateral delta-loop arrays as spacing and parasitic size is changed. In all, the two loops are the same size. Spacings vary from 0.05 to 0.25 wavelength in steps of 0.05 wavelength. Loop perimeters vary from 0.90 to 1.20 wavelengths in steps of 0.10 wavelength. The number below each pattern is the gain of the largest lobe; the pattern is normalized to that value. Arrays with 1.0-wavelength radiators will show slightly more gain if the parasitic element is smaller, and slightly less if it is larger.

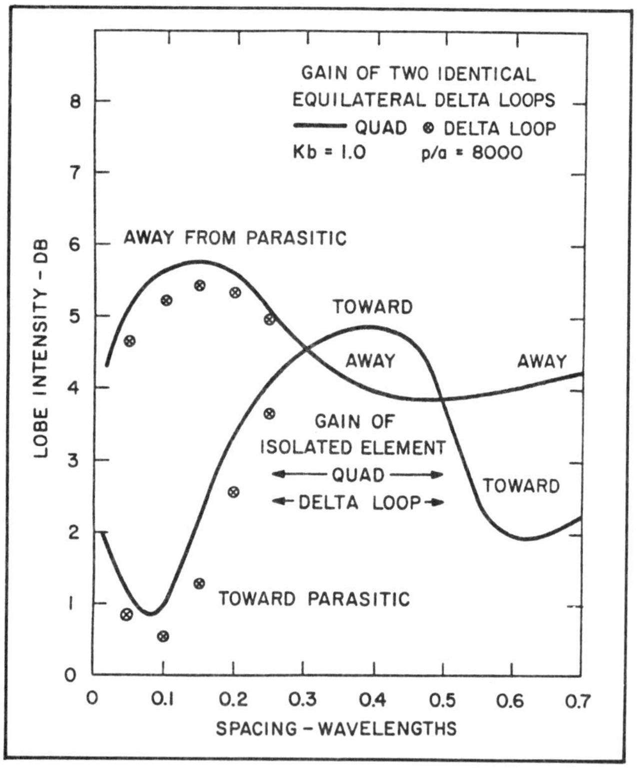

Figure 19. Lobe intensity of two-element equilateral delta-loop arrays with a perimeter of one wavelength. The delta loop values are shown as points, with the quad curves for comparison. Curves will converge to the isolated loop gains at very large separations.

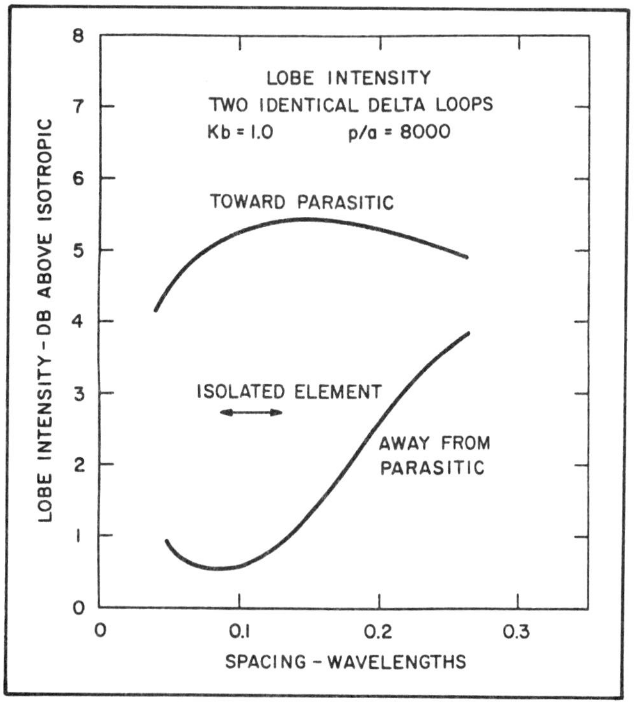

Figure 20. Lobe intensity for two identical equilateral delta loops.

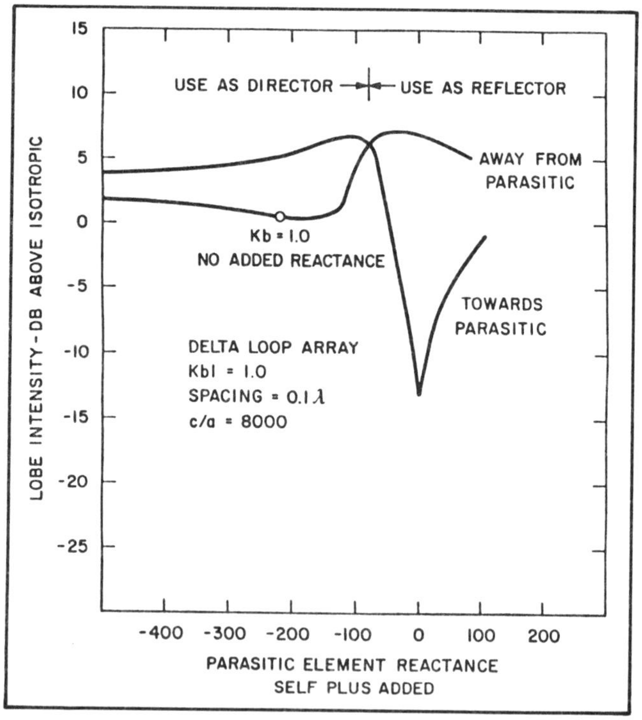

Figure 21. Effect of parasitic element tuning on a delta loop array of 0.1-wavelength spacing. This and the following curves permit design for specific performance, such as maximum gain, maximum F/B ratio or various compromises.

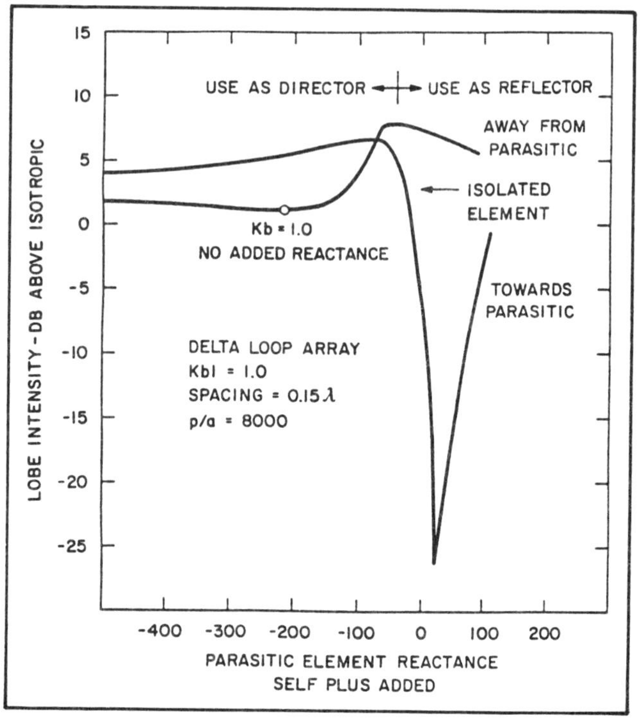

Figure 22. Same as Figure 21, but for 0.15-wavelength spacing. This represents the optimum delta loop (or very nearly so).

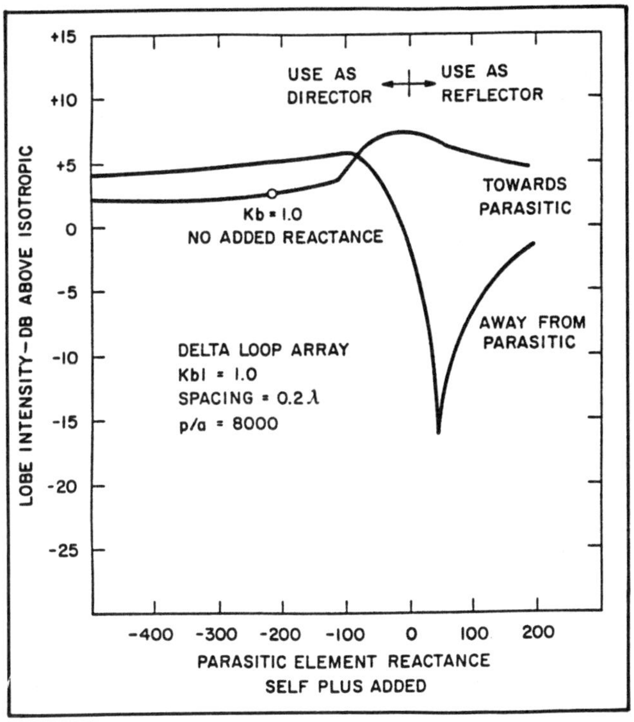

Figure 23. Same as Figure 21, but for 0.2-wavelength spacing.

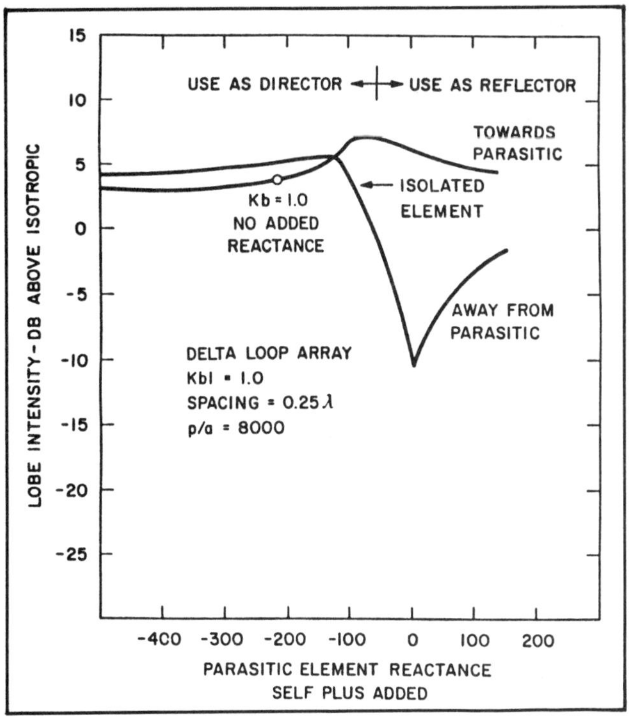

Figure 24. Same as Figure 21, but for 0.25-wavelength spacing.

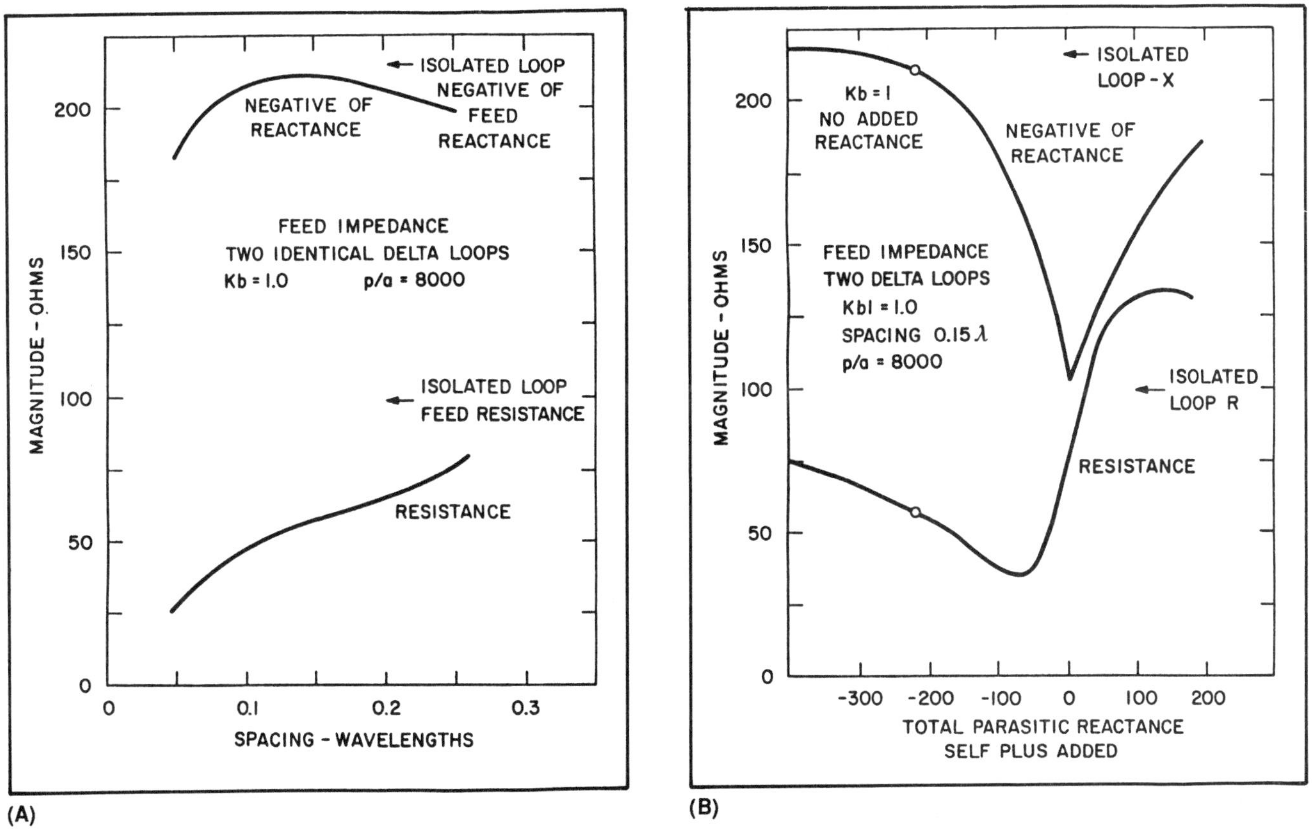

Figure 25. (A) shows drive resistance and reactance of a 0.15-wavelength delta loop array as a function of parasitic tuning. Curves for other spacings are similar, with closer spacing reducing the drive resistance. This is seen in (B), showing the drive resistance of a two-element delta loop with 1.0-wavelength loop circumference as a function of spacing.

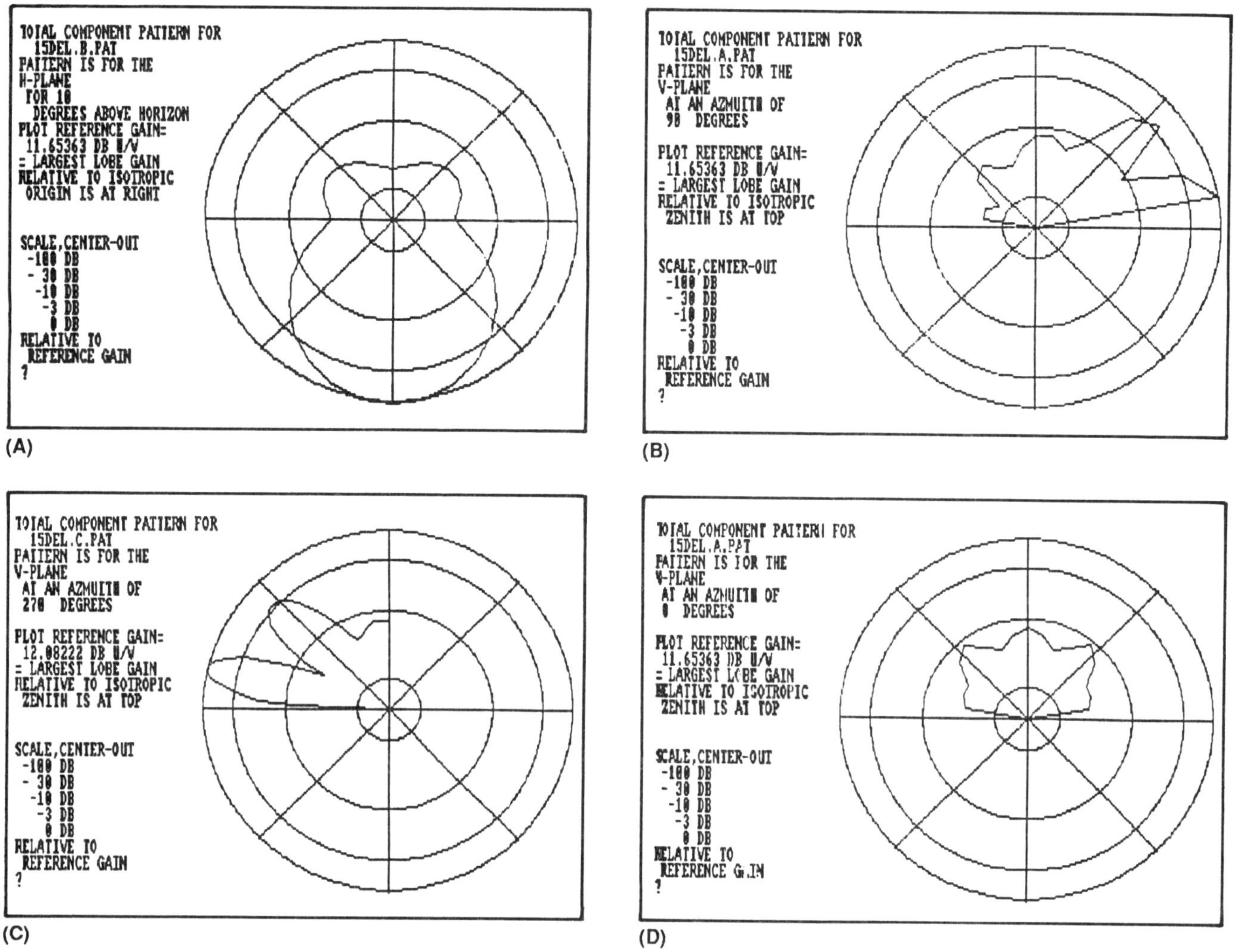

Figure 26. Patterns for a prototype 15-meter delta loop array, with tubing side arms and wire top. (A) is the total component for the cut 10 degrees above the horizon. (B) is the vertical-plane pattern at right angles to the loop plane. (C) shows the main lobe at higher resolution. (D) is the total component for a plane parallel to that of the loops. Feed is at the top center.

was set at 0.078 inch. Number 12 copperweld would be necessary.

The loop length corresponds to a loop of 1.027 wavelengths at 3.6 MHz or 1.12 wavelengths at 3.9 MHz, in the range of reflector performance. Although the effects of variable spacing and closeness to ground would affect performance, it appeared that some directivity would be obtained.

The calculated results are best summarized by pattern plots. Figures 27A and B are vertical-plane plots, bottom-center fed at A and side-center at B. The one at B is the most interesting for DX work because the radiation angle is lower, and there is some directivity. The extent of this is shown in Figures 28A and B, the horizontal cut at 30 degrees elevation. Figure 28A shows the total component, B the vertical. Side-center feed gives primarily vertical polarization. While bottom feed gives the greatest gain (8.0 dB), vertical polarization gives better low-angle gain, although the maximum gain is reduced to 3.5 dB.

Of the additional exploratory patterns made, those of Figures 29 and 30 are the most promising. Figures 29A and B show the total radiation in the vertical and 30-degree elevation-cut planes for bottom-center feed at 4.14 MHz. The gain has increased to a maximum of 10 dB. Directivity has also improved to 6.3-dB F/B ratio. The gain at 30 degrees is slightly below maximum, at 8.6 dB.

Figure 30 shows another possibility. This is also for 4.14 MHz, but the feed point has been moved to the

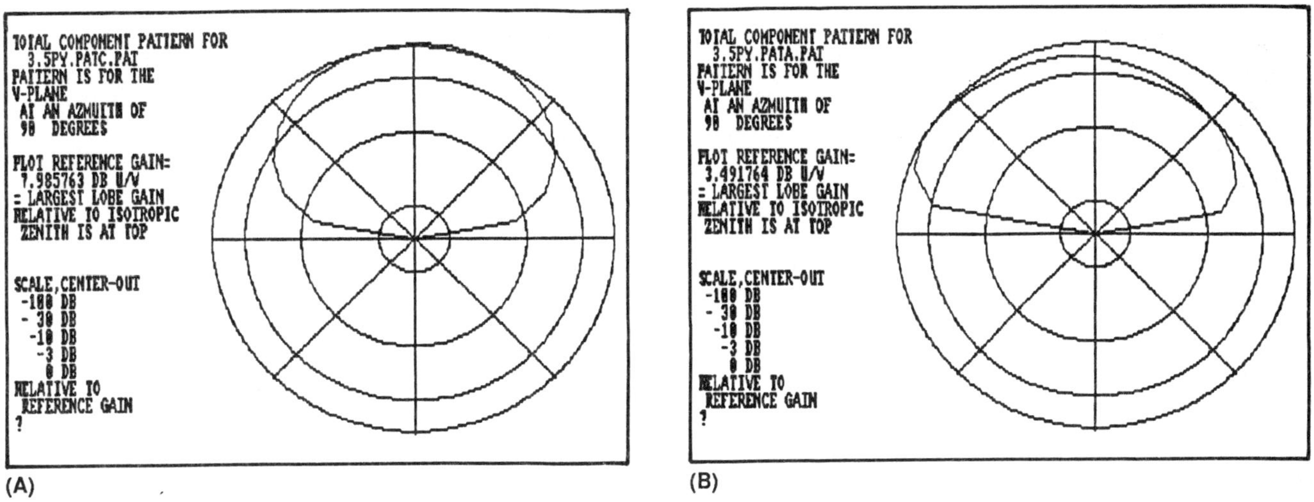

Figure 27. Patterns for a pyramid of two slanted equilateral triangular loops supported from a single tower. See text for dimensions. Feed is at bottom center for (A) and side center for (B). Both are for the vertical plane at right angles to the bottom wires.

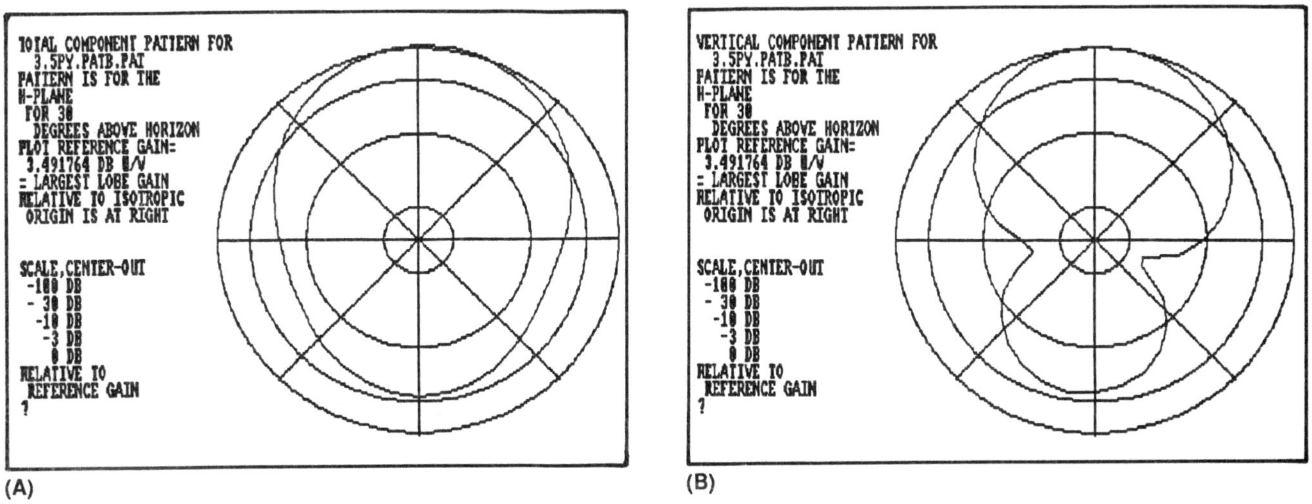

Figure 28. Thirty-degree elevation cuts for the pyramid with side-center feed. (A) shows the total radiation, (B) the vertical component. The small pattern skew is due to current asymmetry in the two sides. It would disappear if the feed were placed one-quarter wavelength from the bottom center of the loop.

corner of the array. This has no effect on the vertical pattern shape, but reduces the maximum gain, to 6.1 dB. The accompanying change in current distribution shifts the 30-degree-cut pattern by about 20 degrees.

A limited number of calculation runs were made for a frequency of 3.6 MHz, using reactive loading of the parasitic element. Two conditions found were:

loading	180 ohms
30 degrees gain	6.7 dB
30 degrees F/B ratio	12.8 db

and

loading	200 ohms
30 degrees gain	10.2 dB
30 degrees F/B ratio	2.9 dB

These pyramid studies aren't exhaustive enough to permit design recommendations. More analysis and experimentation is needed. A suggested approach is a pyramid with feed provisions at the four corners,

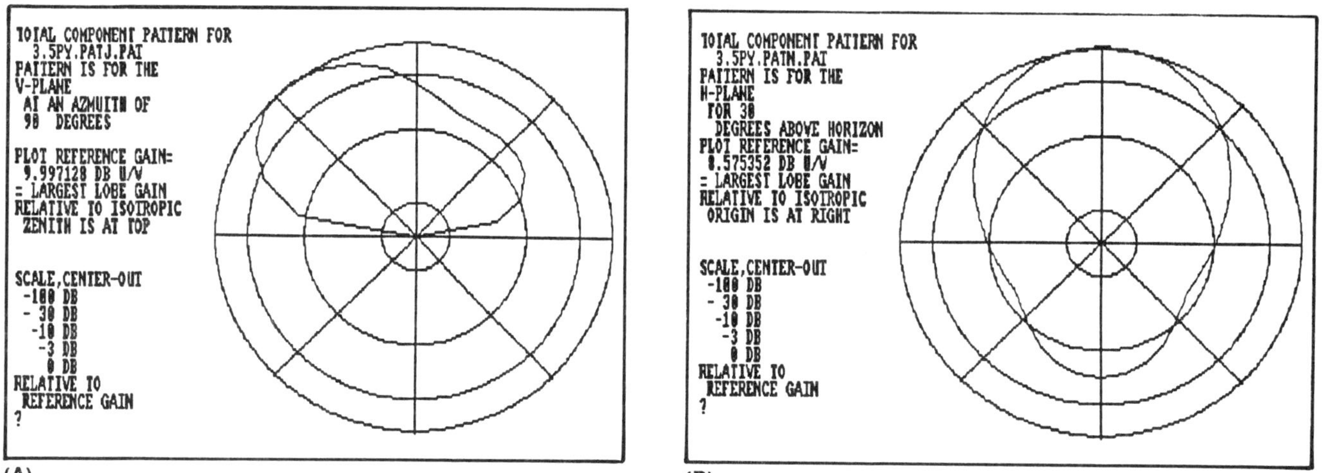

(A) (B)

Figure 29. Patterns of the pyramid array with bottom center feed. (A) is for the vertical plane at right angles to the feed wire; (B) is the 30-degree elevation cut. Symmetrical feed gives pattern symmetry.

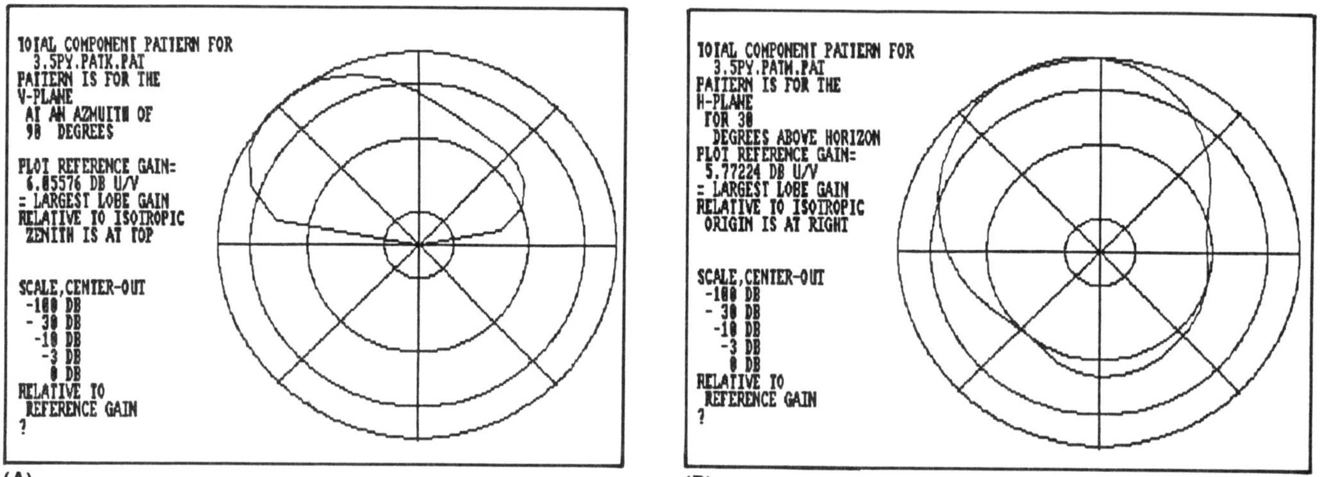

(A) (B)

Figure 30. Patterns for the pyramid with feed at a bottom corner. (A) is the pattern in the plane at right angles to the fed wire. It is nearly the same as for center feed but with lower gain. (B) is the cut 30 degrees above the horizon. The skew due to asymmetric feed becomes sizable.

with three-position relays to allow selection of feed point, shorting the other corner of the fed element, and stub loading at the parasitic element corners. The resulting bi-directional steerable antenna promises to be a good DX performer. One of the super-tower stations could try this on 160 meters; others could experiment on 40 or 20 meters. Even 2-meter scale tests would provide guidance, but ground effects are probably important, so results would be for guidance only.

Concluding Notes

In conformity with the relatively limited use of delta/triangle antennas, this examination has been much less detailed than that for quads. I hope that it will be sufficient to give good design data on the standard delta array, and to open some new approaches for triangle arrays.

Chapter 9

Multiband Quad Feed Systems

Practical aspects of transmission-line problems

It's necessary to consider the methods of applying power to the quad radiator on several bands as a unit. One reason for this is that there are feed-related problems in maintaining good performance with multiband operation.

Single-band Feeds

Figure 1 shows six of the methods that have been used in feeding quads. Open line (ladder line, 300-ohm ribbon, etc.) is shown in Figure 1A. In early quad days, this type of feed was universal. However, there was a variation from current practice in that most of the early quads were made of twinlead, connected to give two close-spaced loops in series. This method was considered important to reduce line VSWR. However, it increased the conductor size, making the loop better as a broad-band radiator. Data previously given allows adjustment of loop dimensions to compensate for the larger conductor size.

A stub may be added as shown in Figure 1B to eliminate reactance at one frequency. Unless the ultimate in performance is the goal, a stub isn't really necessary with open line. However, it does make the line/transmitter coupling somewhat easier. A stub is recommended if coax of small size or appreciable loss is to be used.

The arrangement in Figure 1C uses an RF transformer and reactance-canceling stub to get a good coax match. Sevik[28] shows that the normal quad feed impedance is high for best performance with magnetic core baluns—the reason for using a trans-

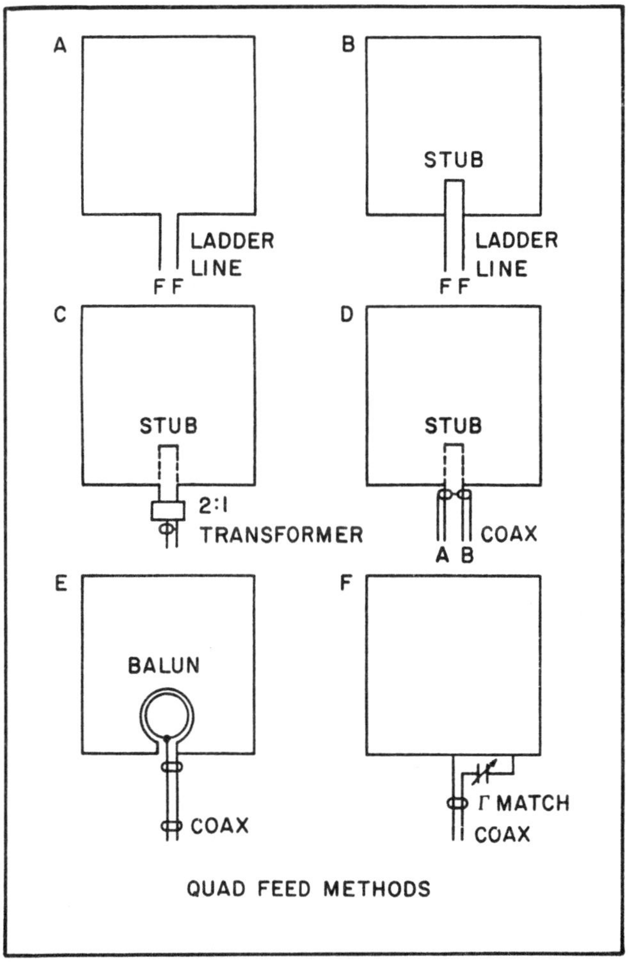

Figure 1. Feed systems for quads. (A) Open-wire feed. (B) Same as (A) with reactance-canceling stub. (C) Coax with step-up transformer. (D) Balanced feed using two coax lines. (E) Coax with balun. (F) Gamma match.

former. (However, Fanckboner[29] shows construction techniques for high-impedance baluns.)

In Figure 1D, two coax lines are used to obtain some of the simplicity of coax. The match is usually better, and voltage puncture and line-heating problems are reduced. A major advantage is that it's no longer necessary to keep the line away from the metal of the tower and boom.

Figure 1E shows the use of a balun. This particular balun is the half-wave line type, which eliminates RF-core problems. For best results, a reactance-canceling stub should also be used. The 1:1 balun developed by W2DU can be used; it's simply a few ferrite beads slipped over the coax. If some pattern distortion is acceptable, the balun can be omitted.

The gamma match at Figure 1F has been used in commercial quad designs. There are rumors of two gamma units fed from a common point to improve broad-band operation, but no data on this was found in the literature.

Any of these methods can yield perfectly good results. There will be differences in performance (usually small) and differences in installation and maintenance methods. Overall, the open line seems to be the best choice, followed by its paired coax version.

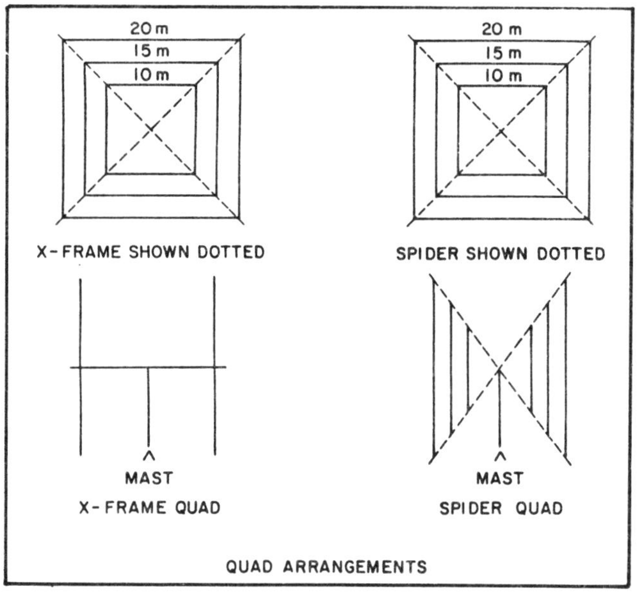

Figure 3. Quad arrangements showing concentric loops for multiband operation. X-frame and spider-type supports are shown. The bands indicated are typical.

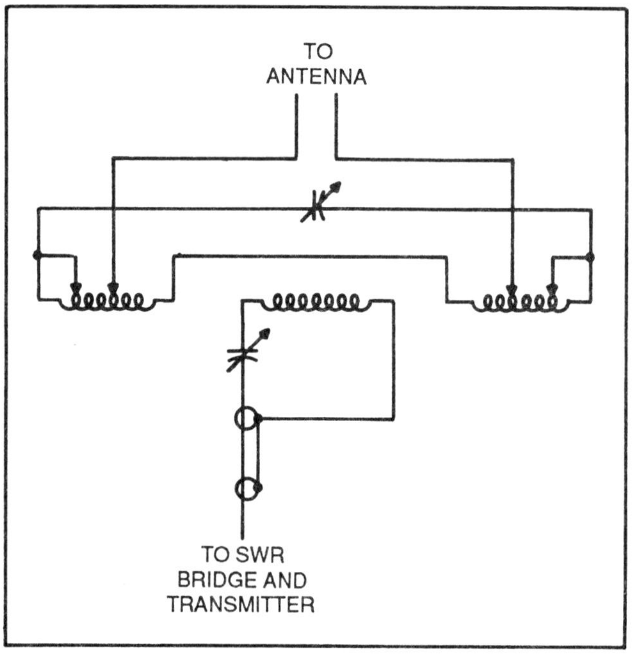

Figure 2. Wide-range balanced-to-unbalanced match unit. For some combinations of impedances and line lengths, it may be necessary to convert to series tuning by moving the parallel capacitor to the series position shown by the asterisk.

Feeding the Line

If the broad-band capability of the quad is to be used, some form of line matching is necessary. Any of the popular designs that can handle a VSWR of 8:1 or so are satisfactory. However, I recommend that you do not use a matching unit that lacks (a) dials capable of being preset to about one degree of capacitor angle or 0.1 turn of inductance, and (b) a unit without a noise bridge or dummy-load tune-up capability.

The current crop of matching units tends to be short on provisions for wide-range balanced line feed. The design sketched in Figure 2 can handle a very wide range. Switching for multiband operation does get complicated, but once set up, operation is no more difficult than for coax designs.

Multiband Quad Arrays

While the basic quad loop is inherently broad-band, the reflector and director must have specific reactance to function properly. Tuning by switched stubs (Gullstad,[30] for example) or by lumped reactance is possible. Except in special situations, the current trend is to use individual elements in multiband designs.

Figure 3 shows the two most common arrangements for the 10, 15, and 20-meter bands. The X-frame allows the electrical spacing to change from one band to another. The spider eliminates the boom, supporting the spreaders at angles to give constant electrical spacing. Designs with more elements are (of necessity) the boom type, sometimes including added intermediate elements for the higher bands. Most designs are for two or three bands, although a seven-band design has been described by James.[31]

Interactions between the loop elements have been examined as part of this study. The results are summarized by Figures 4 and 5 and Tables 1 through 3. Figure 4 shows the current in each loop with a 1000-volt pulse feeding that loop. All show the current increase resulting from the change in impedance at resonance. In addition, the 20-meter loop shows a current increase at two-wavelengths circumference, which corresponds to the second resonance.

Currents in the other loops were also calculated. Except at resonance points, the coupling between loops is small. Second loop currents are typically less than 1 percent of the driven-loop current. Table 1

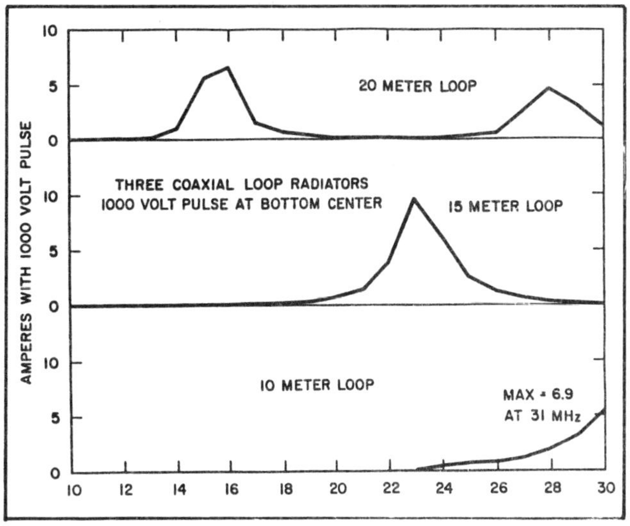

Figure 4. Feed point current calculated for a 1000-volt pulse feeding each of three coaxial loops separately at bottom center. No parasitic elements are present. The response at 28 MHz in the upper curve is due to the two-wavelength resonance of the lowest-frequency element. These values involve calculation approximations.

at a frequency of: (MHz)	**CURRENTS IN ELEMENTS FOR: 10 meters**	**15 meters (amps)**	**20 meters**
16.00	−0.4 + j0	−1.8 + j0	6.5 − j4.3
23.00	−1.3 − j1.5	9.5 + j2.2	0.2 + j1.1
31.00	6.8 + j0.4	−2.7 + j1.5	0.3 + j0.4

Table 1. Currents in each element of a 10/15/20-meter coaxial quad loop array with each element, in turn, fed at the frequency producing highest current with a 1000-volt exciting pulse. These values represent the greatest coupling between loops. Feed is at the bottom center.

at a frequency of: (MHz)	**CURRENTS IN ELEMENTS FOR: 10 meters**	**15 meters (amps)**	**20 meters**
15.00	0 − j.3	−0.2 + j0	5.3 + j4.5
23.00	−1.3 − j1.6	9.6 + j3.3	0.3 + j1.4
28.00	4.6 − j0.2	0.46 − j1.9	−1.2 + j0.83

15/20-meter elements fed.

Table 2. Same as Table 1, but with simultaneous 1000-volt pulses in both the 15 and 20-meter loops.

at a frequency of: (MHz)	**CURRENTS IN ELEMENTS FOR: 10 meters**	**15 meters (amps)**	**20 meters**
15.00	0 − j0.3	−0.2 + j0	5.1 + j4.2
23.00	−1.1 + j0	8.3 + j1.7	0.3 + j1.3
28.00	7.0 + j2.3	−0.1 + j1.0	1.0 − j0.9

All elements fed.

Table 3. Same as Table 2, but with simultaneous 1000-volt pulses in each of the three loops.

shows the currents in each element at the loop resonance points. Coupling at and near resonance increases, with second-loop currents reaching 10 to 20 percent of the driven-loop current.

Coupling Effect

The effect of these resonance-point couplings can be seen in Figure 5 which shows swept-frequency pattern values for each element fed separately. The values shown are for the "on-axis" gain, at right angles to the plane of the loop.

The 20-meter element follows the isolated-element gain curve until about 28 MHz, essentially the point where the contribution of the 10-meter element becomes appreciable. The loss in gain is due to lobe splitting, as for the isolated loop.

The 10 and 15-meter curves show loss of gain near the resonances of the lower-frequency elements. The effect extends some 10 to 20 percent on each side of lower-frequency element resonance. The interaction at driven-element resonance is small to insignificant.

While the general conclusion is that the coaxial-driven element retains full or essentially full performance, there is one caveat. The effect of transmission lines connected to unused elements hasn't been studied. The possibility of accidentally induced resonances does indicate that the best method of feeding these coaxial arrays is by using relay-switched coax, arranged to short circuit the unused element feeds. (However, practical experience with several coaxial-loop quads suggests that the problem is relatively insignificant. This may be because the unused coax has no influence on the parasitic elements. Note that the overall properties of coaxial-loop arrays have not yet been studied.)

James[31] reported no problems with his seven-band spider quad. However, the fact that the element interaction covers a range of 10 percent, or so, suggests that there may be pattern problems as the number of bands is increased. The limits haven't been studied. For multiple-frequency operation, it may be better to use relay-switched stubs to change element tuning. The reactance/frequency and gain/frequency curves of earlier chapters give data for stub design.

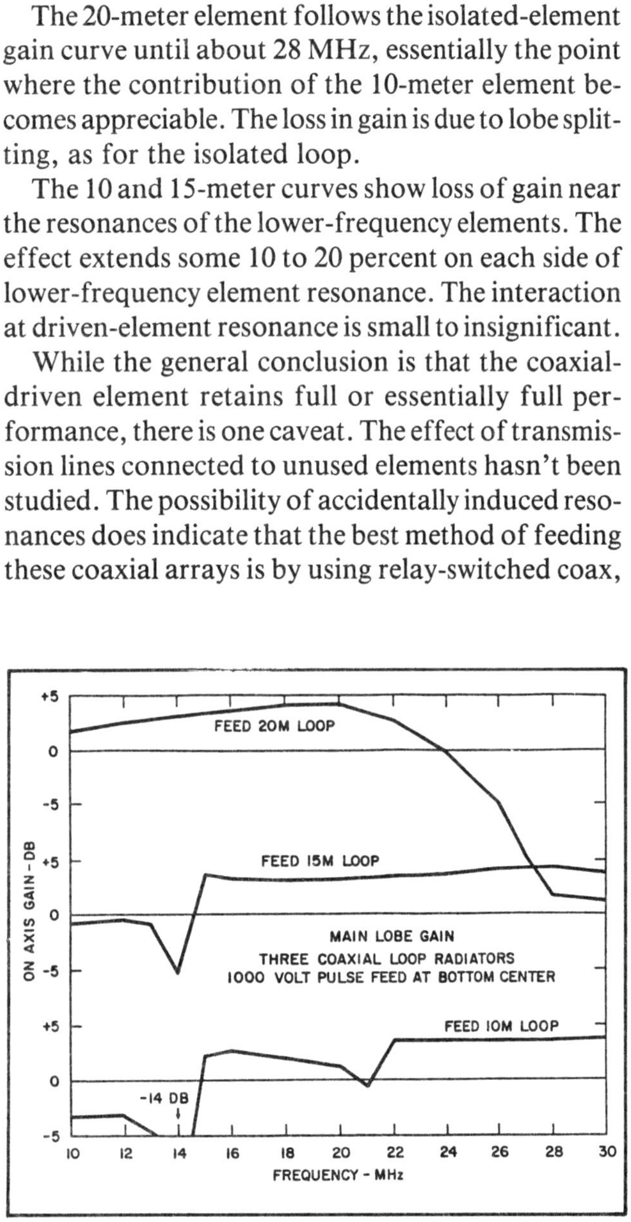

Figure 5. On-axis lobe intensity for the three loops in Figure 4. The fall-off of the upper curve is due to lobe splitting as the frequency approaches the two-wavelength resonance. Note the effect of resonances of the other loops.

Single-line Feed to Multiple Loops

A number of years ago, KP4BCL (now N5IE) pointed out that the way in which quad-loop impedances change allows loops to be connected in parallel if they are not harmonically related. Some conditions for this have been investigated using MININEC.

Figure 6 shows several possibilities. The first two are for two and three-loop feeds. The other is a combined two-loop/single-loop scheme as originally used for 10, 15, and 20 meters by KP4BCL, and used at W4MB for the past 14 years.

Interconnection Test Method

There are computation problems with these multi-loop arrangements because the pulled-together sections act as a transmission line for part of their length. This analysis uses a simple approximation to avoid this problem. The loops are assumed to be the normal coaxial type with no physical interconnection. Each loop to be interconnected has a 1000-volt pulse in place of the actual physical connection. This isn't a perfect substitute, but it will give an indication of interconnection results.

Figure 7 shows the calculated total power with two 1000-volt pulses in the 15/20-meter and three in the 10/15/20-meter loop combinations. The resonances of the individual loops are clear. There does seem to be some interaction in the three-band feed, which is noticeable around 24 and 30 MHz.

Figure 8 shows the on-axis gain for the two conditions. There's marked gain reduction at some frequencies for the three-loop type and some for the two-loop

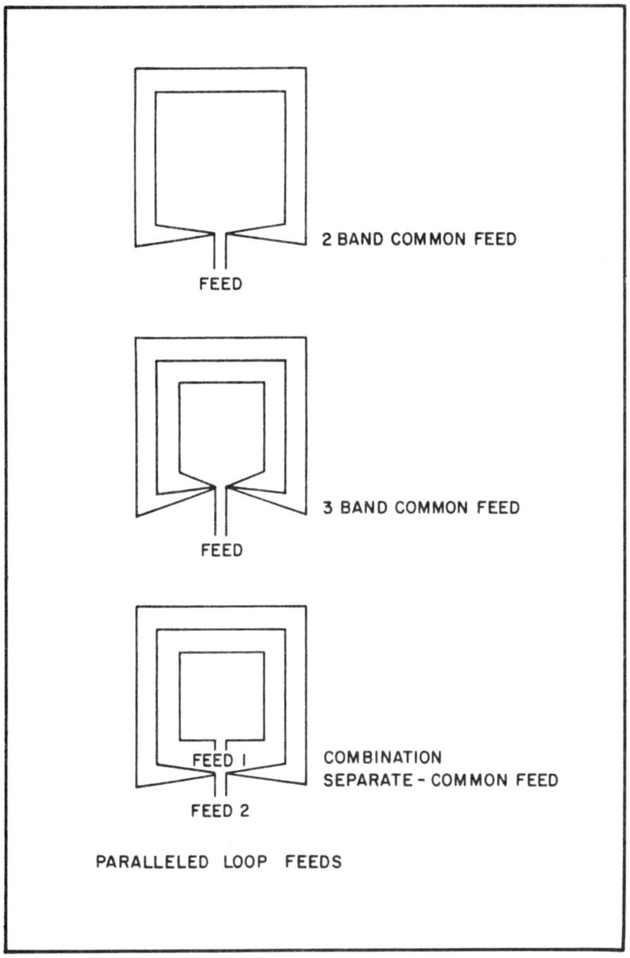

Figure 6. Possible methods of paralleling loops for multiband operation. See text for reasons the combination type is preferred.

version. The reduction is relatively small at resonant frequencies and large just below resonance.

The first two patterns of Figure 9 show the main features of the degradation in performance of the three-loop interconnection. This figure is for the three loops elevated 60 feet above a perfect ground at 28 MHz. The curves of A and B are for the three-band common feed; C and D are for the separate-common feed of Figure 6.

Comparison of 9A and B shows that the lowest-angle lobe lies in the plane of the loop, where the next higher one is at right angles to the loop, as is normal. The lower lobe is primarily from the 20-meter loop, acting as a 2-wavelength radiator.

Figures 9C and D, for separate 28-MHz feed, show the normal on-axis shape for the lower-angle lobe. There's some increase in plane radiation, again due to radiation from the 20-meter loop. The fill-in is less

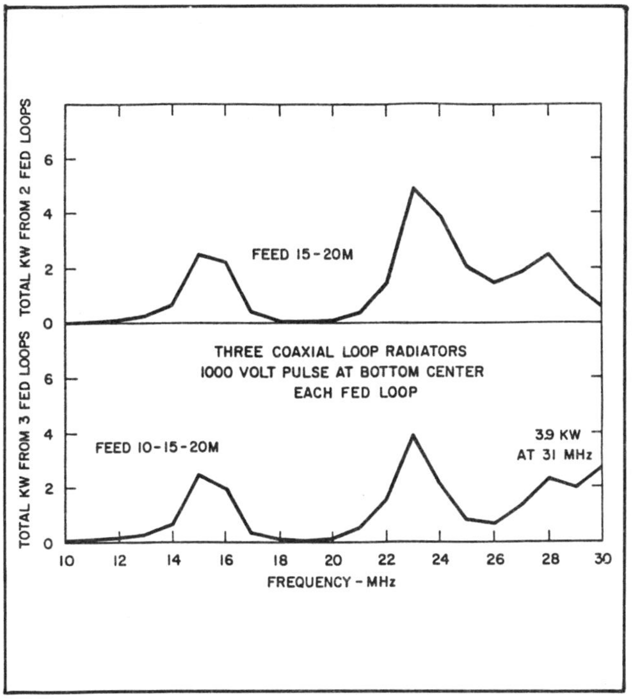

Figure 7. Results of approximate analysis of paralleled loops by simultaneous feed of separate loops. Total radiated power is shown. The resonances of the three loops are obvious.

because the induced current in the 20-meter loop is lower than that resulting from direct feed.

Note that the angles weren't chosen for the peak of the lobe, and that there is an effect on gain for the separate-loop feed. This becomes small at and above resonance. In all cases, the loops are good radiators.

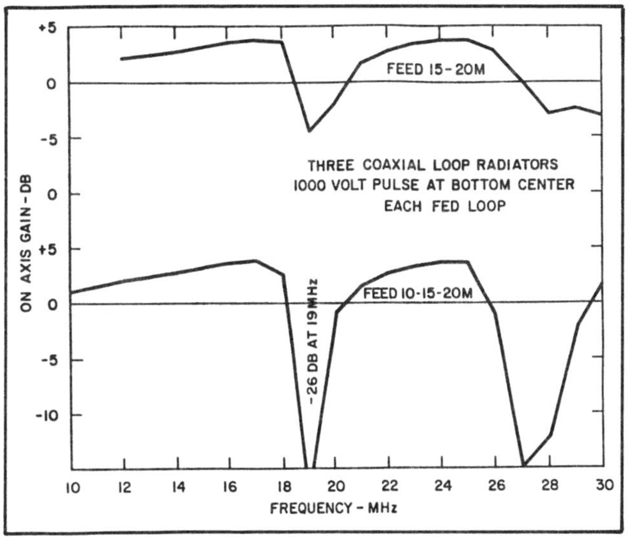

Figure 8. Main-lobe intensity (approximate analysis) of the two and three-paralleled loop-feed conditions. On-axis gain loss is distinct at frequencies below resonance.

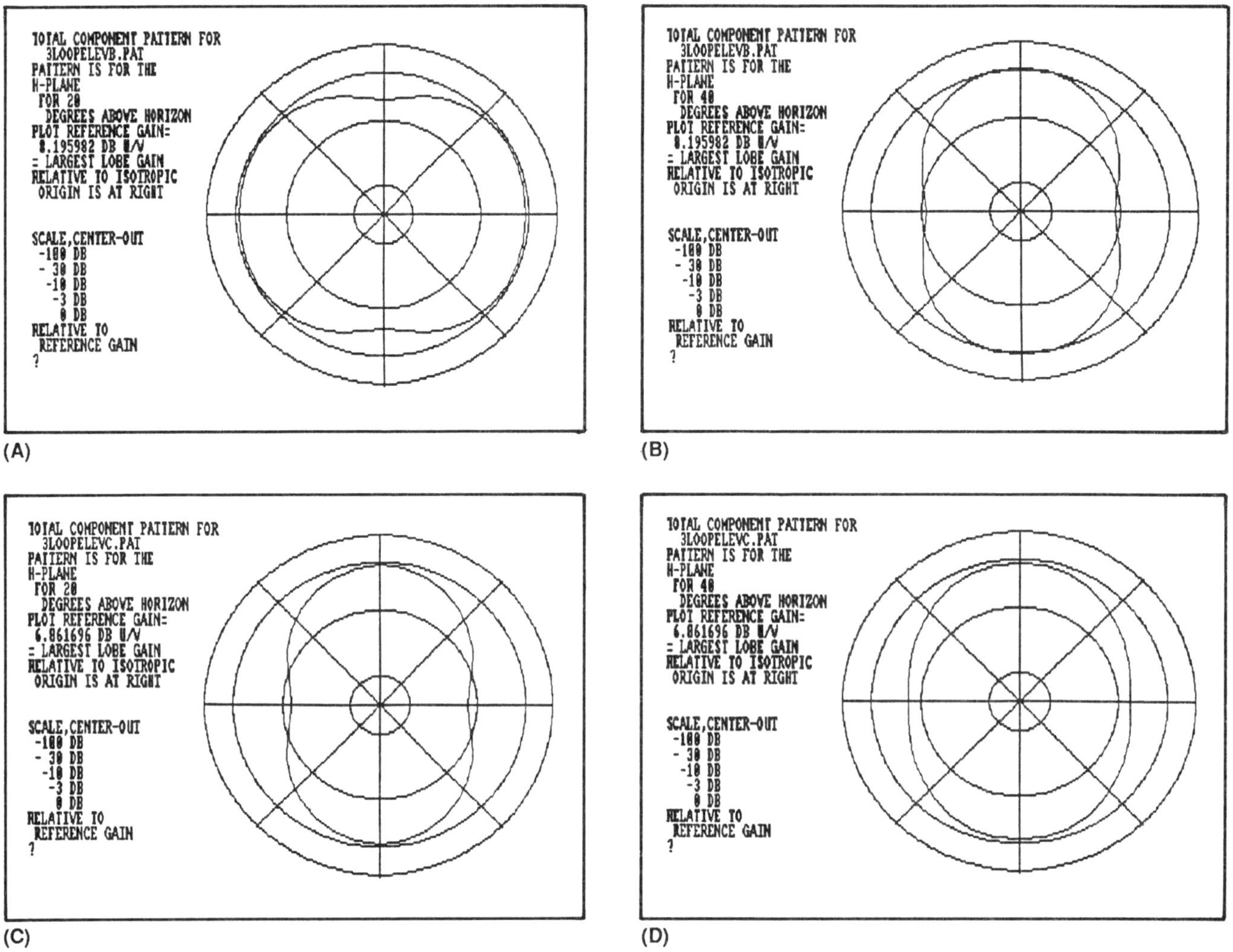

Figure 9. Patterns for the three fed-loop conditions. The analysis is approximate. The antenna is 60 feet above ideal earth. (A) and (B) are for the three fed-loop conditions at 20 and 40 degrees elevation. The low angle directivity is toward the loop plane, rather than at right angles to the loop. (C) and (D), for two-loop feed, show the normal pattern. Elevation angles weren't selected for maximum lobe intensity, which is indicated on the figures.

It is the pattern that changes, not the ability to radiate.

The overall conclusion is that multiple-element feed from a single transmission line is best confined to non-harmonically related loops. Attention should be given to radiating-loop size, accepting the penalty of reactance introduction to minimize pattern distortion. To the accuracy of this analysis, the 1.1-wavelength radiator suggested for gain reasons also appears good for pattern. The harmonically related loop interconnection does radiate perfectly well, and can be used if pattern change isn't important.

Mast, Boom, and Spreader Interaction: The X-loop

The triband quad at W4MB has good forward-lobe performance, but the high-angle back lobe on 20 meters is much larger than on 10 and 15 meters. Several different parasitic-element sizes didn't reduce the back lobe appreciably. This raised the possibility that the back lobe was being increased by radiation from the mast at 60 feet height—nearly a full wavelength.

Analytical Model

To investigate this, a model system was analyzed. The loop was a single bottom-fed one-wavelength element, spaced 1/16 wavelength from the mast. The mast was modeled for analysis as having an effective diameter of 1/80 wavelength. Mast height was 60 feet.

The analysis showed that the mast-to-loop coupling was totally negligible. Maximum induced mast current was about 10^{-6} the loop current. There was no effect on the pattern.

The W4MB quad is a size-compromise design, with the boom restricted to 21 feet to meet turning-radius needs. This is close to 15-meter dipole length. Mast-to-boom interaction was modeled with the same loop and with a half-wave boom. As for the mast, interaction was totally negligible. Current in the boom was even lower than for the mast. There was no effect on pattern.

As a check of multiband possibilities, the combination was analyzed with the boom fed as a dipole. Coupling to the loop was again negligible, and the pattern was that of a normal dipole. A rotary dipole-quad combination will work well.

The conclusion is that the mast and boom normally present no problems. However, this conclusion may not hold if the system isn't reasonably symmetrical, a factor not modeled. Though not investigated in this analysis, the interaction shouldn't be severe with loop side feed. At some angles there is appreciable cross polarization, so the modeled design would have shown any significant coupling.

X-frame/loop Coupling

The degree of X-frame-to-loop coupling is quite different. Figures 10 and 11 show the variation in drive resistance and reactance near resonances for a one-wavelength isolated loop and an X-frame with one-half wavelength overall. The loop and the two X-frame arms are separated 0.004 wavelength, and are assumed to be totally unconnected. Isolated loop values are shown for comparison. Calculated resonances for the loop and arm are shown.

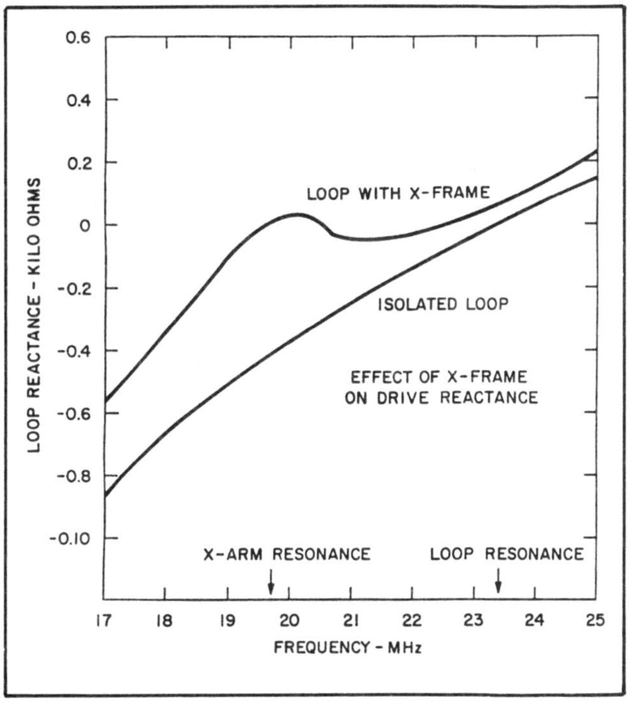

Figure 10. Effect of an X-arm spreader on loop-feed reactance. The resonances of the loop and spreader arm when isolated are indicated. The isolated loop-feed reactance is shown for comparison.

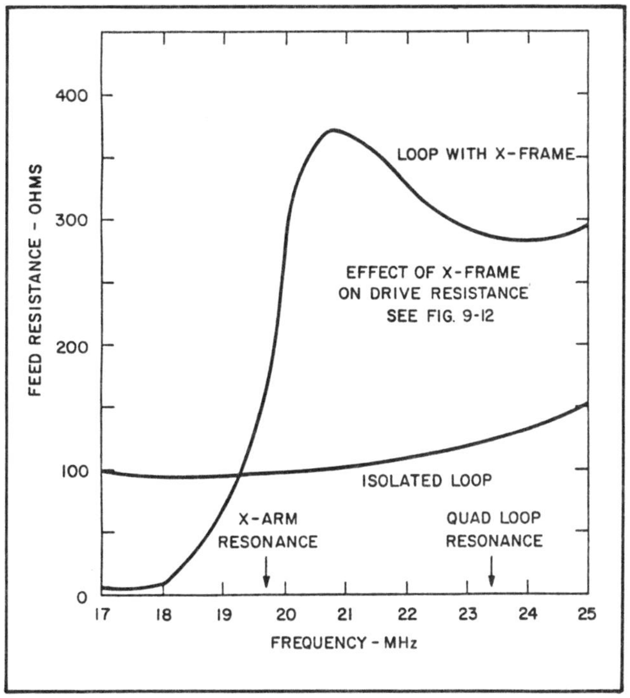

Figure 11. As in Figure 10, but for feed resistance. (See Figure 12.)

This high degree of coupling isn't altogether bad. Figure 12 shows the calculated on-axis gain for an isolated loop and the loop with the X-frame. The major adverse effect is confined to the frequency range below dipole resonance, where the dipole is acting as a director. Essentially, the X-arms are taking energy from the field at right angles to the loop, and radiating it in the plane of the loop. This effect continues above dipole resonance, but is much less evident.

The major pattern features of this X-frame/quad combination are shown in Figure 13. The pattern at A is for the quad-element-fed condition. Except for the slightly lower gain, this is the same as for an isolated loop. The pattern at B is for one arm of the X-frame fed at the center, and shows the contribution of the frame. The pattern is skewed toward the loop plane; the skew would be in the opposite direction for the other arm. Note the magnitude of the total gain.

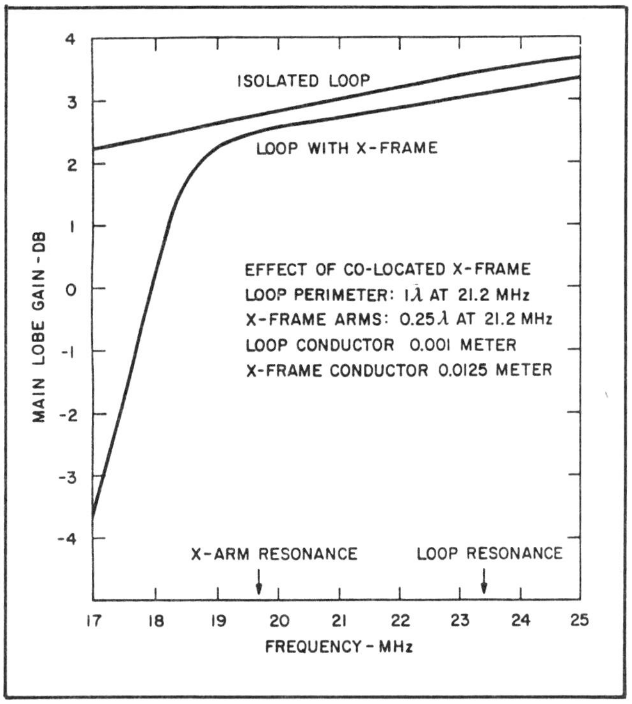

Figure 12. Comparison of lobe intensity at right angles to the loop for an isolated loop and for the loop/X-frame combination. Note that the values in Figures 10 through 13 involve geometry which may increase approximation errors in the calculations.

This interaction is due to the loop: Dipoles crossed at 90 degrees are independent.

The primary conclusion is that it's best to avoid metallic spreader arms and use fiber glass or even the classic bamboo spreaders. If metal arms must be used, resonance effects must be taken into account.

X-frame/Loop Tuning

The strong coupling suggests that there is an interesting possibility if metallic arms are used. In this case, it's possible to tune the X-frame elements to conform to the tuning of the loop. In other words, an X-frame supporting a quad reflector would be made longer than resonance, to act as a Yagi reflector element, etc. Figure 14 shows typical loop and arm dimensions.

This technique has been used for two-and three-element quads at W4MB for many years, always for the 15-meter loop. The 20-meter loop is supported by wooden extensions to the 15-meter tuned-X elements. The claimed gain improvement couldn't be verified given the accuracy of the measuring equipment available, but the pattern shows definite lobe narrowing as compared with a design having nonmetallic arms. This implies some increase in gain. This tuning is recommended if metallic spreader arms are to be used.

Interactions between elements in multi-element quads haven't been investigated due to lack of adequate computer capability and time. The entire area of incidental interactions, and the use of these, needs more work.

Spreader-fed Quad Loops

One of the sights of CB country is a six-element, 11-meter quad array. It appears that all of these use the spreaders for feed. This is a definite advantage in this service, as either the horizontal or vertical spreaders can be fed to give corresponding polarization, as shown in Figure 15. For some reason the idea isn't

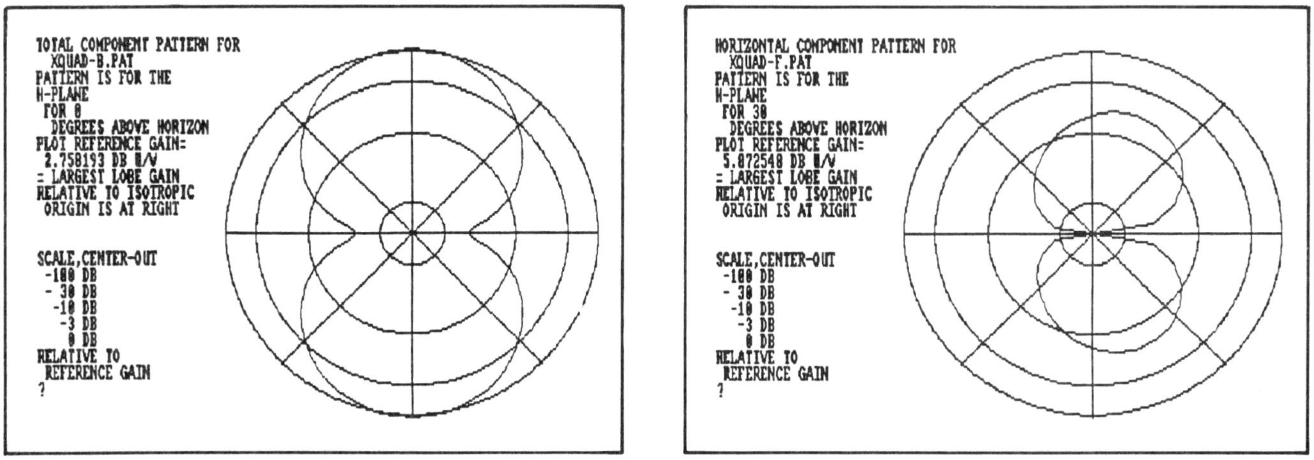

(A) (B)

Figure 13. Patterns for the X-frame/loop combination. (A) with loop feed; (B) with one spreader-arm fed.

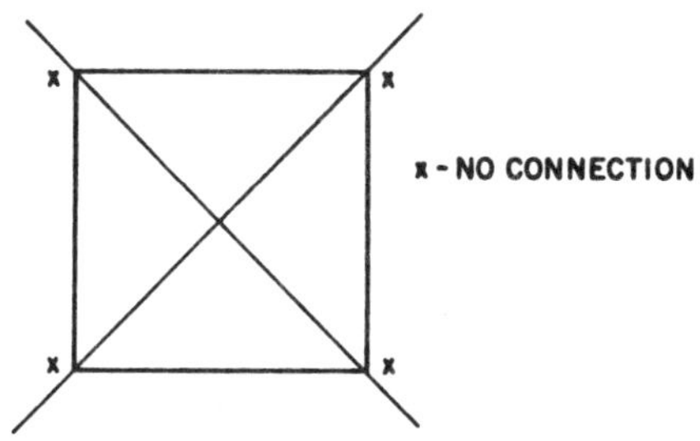

LOCATION	TYPICAL LOOP	TYPICAL ARM
REFLECTOR	1.1λ	0.49 - 0.50λ
RADIATOR	1.0 - 1.1λ	0.47 - 0.48λ
DIRECTOR	0.9 - 0.95λ	0.45 - 0.46λ

LOOP WITH TUNED X-FRAME SUPPORT

Figure 14. Arrangement for spreader arms tuned as Yagi elements. Typical spreader-arm lengths and loop perimeters are shown in wavelengths. Other loops are not shown.

common in amateur use, although both polarizations are used—especially on 2 meters.

This antenna feed has been modeled for this study using typical wire size for the loop and tubing size for the spreaders. It was assumed that the feed point was at the center of one spreader arm. The arm 90 degrees to the feed arm was checked for both open and shorted conditions. No difference was found, so the treatment of the other feed possibility can be as is convenient.

The feed point resistance and reactance are shown in Figures 16A and B. The radiator has good broadband performance, with nearly constant drive resistance over the range 0.9 to 1.5 wavelengths perimeter. Reactance is relatively high and variable. At the feed point, the apparent resonance is at 1.35-wavelengths loop perimeter. This change from the 1.12-wavelength perimeter at resonance of the loop alone is the contribution of the spreader. It appears that this acts partly as a radiating component, and partly as a feed transformer. Gain for the lobe at right angles to the spreader-fed loop is similar to, but higher than, that of a loop alone. The difference amounts to nearly 1 dB around 1.3 to 1.4 wavelengths perimeter.

In CB practice, there are sections of transmission line between the spreader ends and the actual feed point. These are often formed into an eggbeater shape. For a single radiator, the dimensions of these matching sections can be calculated from Figure 16. This feed scheme was tried with both 10 and 15-meter elements present, with both connected to the arms of the fed spreader. The results are summarized in Figure 17. The gain performance is quite good. Matching will be required, of course. No attempt was made to find the best dimensions, nor to determine the effect of loop connection to the unfed spreader.

No work has been done on arrays using these types of elements, and no design data was found in the literature. The feed concepts appear to merit more use as a driving radiator, in 2-meter beams in particular, and possibly in the lower two bands of a dual or triband array. Their value as parasitic elements needs to be studied further.

Quagi Use: Quagi Design Data

In multi-element multiband arrays, the spacing between directors for good gain on the lowest band is too great for best performance on the highest one. *The ARRL Handbook*, for example, shows an additional highest band director loop between the main loops, to obtain better spacing.

In earlier chapters, it was noted that Lindsay[12] and others had shown that the director elements can be loops, plates, or rods, with little effect on gain and lobe shape. This suggests that the additional element

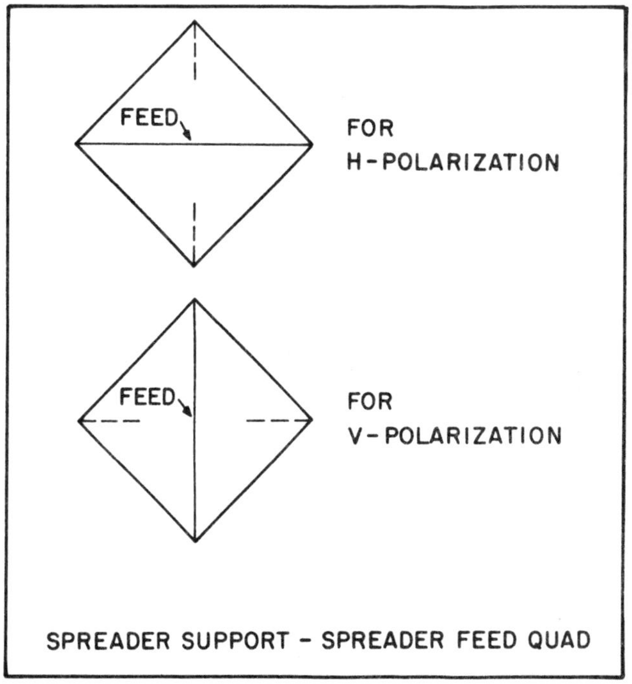

Figure 15. Schematic showing use of spreader arms as loop feeds. Two polarization possibilities are shown. Simultaneous feed with 90-degree shift may be used for circular polarization. The unfed arms may be open or shorted together.

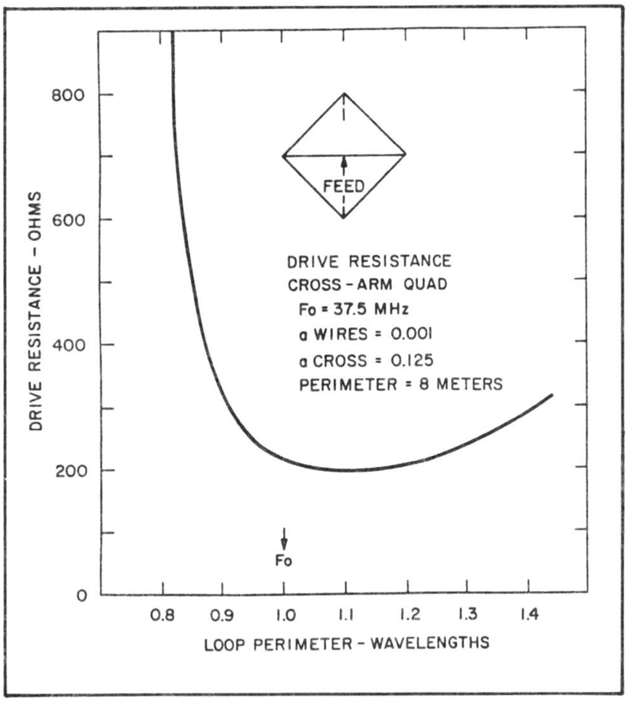

(A)

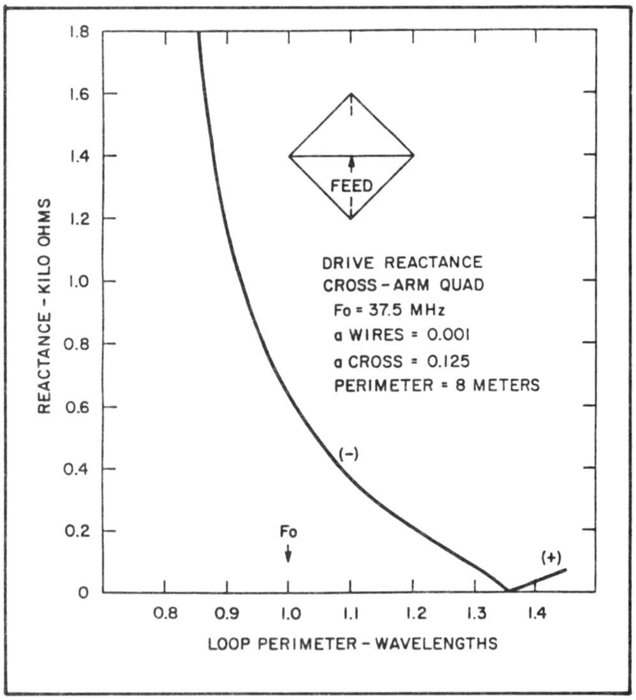

(B)

Figure 16. Drive impedance for the arm-fed diamond loop. (A) Drive resistance. (B) Drive reactance. The isolated loop resonance point is shown. The values are the same with the cross arms open or shorted together.

can be a Yagi rod, and can give as good performance as an additional loop. Figure 18 shows such an antenna, a 10/15/20-meter design with a 10-meter Yagi fill-in. Also shown are the spreaders tuned to 15 meters, as already described. This is the antenna currently in use at W4MB. It was also used in the WARC band experimental work of KK2XJM. On the WARC bands, the antenna functions as a rotary loop, with zero dB front-to-back ratio.

The length of the added Yagi element and the length of quagi elements can be determined generally from

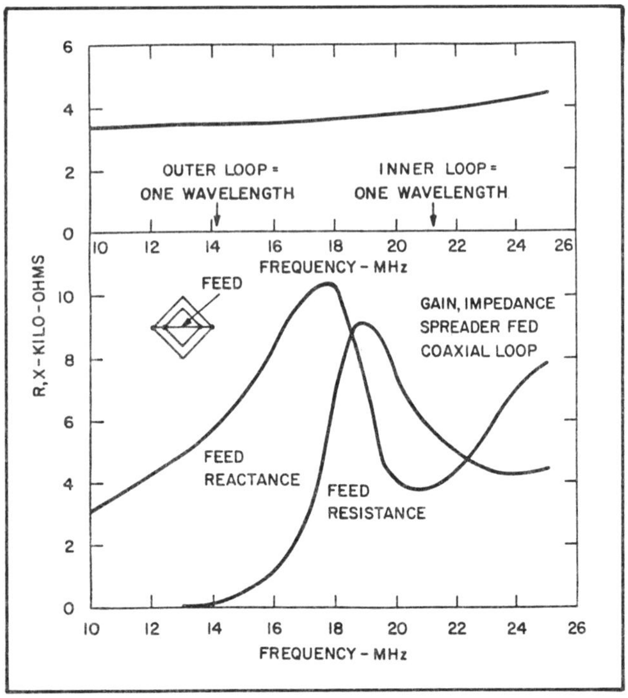

Figure 17. Gain and feed impedance of two concentric diamond loops connected to a common spreader arm. The 90-degree arm is assumed to be of nonconductive material.

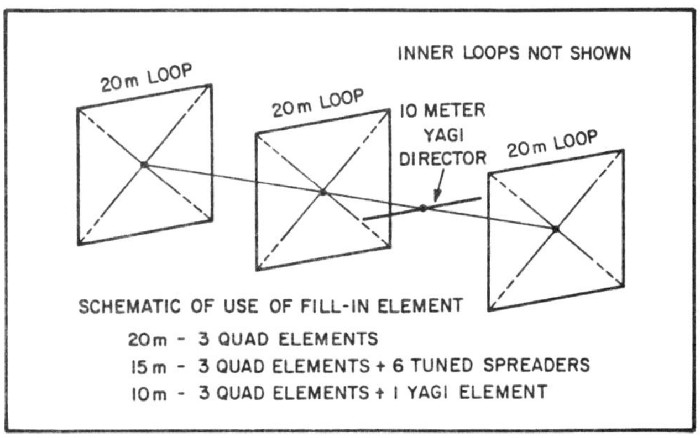

Figure 18. General appearance of a triband concentric quad array using a high-band Yagi director, and with cross arms tuned to the intermediate band. Inner loops are not shown. Insulated spreader parts are indicated by dashes.

the design curves of Figures 19 through 21. These are for spacings likely to be used from 0.05 to 0.15 wavelength. Interpolation may be used for other spacings. Only one rod size is shown; the curves and scaling factors in Lawson[19] may be used if another rod size is desired.

To ascertain rod length for a given design, determine the spacing, 0.15 wavelength, for instance. Select the design criterion, say maximum gain. The rod should be shorter than a half wavelength by the ratio 25.5:28.5 (see Figure 21). It should be 0.447 wavelength at the frequency of maximum gain. These curves can be used for general quagi design. No corrections are needed if all dimensions are scaled by the frequency ratio of new bands.

It isn't necessary to use a quad loop as the other director element. For example, G3NVA (now a VE7) used a pair of trap-Yagi directors with a quad radiator and reflector to produce an outstanding signal from G-land.

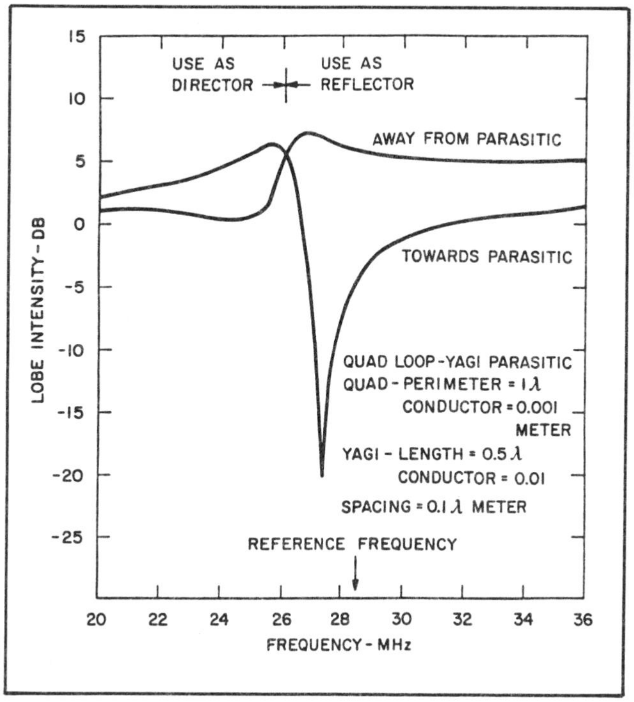

Figure 20. Same as Figure 19, but for a spacing of 0.1 wavelength.

Open-quad Loops

As noted in the conceptual approach of Chapter 1, the lobe splitting that occurs as the loop approaches 2.0 wavelengths perimeter can be avoided by opening the loop. The general scheme is shown in Figure 22. This also shows the practical problem that arises.

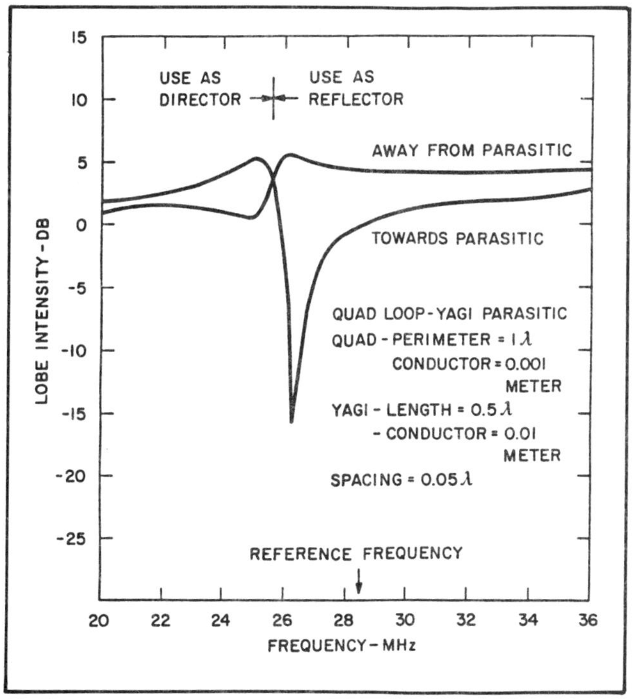

Figure 19. Lobe intensity for a quad-loop/Yagi parasitic element with spacing of 0.05 wavelength. See text for use, and for scaling factor.

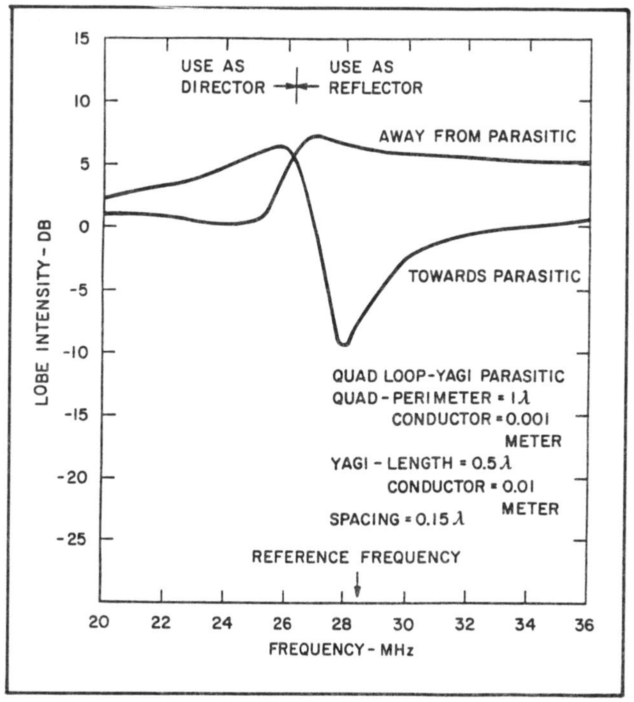

Figure 21. Same as Figure 19, but for a spacing of 0.15 wavelength.

In a closed one-wavelength loop, the feed point is a high-current point. But when the loop is opened symmetrically, the feed becomes a high-voltage or high-impedance point. Current feed of the open loop requires asymmetric feed.

A single-band open loop can be voltage fed, say, by a quarter-wave stub, as in the J-Pole. For two-band operation, a solution to the feed requirement is to use the two-frequency network pioneered by Pichitino[32] and shown in Figure 23. This uses a combination of series and parallel resonance to meet the impedance requirements. (See Pichitino's original article for design details.) The antenna opening in a two-band design can be done by relay or by another series-parallel network.

In the 1978 addition of *The ARRL Handbook*, VK2AOU took a different approach. Instead of opening the loop, he accepted the gain penalty of operating with a smaller loop, and combined a matching net-

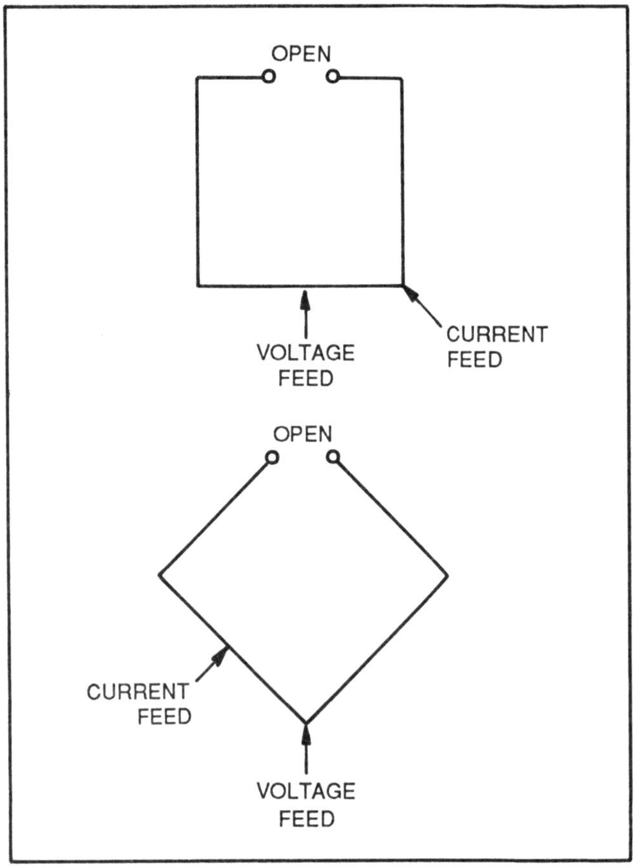

Figure 22. Two wavelength-perimeter open-loop quad and diamond loops. Points for voltage (High-Z) and current (Low-Z) feeds are shown.

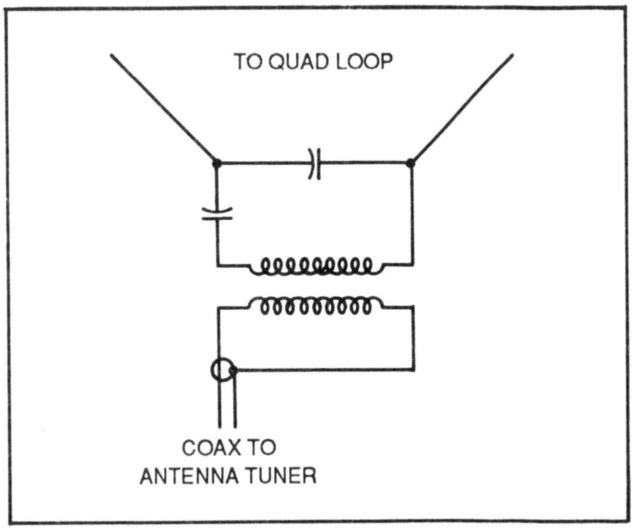

Figure 23. Use of a network having two resonances for closed-open loop feed. Typically 20 meters and 10 meters are series and parallel resonant, respectively. Loop opening may be by a relay or by another two-resonant network, omitting the feed. (After Pitchitinio.[7])

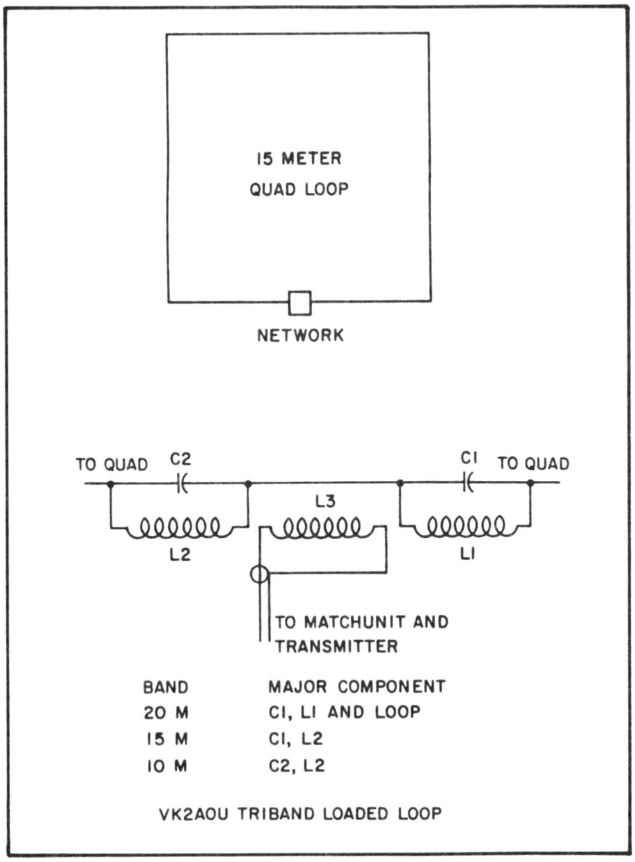

Figure 24. Three-resonant-point network used by VK2AOU to change the resonance of a loop to an added higher-frequency and an added lower-frequency band. Typical use would shift the resonance of a 15-meter loop to 20 and 10 meters.

work with reactance-compensating elements. The essential parts of this are shown in Figure 24. The network for the parasitic elements omits L3 and the feed line.

The performance of the open-closed loop system can be seen in Figure 25. At 2.0-wavelength circumference, the gain is higher than for one wavelength by 2.5 dB, a worthwhile increase. In a wideband loop design, the loop should be closed below about 1.3 wavelengths perimeter, and should be opened above this point. Gain will be at isotropic or above for the range 0.7 to 2.5-wavelengths perimeter.

Drive resistance and reactance for the opened quad are shown in Figure 26 and for the diamond in Figure 27. Some form of open-wire feed is indicated because of the rapid change in feed values with frequency.

Two pattern features of the open loop are shown in Figures 28A and B. Figure 28A shows the vertical plane half pattern in the loop plane and 28B the pattern at right angles to the loop. As compared to the closed loop at one wavelength perimeter, the in-plane radiation is smaller, and the main lobe has less vertical radiation. Both are desirable for arrays.

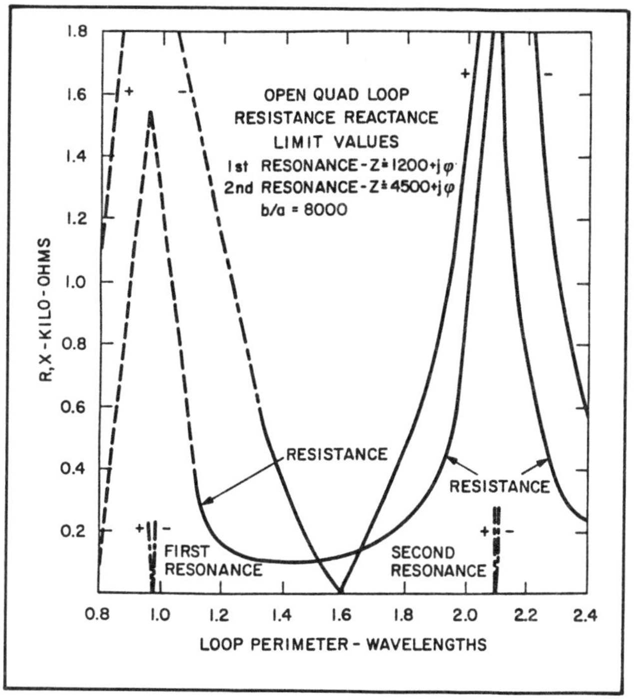

Figure 26. Drive impedance of an open-quad loop. Because of the rapid change in values, the reactance curves are not fully plotted: the resonance perimeters are indicated by the zero-reactance points.

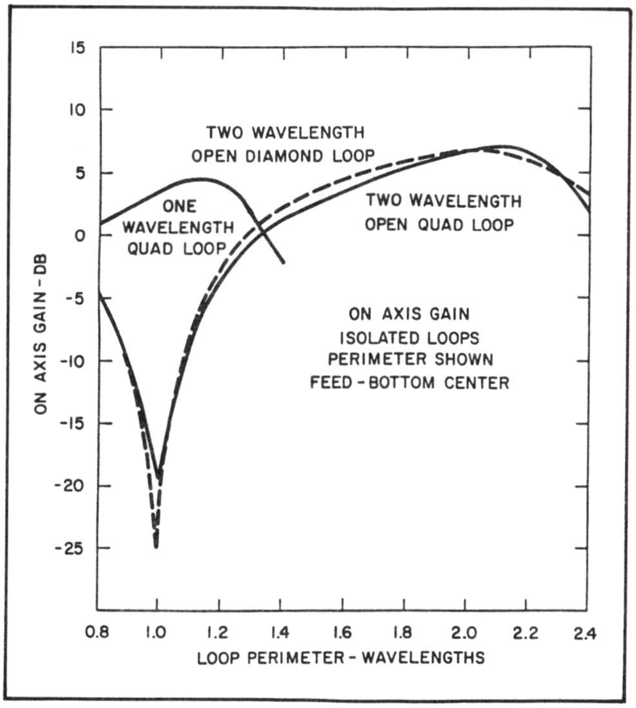

Figure 25. Gain of the lobe at right angles to the loop for open quad and diamond loops. The gain of a one-wavelength closed loop is shown for comparison. Feed is at bottom center.

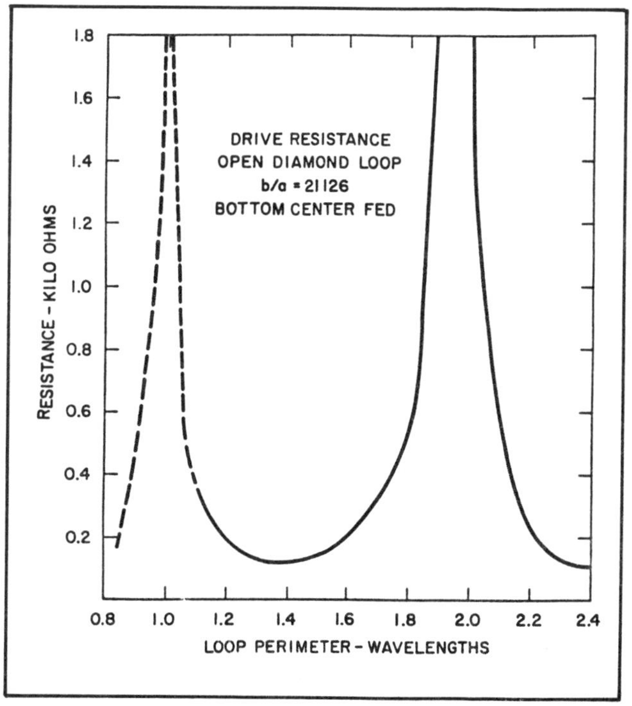

Figure 27. Drive resistance of an open diamond loop. The reactance curve is very similar to that for the quad. Feed at bottom center.

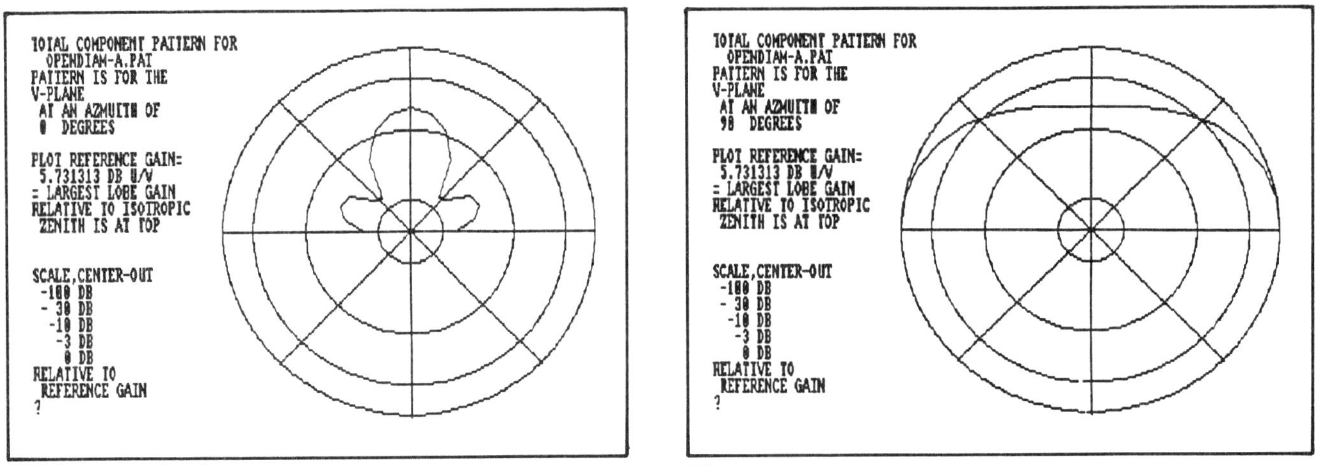

(A) (B)

Figure 28. Two half-patterns for an open-diamond loop. (A) Total vertical plane intensity in the plane of the loop. (B) Total vertical plane intensity at right angles to the loop.

Except for Orr,[22] no data on arrays of open loops is available. If MININEC analysis is undertaken, about 32 segments per loop will be needed for reasonable accuracy.

Chapter 10

Mechanical Design

A look at hardware and maintenance

When quads are discussed on the air, it's likely that one or both operators will mention having had a quad, "but it got damaged in a storm." Other conversations will bring up the matter of size and of the difficulty in handling such a large antenna. Often there will be a wish to have a quad expressed, most probably because of its reputation for good performance at low height.

This chapter takes a look at construction methods that give long life and failure-free operation. Included are handling techniques to make the job of erecting the antenna relatively easy, and notes on maintenance. The techniques are biased toward high-frequency designs.

The Quad Conductor

While there's a bandwidth advantage in using large conductors in quads, factors such as weight, wind forces, and cost dictate that high-frequency quads use small conductors; that is, wire elements. This doesn't mean that the gain or F/B ratio must be seriously compromised, but mostly that change of VSWR with frequency is relatively large. However, this isn't a problem with a good transmission line and a wide-range antenna tuner.

Of the various types of wire for elements, it appears that copperweld is by far the best. It's strong and far less likely to fail because of a sharp bend or wind-induced vibration. These factors mean that very small wire can be used.

In principle, the increased resistance of small wire is a disadvantage. Practically, however, this seems to be unimportant—a consequence of the high radiation resistance of quad loops and arrays. Number 18 copperweld has proven satisfactory for 10 to 20 meters. But it appears that no. 14 or no. 12 is better for 40 and 80 meters, because of the longer span required.

Copperweld is made in various grades, corresponding to the percentage of conductivity as compared to pure copper. Because RF conduction occurs along the wire surface, low-conductivity grades are as good as the higher ones. Grades with 30 or 40 percent conductivity have worked well in tests. A good source of copperweld wire is "electric-fence wire" sold by some hardware stores and mail-order houses. The minimum order is usually a quarter mile in length—an amount sufficient for a lot of experimentation!

Aluminum-coated wire is an alternative to copperweld and has been supplied with some commercial quads in the past. It is lighter and a little less noticeable, but it's necessary to use mechanical clamps at the joints instead of solder.

The other option is pure copper. Only hard-drawn seems to be satisfactory, and particular care should be taken to avoid sharp corner bends and kinks in handling. The size needs to be large, say no. 12 for a 20-meter quad. An alternative is to twist three smaller wires together, which reduces the risk of failure due to bending and vibration.

W6SAI and others warn that the presence of wire insulation of any thickness changes the required loop dimensions appreciably. If you use insulated wire, build a small loop first, determine its resonant frequency, then apply the scaling factors found in the preceding chapters to get the full-scale dimensions.

Use of twin-lead for a conductor, as in the first quads, was mentioned earlier. In principle, this increases the effective conductor diameter, and makes the loop better for broadband operation. Full-scale tests of a variation of this have been made, using two reflector wires spaced 6 inches apart, and centered on the design dimensions. There was some change in the back-lobe variation with frequency, as compared to a single wire. However, the main effect was a shift in frequency of best F/B ratio. The overall conclusion was that the improvement didn't justify the complexity or added weight and wind loading.

Supporting the Wire

The weak point of the element is the corner bend at the wire support point. This is 90 degrees in a conventional quad. *The ARRL Handbook* shows several methods for making these support bends, and a number of articles have shown other techniques. But for copperweld wire, experience has shown that a piece of string is a perfectly adequate corner support. The reasons why this works include the elasticity of the string and its vibration-damping effect. Of course, there are points to watch. The string must have adequate strength, and still be elastic. One-sixteenth inch nylon "fishing line" or "mason's cord" is satisfactory for a no. 18 wire 20-meter loop and its 10 and 15-meter companions. A 40-meter loop could use 3/16-inch cord.

The best way to fasten the cord to the wire seems to be the sailor's anchor bend, invented to fasten soft line to the metal of an anchor ring. This knot is made by passing the cord around the wire several times, then passing the free end through all the loops, as illustrated in Figure 1. The assembly is pulled tightly around the wire, and finished off with two half hitches around the main part of the cord. Instead of cutting off the free end with a knife, use a small blowtorch, leaving the strands fused together to prevent unraveling and slippage.

The cord is best fastened to the support arm by a clove hitch, backed up with two half hitches. In initial setup, the clove hitch can be slid along the support to square up the loop. Then, to prevent further slippage, wrap several layers of friction tape around the support, immediately below the cord. There should be 6 inches of cord between the wire and the support to form a shock absorber.

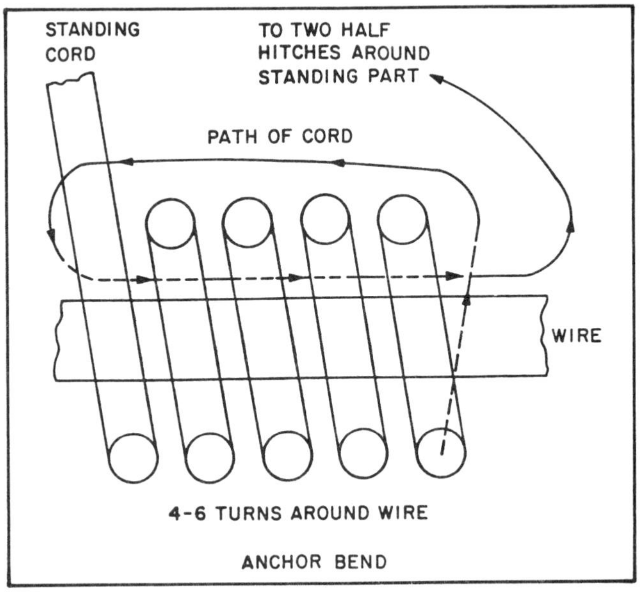

Figure 1. Tying an anchor bend, used to fasten the loop conductor to the spreader arm. The line with arrows shows the path of the free end after forming the multiple turns around the wire. Pull tight, then finish off with two half hitches around the standing part.

Note that friction tape is specified. Black plastic tape is unsatisfactory, because the adhesive softens with time, allowing the cord to slip. Unprotected friction tape will eventually dry out and fall off, but you can prevent this from happening by covering the tape with Rustoleum® aluminum paint. Rustoleum is recommended because its base is compatible with friction tape adhesive, and because it doesn't dry completely. This friction tape fastening is also recommended for attaching the coax feed to the boom, and the coax and rotator cables to the tower.

If copper rather than copperweld is used, this simple support may not be good enough. One suggestion for a sturdier support is to form copper tubing into a 90-degree bend of about 1-inch radius. The tube ends should be flared and rounded as shown in Figure 2. The wire is passed through the tube, which is then supported by cord from the spreader. The flare prevents a sharp-corner contact. I have never tried filling the tube with silicon rubber, but this would seem to help damp out wind-induced wire vibration, which is often the cause of wire failure.

The antenna element must not touch metal parts at any time. Such contact makes noise on receive and damages the wire due to arcing on transmit. An easy way to prevent contact is to split an 8-inch segment of PVC pipe in half lengthwise, and fasten one piece

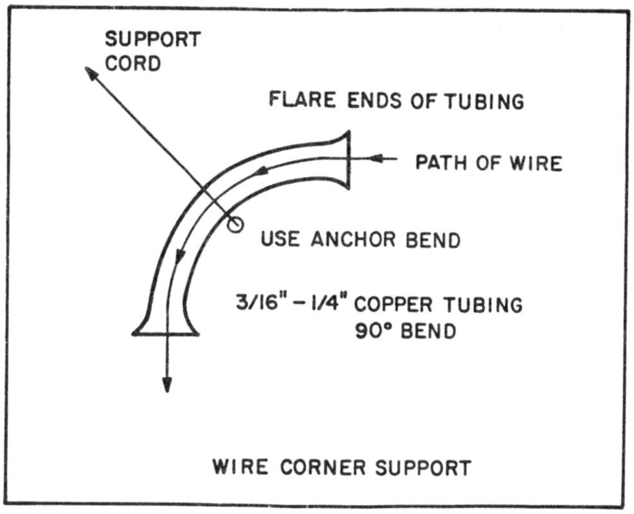

Figure 2. Suggested method of forming a length of copper tube as a corner support if copper wire is used. The flare prevents the wire from rubbing on sharp corners.

where contact would otherwise occur. This pipe can also be made long enough to keep the support string from slipping. Use painted friction tape for fastening.

Loop-support Frame

For either the X-frame or spider quad, there are three choices for the support arms or spreaders: nonmetallic arms; metal arms, insulated to avoid resonances; and resonant metal arms.

Original quads used nonmetallic arms of bamboo, still a satisfactory material if well protected. Varnish and plastic resin coatings are good for a few years. A layer of fiber glass and resin, painted to stop sun deterioration, and completely covering the bamboo, will give long life. A friction-tape wrapping painted with aluminum Rustoleum will last a long time if repainted every few years. Recently, shrink tubing has been used; it should be protected from sunlight.

The best nonmetallic arms are the tapered fiberglass tubes made for the purpose. They are light, strong and long lasting. They are available with manufactured quads and on separate order, and are often found at hamfests. All of the nonmetallic spreaders I've seen expose the plastic to the sun. This reduces spreader life. A better practice is to paint the arms with a good grade of white marine epoxy, formulated to withstand sun exposure. A coat may be needed each 3 to 5 years.

Wood is another material used for nonmetallic arms. It's sometimes used for full-length spreaders, but more often for end inserts. Wooden arms must be protected from rain, sun, and insects. The old technique of boiling in paraffin works. Coating the arms with epoxy, or wrapping them with fiber glass or tape, is even better. Be sure to provide drain holes just above the wooden inserts for the lower arms.

You have two choices if conducting arms are to be used. One is to cut the arms to lengths well away from operating frequencies. For a 10/15/20-meter quad, this is easy if the spreader is cut and insulated at the center; the outer part may be a nonmetallic material. A typical set of dimensions is shown in Figure 3. Note the use of an insulating barrier between the inner ends to avoid contact if a holding clamp should loosen. The insulating pipe shown should be split to give good clamping action. It can be replaced with an insulator made of about 6 to 8 layers of friction tape.

This figure shows a wooden end—in this case, a piece of dowel. The nonresonant spreader's length is chosen to avoid resonance, or, if this is no problem, to allow use of standard tubing lengths.

If the spreader is to be resonant at an operating frequency, there will be no center cut, and it isn't necessary to insulate the spreader where it is supported. In fact, a good connection from the spreader to boom, to mast, to ground is a benefit in lightning-prone areas, and in dry climates where wind-blown dust is a static producer.

The resonant spreaders can be used in either of two

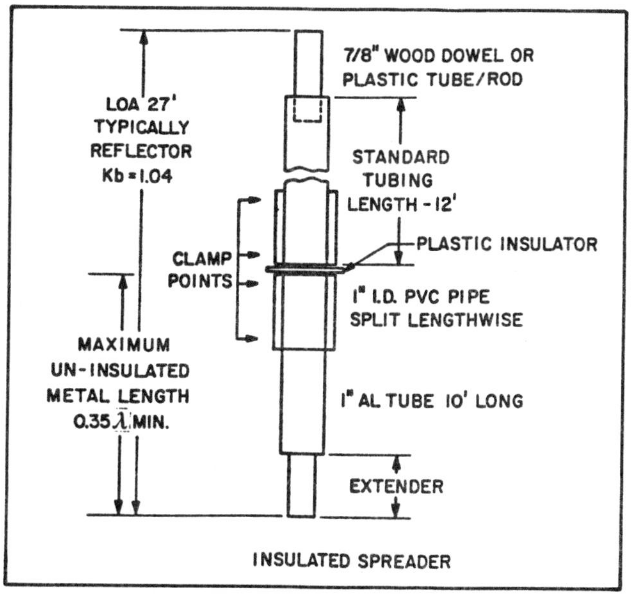

Figure 3. Typical spreader using a metal (usually aluminum) tube, sectioned to prevent resonances in the operating range. Dimensions shown are for a 10/15/20-meter quad. Insulators to prevent element-tubing contact aren't shown.

ways. One is to tune the spreader to the frequency of one of the loops carried by the spreader arms. That is, the spreader supporting a director loop would be tuned as a director, as discussed in Chapter 9. The tuning can either be for broadbanding or for gain.

The second way to use resonant spreaders is to increase the performance for bands away from the nominal loop resonance frequencies. For example, tuning the supports to 18 MHz and feeding an 18-MHz signal to the 20-meter loop would give reasonable 18-MHz performance as a quad/X-frame quagi. With a three-element quad, the director tuning could be to 24 MHz and the reflector to 18 MHz, giving good performance on the two added bands. Or, the parasitic element arms could be tuned to 10 MHz; even 7 MHz would be possible if a loading coil were used at the spreader-arm center.

At W4MB, the tuning has been to improve 15-meter performance. Spreader dimensions that have worked well are shown in Figure 4. The tubing and extender sizes are based on 85-mph winds. They have given no trouble in storms with wind gusts to 75 mph. Currently, I crank the tower to its lowest height on receipt of a hurricane warning (always a possibility in Florida). Wind forces are lower near ground, due to friction.

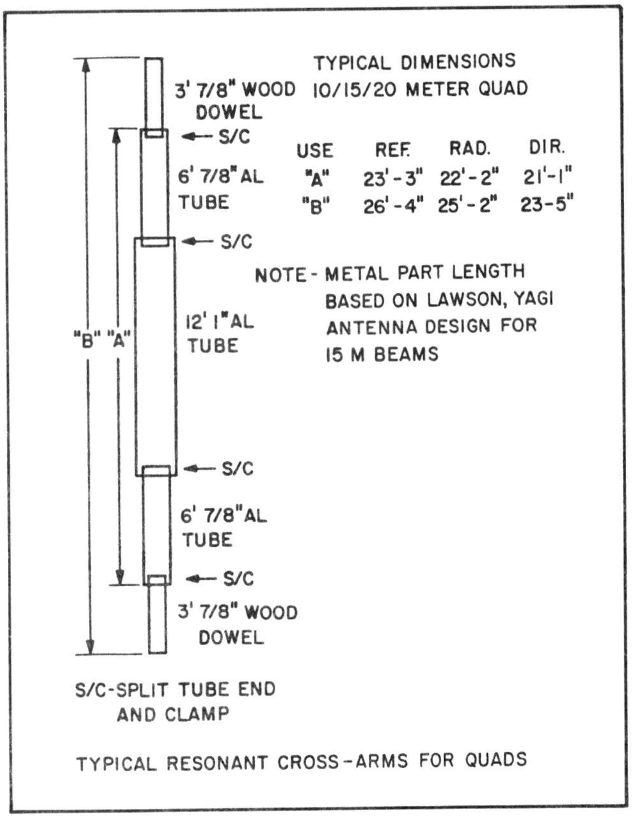

Figure 4. Typical spreader using metal (aluminum) arms tuned to resonate in the operating range, in this case 15 meters. The arms carrying the reflector are tuned as a Yagi reflector.

A stronger design would be needed for hurricane-area antennas that must remain at full height during storms. (See the following discussion on towers.)

Boom Design

Most amateur antennas use a tubing boom of sufficient diameter and wall thickness to withstand the loads imposed. In almost all cases, less windage and weight result from a design using boom guys. To get a clear understanding of the range of possibilities of tension-compression design using tube in compression and wire in tension, look carefully at a model of a full-rigged ship, or even a drawing like those found in the better dictionaries. It will take a little time, but try to understand the purpose of each line. Then see how this applies to your antenna problem.

Figure 5 shows a design for booms up to 21 feet in overall length which is suitable for two elements on 71 MHz, three elements on 14/21 MHz, and four elements on 24/28 MHz. This diamond-stay principle is readily extended to greater length. Figure 6 shows the key elements. In both these layouts, design emphasis has been on use of commonly available material and on construction simplicity. Greater strength is possible using the components and techniques found in sailboat rigging. If a large antenna for permanent mounting on a high tower is needed, these improved techniques should be adopted.

Check the lengths of the stays—the diamond is really a quad loop. There's a possibility of additional antennas, say for 2 or 6 meters, by feeding one or more diamonds.

The key to these two boom designs lies in the jacks used to carry the diamond stay arms, and also the spreader arms. A separate jack also joins the boom to the rotating mast. Basic construction is shown in Figure 7. The materials are for the short-boom design, with 10/15/20-meter elements. These should be suitable for such elements on the longer boom. With larger elements, or for the center jack of the long-boom design, the dimensions and weights must be increased.

A good way to fabricate these jacks is to cut the pieces to length with a saber hacksaw blade. Use plenty of oil. Drill a single 1/4-inch hole in one web of each spreader arm, at the center. Clamp the arms together, and drill them through the center arm holes. Fasten the arms together, using lock washers under the nuts.

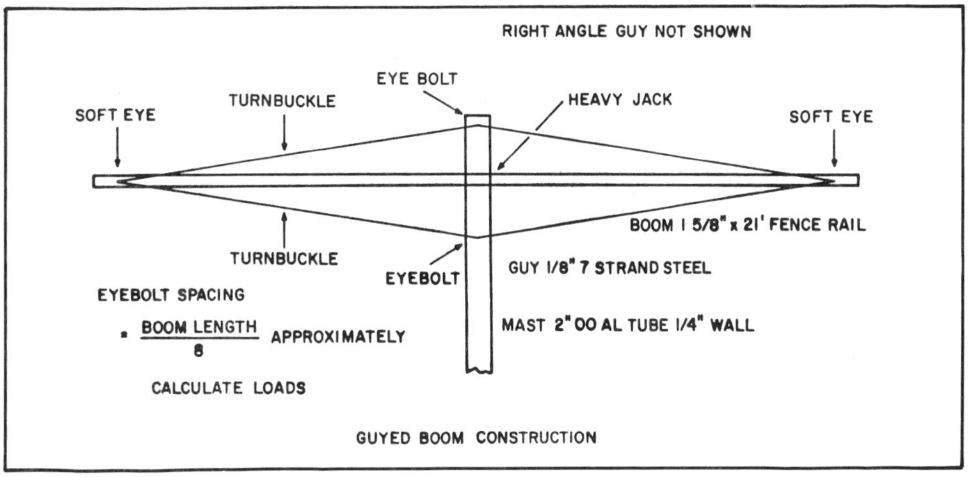

Figure 5. Use of diamond stays plus a small tube for the boom, giving a low-weight, low-windage design. The design is good for a close-spaced three-element 10/15/20-meter quad. Fitted with torsion guys (discussed later), it will support a light-construction, 40-meter unit.

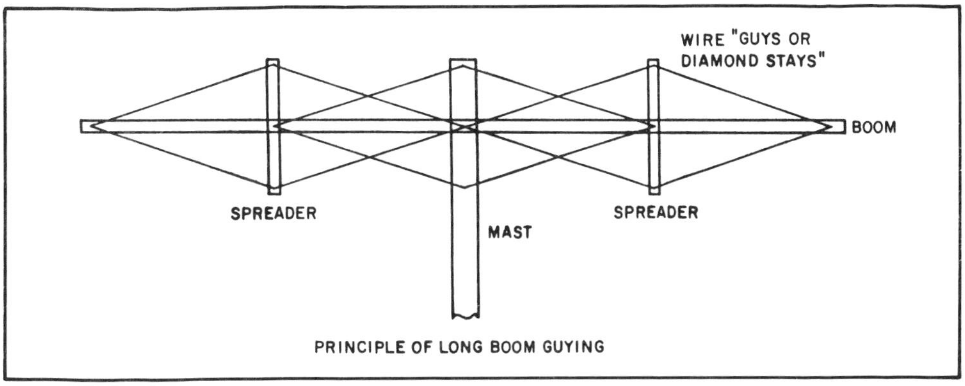

Figure 6. Use of multiple diamonds in long-boom construction. Tubing and wire sizes should be determined by a complete load/strength analysis. Note that this construction is suitable for Yagi elements and for quad spreaders, as well as for booms.

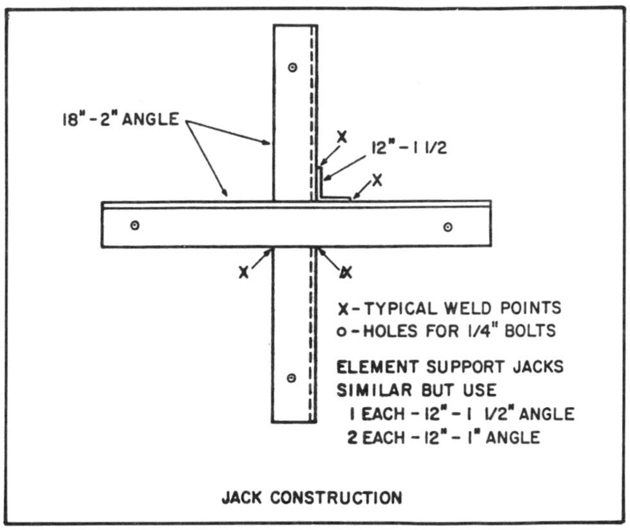

Figure 7. Construction details of the jacks used for spreader-boom and boom-mast attachment. Angle size and thickness vary with the application. Some weight reduction is possible using aluminum angle rather than steel, but the cost is much greater.

Now the welder need only run beads along the mating surfaces. The holes can be used to fasten the assembly to the boom.

If a job shop is available, the assembly may be hot-dip galvanized. However, two coats of Rustoleum aluminum paint are nearly as good (touch up after final assembly). If all hardware is also painted, steel is usable instead of stainless steel. Because this paint never really sets up, paint the screw threads too, making the same after-assembly touch ups.

For loops up to 20 meters, the spreaders can be held in the jacks by a hose clamp at each end of the jack arm. Insulated spreaders would need two clamps per tubing end. Larger loops with heavier arms should use through-bolt fastening, with a short doubler plate on the side of the tubing away from the jack arm. Note that the common stainless-steel hose clamp often has a plain iron screw cam, so these should be painted with Rustoleum.

Experience with the short-boom type indicates that no torsion control is needed. Unless the boom has a heavy wall, 40-meter elements or long-boom types will need protection against wind-induced twisting instability. The principle is illustrated in Figure 8.

Mast

In the usual design, there will be a vertical mast section between the boom and the top of the tower. Typically, this is of heavy-wall tubing, and extends through a lateral rotating bearing at the top of the tower to a thrust bearing some distance down the tower, as shown in Figure 9. For the 21-foot boom, a mast of 2.5-inch aluminum with 0.25-inch wall has been found to work well. The mast must carry all wind and weight forces to the tower, so adequate bending strength is needed.

If the quad elements aren't too close to the mast, the mast can be fitted with diamond stays to reduce weight and windage. A single mast-to-boom stay can be used if the loads aren't the same on both sides of the boom. Mast-to-boom stays are also useful for torsion-load control.

Guy Wires and Fittings

In the 21-foot boom design, the 1/8-inch, 7-strand steel cable sold as TV mast guy works well. However, the galvanizing isn't very good, and the cable should be painted with Rustoleum after installation.

A version of a soft loop is satisfactory for fastening the guy to the boom. It's made by taking two turns of the guy wire around the boom, pulling it snug, then wrapping the free end around the standing part for a distance of about 2-1/2 inches, or about 20 turns. A 1/4-inch bolt extending through the boom keeps the guy from slipping. It's best to keep this anti-slip fastening separate from those attaching the spreaders to the mast, so a loop assembly can be removed easily as a complete unit.

Each guy should have a turnbuckle to set the initial tension and to allow removal of boom bends. There are devices and rules for setting the proper amount of tension. A simple rule used in sailboat rigging is that unloaded guys should look tight and feel loose. It's neither necessary nor desirable to set the tension so high that the wire makes a noise when plucked. The guy shouldn't make a sharp bend at its center. For tube spreaders, two parallel saw slots cre-

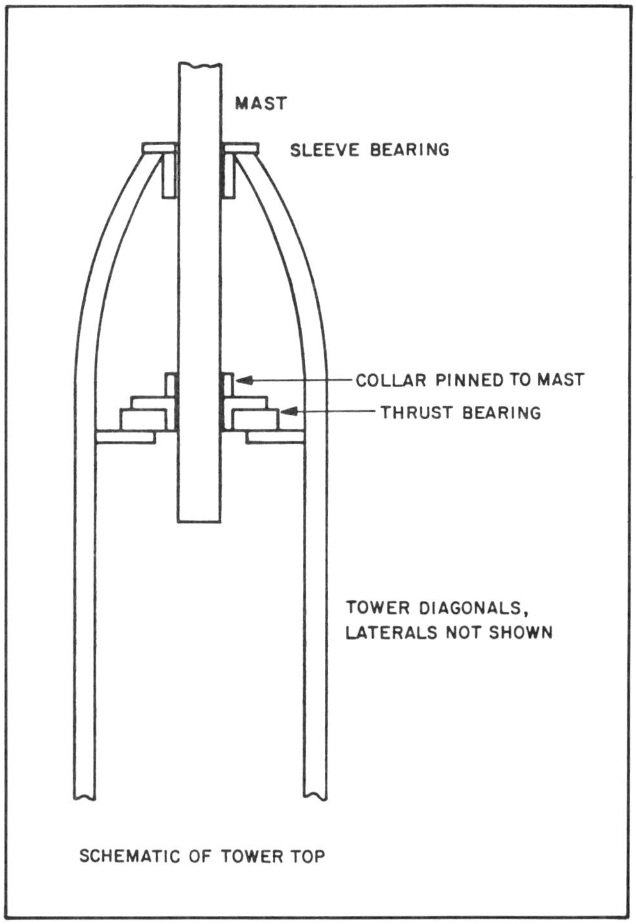

Figure 9. Schematic of a tower top assembly, showing separate bearings for lateral forces and for weight loads. Having the thrust bearing take the load reduces load on the rotator and simplifies service operations. (Tower diagonals, laterals not shown.)

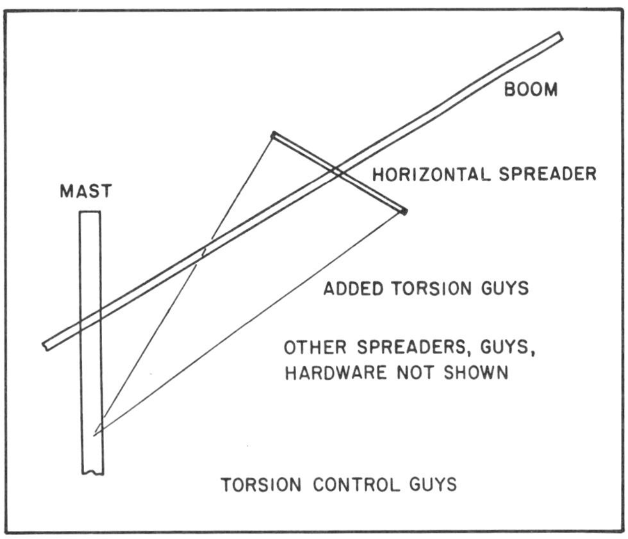

Figure 8. Principle of guys for torsion control. If cross-arm strength is sufficient, these guys can also replace part of the diamond stay assembly. (Other spreaders, guys, and hardware not shown.)

ating a gap just wider than the wire is good, providing the metal tab formed is bent toward the tube center. If the tube must extend beyond the point of guy fastening, an eye-bolt or an eye-nut can be used.

Standard sailboat design uses tangs and clevis fittings at guy-cable ends. Several types of cable fittings are available, like the "Norseman" and "Nicopress." This hardware is available in a wide range of sizes and weights, and is recommended for large antennas of all types.

Ice Proofing the Quad

The greatest load on antennas is often due to icing. Any design will fail in some way if the ice really builds up. If you review the published photos and accounts of quad failure due to ice, you'll see that the problem is primarily due to support-arm bending. In many cases, especially with fiber glass spreaders, the quad recovers shape as the ice melts.

This observation shows the way to reduce ice-storm risk. Simply add additional guys to take the bending loads of the ice. An inexpensive way of getting modest protection is to prebend the element support spreaders in one direction, then provide guy support against further bending. This prebending occurs automatically if the tension on the quad loops is made high by sliding the support fasteners (string, etc.) farther out on the spreaders. After the bend is established, the tips of the spreaders are guyed to the corresponding spreader of other elements, or the boom. Figure 10 is a sketch of a two-element design. The alternative is to extend the boom somewhat, and guy the ends of the spreaders to the end of the extension. The extensions do not need

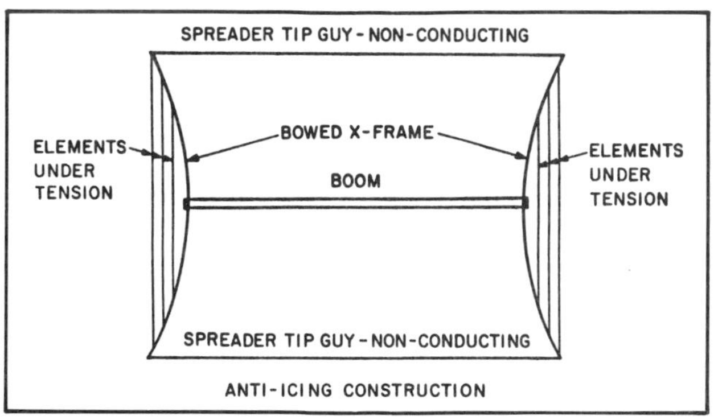

Figure 10. Schematic showing thin-wall extender tube as a torque absorber. The tube may extend to the bottom of the mast section, if necessary. Inspect the center part of the torque tube at intervals to see if excessive load has induced splitting.

to be quite as heavy as the boom. If space is a problem, keep the extenders within the turning circle of the original design.

These additional guys can be of wire with insulators to avoid resonances if necessary. However, it's probably better to use nonmetallic guys, mason's cord or glass-core guy material. Keep the guy diameter small to decrease the ice load, and to avoid excessive strength. In an ideal design, the guys would break first, allowing the spreaders to deflect. This would likely crack the ice, causing some to fall off, thus reducing the load.

The Tower

Despite their higher cost, I must confess a partiality for fold-over towers which allow all antenna work to be done at ground level. Certainly, these look more and more attractive as the years go by and tower climbing becomes more arduous. The quad is easier to handle if there are no guys to work around.

The easiest tower to use is the sky-needle type, because it folds over in one piece. If the entire needle is rotated, there's a marked reduction in antenna support complexity, and weight and windage aloft are considerably reduced. See Reference 33 for a design technique for these towers. Almost as easy to use are the crank-down-and-fold-over types made by several companies, and the fold-in-the-middle variant of standard guyed towers. Check back copies of W6SAI's series of columns in *CQ* and *Ham Radio* magazines for ideas.

For the average installation, with a house and its accessories to consider, and trees to clear, careful layout of the antenna-occupied space is a necessity. This should include ground handling, erection and aloft-occupied volume. If things look tight, it's probably worthwhile to build a scale model and practice handling techniques with it.

The following steps work well with fold-over installations:

- Prepare the foundation.
- Install the standing portion of the tower.
- Install the movable section. Some designs have built-in provisions for this, in others an auxiliary like a gin pole or portable crane must be used.
- Practice the erection process. Make certain that there are no interferences.
- Install the rotator and mast.
- Install the boom; check again for clearances.

• Assemble the loops and spreaders and install them in turn, working from center out. As soon as the radiator is installed and connected, check to make sure the feed system works.
• Touch up all paint, protect the coax connectors, and make certain that all fastenings are secure.

For guyed tower installation, it's usually best to assemble and test the quad assembly on the ground. Use a short temporary mast as a support. At the tower top the bearings are installed, followed by the mast. The latter is fastened temporarily, so that it just projects out of the tower.

This temporary setup provides an opportunity to proof test the antenna and boom. This is done by hanging temporary weights at various points, which have been chosen to exceed the calculated forces at that point by 50 percent, or so. The spreader and spreader jacks can be checked by applying horizontal forces with a scale, or by making a weight test before the spreader is attached to the boom. If you make this test, keep a careful record and have it signed by a knowledgeable test witness (see the section "General Notes On Towers").

Raising the quad/boom assembly to the tower top is relatively easy if a temporary trolley is rigged, as shown in Figure 11. (Hoisting and control lines are shown, but you may want to build a small model and work out a system of your own.) After the mast-to-boom attachment is completed, and the feed is connected and tested, the mast is raised to its final position.

You might want to install a temporary work platform alongside the mast and perform the final assembly from this. Such platforms can be constructed easily from painters' scaffolding, which can be rented for short periods in most cities at a reasonable price. I have used these to a 60-foot top-platform level. Temporary guys are needed at these heights.

Review the tower hints in *The ARRL Handbook* before starting the design, and again before starting the actual installation.

Rotator System

One of the secrets of long-life rotary antenna systems is the use of a shock absorber between the rotator and the antenna. This may be even more important for quads than for Yagis, but it is an important life extender for rotators of all types.

Some rather elaborate shock absorbers, which use shock cord and helical springs, have been described in the amateur literature. Such complexity is unnecessary. A simple extender of relatively thin-wall tube joining the mast and the rotator does the job. This assembly acts as a torsion-tube spring, similar to that used in some automobiles.

The same chain-link fence top rail suggested for the boom will provide a satisfactory tube. A 21-foot length of the 1-3/8 inch thin-wall type is satisfactory for a two-element quad. A three or four-element quad can use the same size heavier-wall tubing, with length extended to 30 feet. Figure 12 shows the general layout of the tower top including bearings. Note that, in principle, many rotators can serve as the thrust bearing. Practically, a separate bearing is better, because it al-

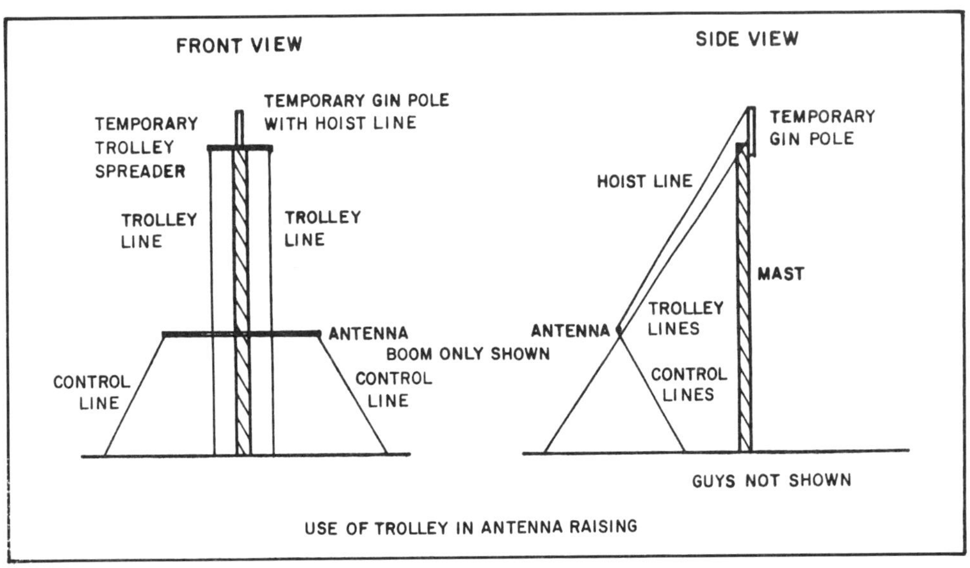

Figure 11. Schematic of a method to make the quad assembly more rigid for ice loading. This is also useful in high-wind areas. An alternative to bowing the elements is a boom extender, plus added guys from the spreader ends to the extender end.

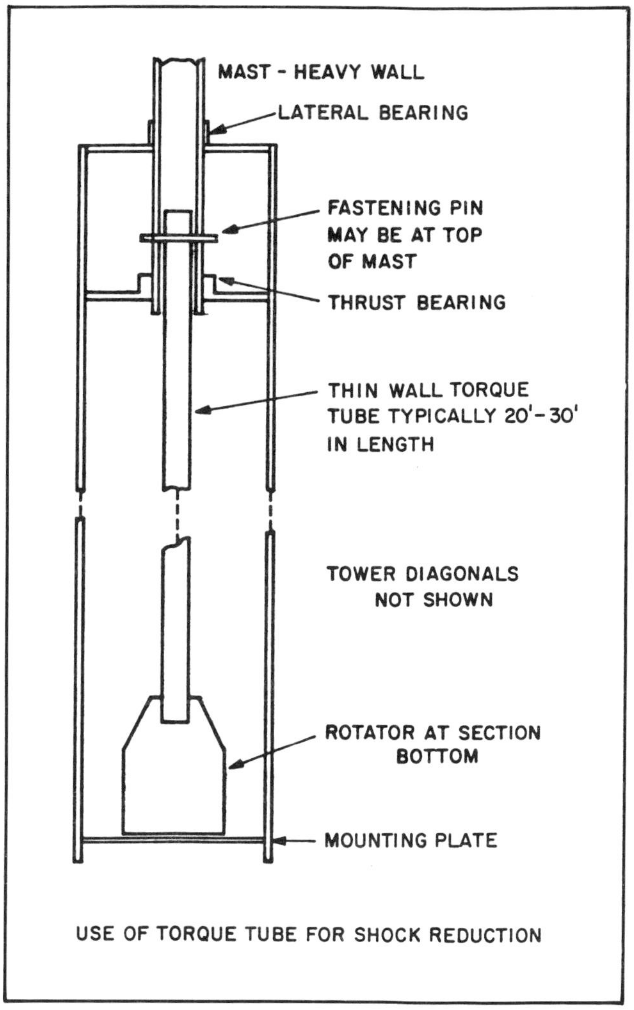

Figure 12. Schematic of antenna erection method using a temporary trolley in addition to the usual temporary gin pole. (Tower diagonals not shown.) With a very large antenna, added control lines may be needed. A practice run using only a boom is recommended. Remember that "kite-eating" trees also like to chew up antennas, so check clearances carefully.

lows rotator removal without introducing the need for a temporary support.

This shock-absorber system really makes a difference in rotator life. A TV rotator of the "Alliance" class will handle a two-element quad. A heavy-duty amateur type will take care of three-element quads and four-element Yagis without excessive wear and tear. Additionally, the absorber reduces the load on the top bearings and fastenings.

General Notes On Towers

More and more, there are city and county-code requirements for tower installations. If you don't work within these requirements, trouble may occur if anyone makes a complaint—no matter how frivolous. Get a copy of the local code and understand it. If you are told that no permit is needed, or that a section of the code doesn't apply, get it in writing: at least write it down and have it initialed.

Many codes have requirements for design certification. You'll need a to find a licensed engineer for this. It's a nuisance and some small expense. But it's a trouble saver, and may even protect you in the event of a tower accident.

Keep a full record of all design data, construction details, and tests, including photos. In these days of litigation, it's sensible to take precautions against charges of improper design and negligence. In this regard, the techniques and material weights and sizes given here have proven satisfactory over years of use. They aren't warranted to meet any building code automatically, nor to be satisfactory for use in your antenna under your conditions of use. If you aren't qualified to perform and check the design yourself, get local expert assistance—for design and installation.

NOTES

Chapter 11

Design Variations

Six quad designs and suggestions for further work

This chapter is devoted to variations of the basic quad. The approach is exploratory rather than definitive. The intent is to establish the general performance of a particular design, and then indicate paths for further analytical or experimental work. Some of the variations appear to have considerable potential in some applications.

Bent-element Quads

In a bent-element quad, one or more of the sides are bent to achieve some stated or claimed purpose. Usually, opposite sides are treated to maintain symmetry about an axis.

Batwing Quad

In the batwing quad, the sides that aren't fed are pulled toward the center of the configuration, as shown in Figure 1A. Note that this amateur antenna resembles another design used in television broadcasting. However, the TV type uses added horizontal and vertical members to connect the sides of the figure, and is regarded as a skeleton of a "fat dipole."[34]

The bent sides bring two high-voltage points closer together, increasing the capacitance. This is a form of loading, so the resonant frequency will be lower than that for a square of the same total conductor length. The bent side also requires that the top and bottom be closer together. Thus, there is a reduction in area, and a resulting reduction in gain. Alternatively, the change can be regarded as moving two dipole sections closer together, which reduces the gain of the pair. A change in feed resistance must be expected as the shape approaches that of a folded dipole with the ends bent toward the center.

It's convenient to identify the particular batwing by the percentage of normal loop height. For example, an 80-percent design would have the wings bent to reduce the height by one-fifth. By geometry, the narrowest part of this loop will have a width of 60 percent of the narrowest part.

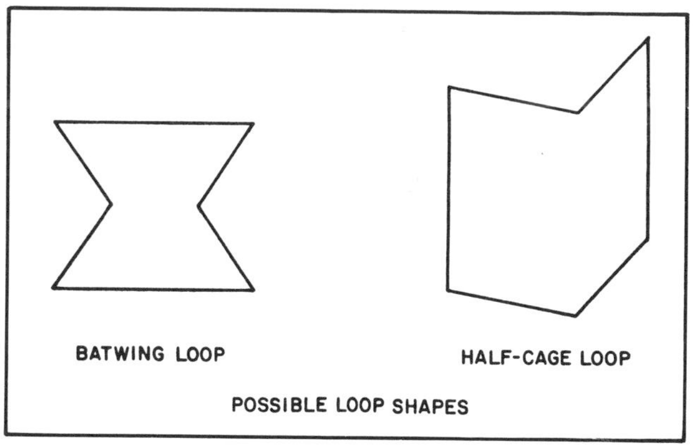

Figure 1. Possible loop shapes, bent-side quads. (A) Batwing quad, with bend in sides and feed at bottom. Height decreases compared with square loops. (B) Half-cage loop, with bend in top and bottom, out of plane containing vertical wires. Feed is normally at center of bottom apex. Usually used with a mirror-image reflector as in the birdcage antenna of G4ZU. Width decreases compared to a square loop.

Figure 2 shows the effect of element bending on gain for two heights. These values should be compared with Figure 8 in Chapter 4, for the straight-sided loop. The loss of area shows as a reduction in gain, from a maximum of just over 5 dB for the full loop to about 3.2 dB for the 60-percent batwing. Because of the loading effect of the bent element, the maximum-gain frequency changes by about 10 percent for the same pair.

The effect on drive resistance is shown in Figure 3, which may be compared with Figures 2 and 4 of Chapter 4. The drive resistance near the first resonance decreases, and the minimum moves higher in frequency by a small amount. The drive reactance near resonance is shown in Figure 4, which may be compared with Figure 6 of Chapter 4. Resonance (zero reactance) moves higher in frequency as the sides approach each other more closely as a result of the increased capacitive loading.

While the effect of the side bend is noticeable, it isn't really enormous. Thus, the performance of a quad assembled from two batwing loops should be good. This is borne out by Figures 5 and 6, for 80 and 60-percent height radiator-reflector quads. The closest data for a full-loop quad is found in Figure 5, Chapter 10. Comparison of these figures indicates that there is a relatively small effect on forward gain. No attempt has been made to establish conditions for good F/B ratio.

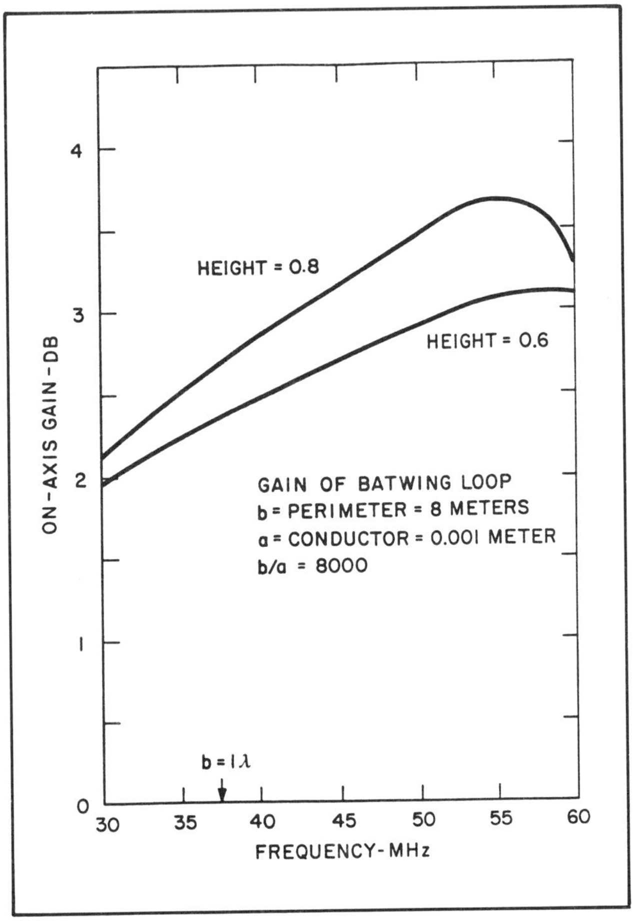

Figure 2. Gain of batwing loops. Gain versus frequency is shown for batwing loops of two heights. Loop perimeter is one wavelength at 37.5 MHz.

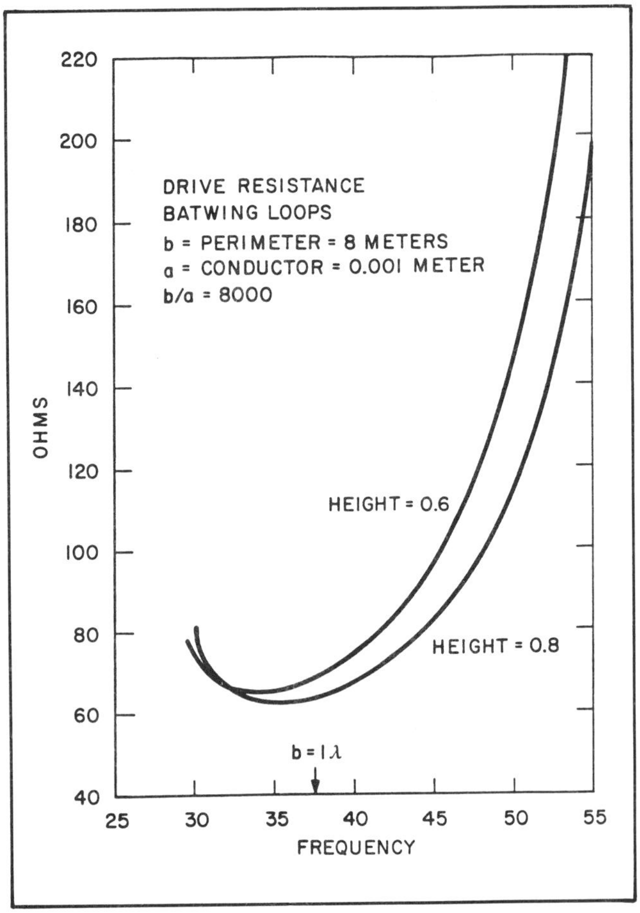

Figure 3. Drive resistance of batwing loops. Drive resistance versus frequency of batwing loops is shown for two heights. Loop perimeter is one wavelength at 37.5 MHz.

The major mechanical effect of the batwing design is a reduction in height. The turning radius, feed, and support-arm problems remain the same. In view of the relatively small mechanical gain and the noticeable change in performance, it appears that the batwing quad will continue to be a rare design.

Birdcage Quad

Instead of the loop side being bent toward the center, the center of inflection may be at right angles to the plane of the loop. The designs in the literature show this configuration in the horizontal rather than the vertical plane. This is to simplify mechanical design problems, allowing one side of the loop to be the primary mechanical support. This eliminates the need for a separate spreader. The general appearance of such a loop, known as the half cage, is shown in Figure 1B. Here the width is reduced compared to the full loop.

The general performance is evident from Figure 7, which is for a width of 80 percent. As for the batwing, there's a reduction in gain and drive resistance, and a small shift toward a higher resonant frequency. Two of these bent-out-of-plane loops may be placed back-to-back, to form a version of the quad. Two variations are found in the literature; they vary somewhat electrically, but more so mechanically. The "G4ZU birdcage"[35] uses 1/8-wave horizontal tubing arms, 1/4-wave vertical wire elements, and a vertical arm spacing of 0.175 wavelength. The originator of the second variation, "The Swiss Quad," is unknown. It

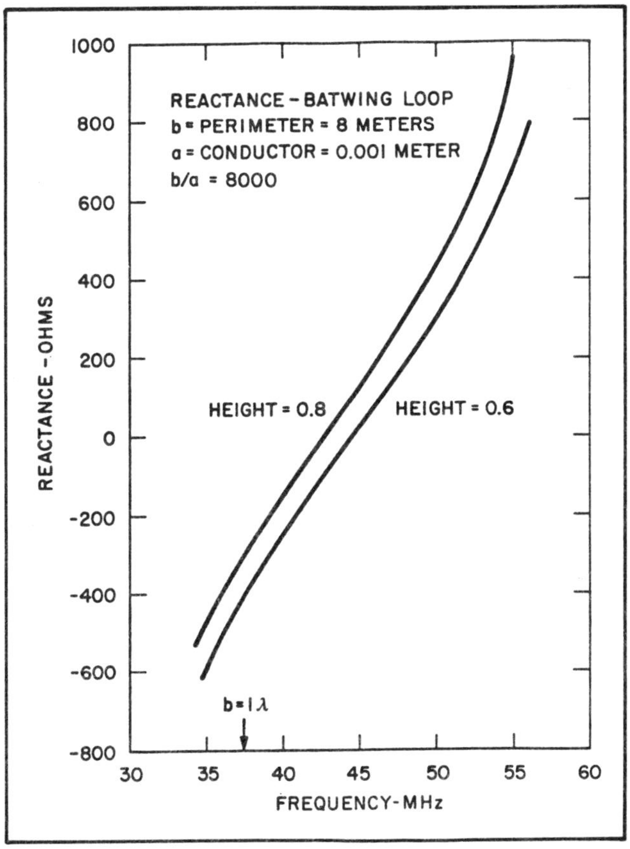

Figure 4. Drive reactance of batwing loops. Drive reactance versus frequency of batwing loops is plotted for two heights. Loop perimeter is one wavelength at 37.5 MHz. Resonance occurs at a higher frequency compared to square loops and is a function of height as well as wire size.

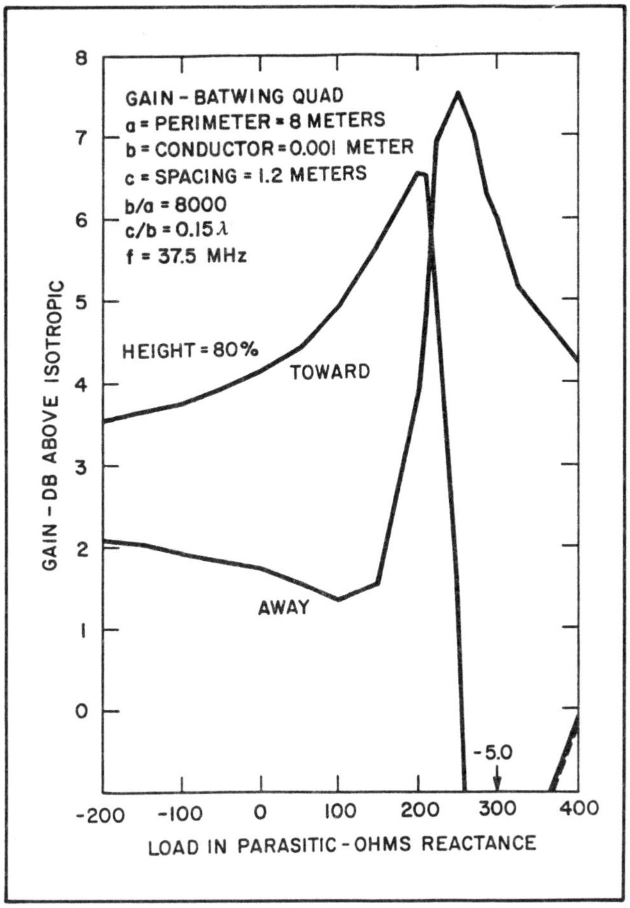

Figure 5. Gain of an 80 percent height batwing quad. Gain versus parasitic-element tuning reactance is shown for two batwing loops separated 0.15 wavelength, operating at 37.5 MHz. As for the standard quad, maximum gain and F/B ratio occur at different tuning conditions.

appears to exist in three versions: one essentially the birdcage; a second with the V-arms touching, used with a gamma match; and a third with both front and rear elements driven. See Orr[36] for a partial discussion.

For this study, the family was modeled by placing two of the 80-percent loops back to back, spaced 3.9 feet (1.2 meters) at the horizontal arm apexes. One loop was fed at the bottom Vee center. A tuning reactance, equivalent to a stub, was placed at the other Vee center. Wire was used for all elements, but tubing shouldn't change performance appreciably. The performance of this antenna is summarized by Figure 8, with gain and drive resistance plotted against stub reactance. Unlike the regular quad, best performance is with director tuning. Maximum calculated gain is 7.5 dB and maximum F/B ratio about 20 dB. Drive resistance is in the range of 18 to 75 ohms around maximum gain; drive reactance is about − 175 ohms. An

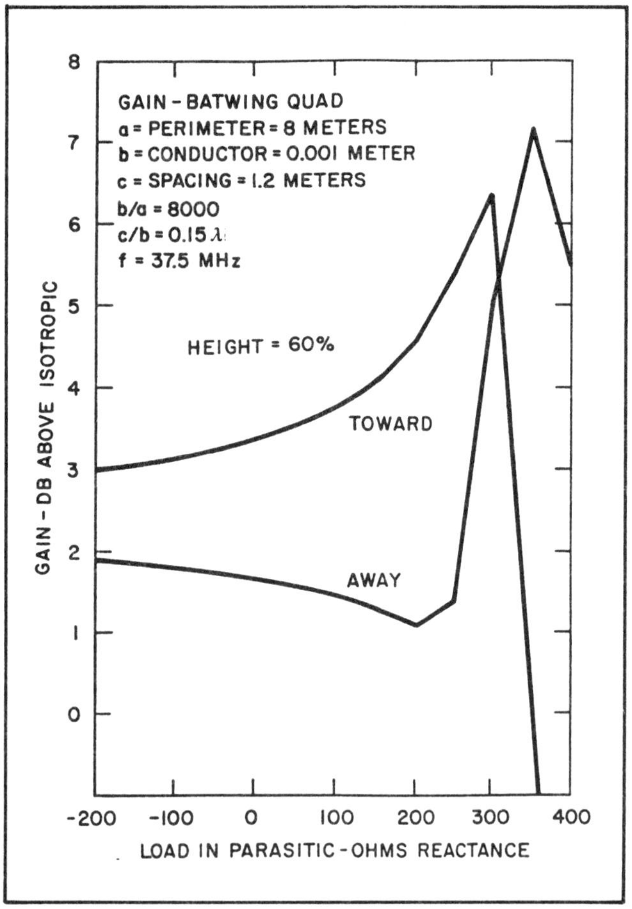

Figure 6. Gain of a 60 percent height batwing quad. Same as Figure 5 except for loop height. Note that these two figures may not represent best design conditions, since effects of other spacings were not studied.

inductive stub across the feed point will often give adequate matching. No attempt has been made to determine the best performance possible, or to study the effects of variations such as tubing supports, spacing, use of loops of different sizes, or different feed locations.

This design appears to combine reduced turning radius, simple construction, and quite impressive performance. Further design studies and experiments are recommended.

Opened-loop Quad

Diamond loops with the top opened were considered in Chapter 9 as a way of securing two-band operation with a single loop. Gain was found to increase to a maximum of about 7 dB. Feed resistance (bottom center) increased, on the order of 1 to 2 k.

A brief investigation of beams using two open-loop elements has been made. Loops were 8 meters perimeter, or 2 wavelengths at 75 MHz. If a two-frequency design is used, element spacings will be relatively wide at the overtone frequency, from about 0.2 to about 0.6 wavelength. Figure 9 summarizes the results for a spacing of 3.9 feet (0.8 meter), or 0.2 wavelength at 75 MHz. There was very little parasitic action around the 2-wavelength resonance point. Instead, the combination acted as a parasitic beam close to the 1.5-wavelength resonance condition (around 60 MHz). The combination of sizes used gave about 5 dB gain with respect to a single loop, and about 15 dB F/B ratio. All conditions studied provided the same general performance.

The effect of loading coils at the center of each arm of the parasitic element was also studied. Normal tuning action was found near the 1.5-wavelength condi-

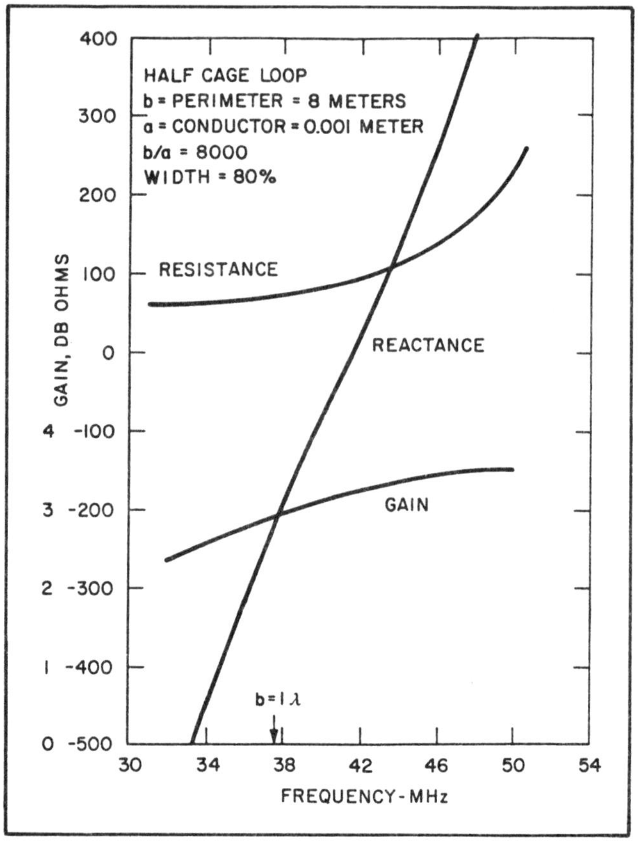

Figure 7. Design data for a half-cage loop. Gain, drive resistance, and reactance as functions of frequency for an 80 percent-width half-cage loop are shown. Perimeter is one wavelength at 37.5 MHz. Resonance occurs at a higher frequency compared to a square loop, and is a function of wire size and loop width.

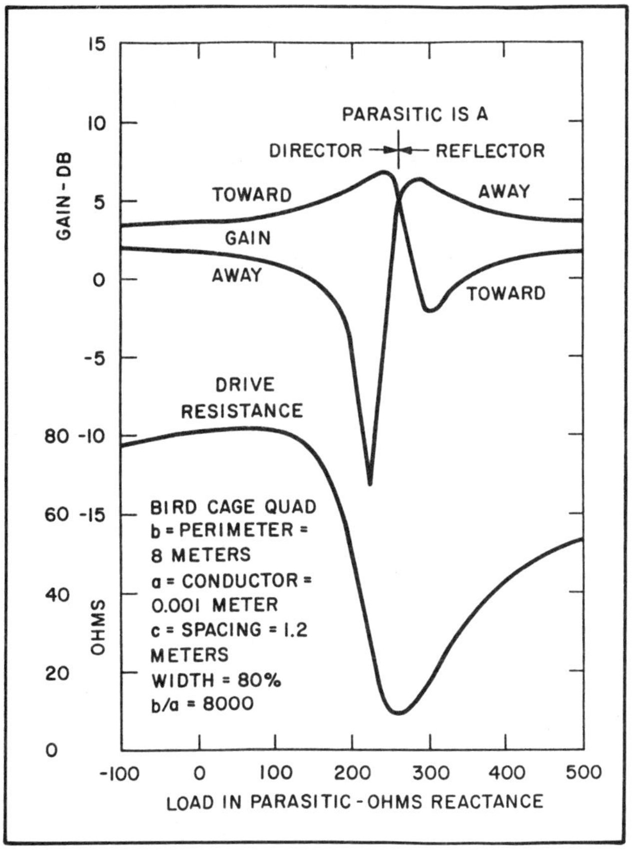

Figure 8. Performance of a birdcage quad. Gain and drive resistance of a birdcage quad consisting of two back-to-back half loops of 80 percent width versus parasitic-element loading. Unlike the conventional quad, best performance occurs with the parasitic element tuned as a director. The data of this figure may not represent best design conditions, since the effects of tubing top and bottom sections and other widths were not studied.

tion, but very little effect was observed around 2.0 wavelengths. This was true for all spacings studied. Incidentally, this performance pattern appears to have been found earlier. The post-World War II use of the single opened loop as the basis of the bi-square beam appears to be a result of the poor experimental performance of the two-loop version. (However, no reference to this has been found in available literature.)

It's evident that more investigative work needs to be done on this type of antenna. The effect of opening the parasitic element at the bottom, and at bottom and sides, are items for study, as is the effect of closer spacing. Capacitive loading of the parasitic element is another potential item. Not enough is yet known about the design to predict results; however, some large curtain arrays—the Chireix-Mesny, for example—are related, and perform well.

Stub-switched Loop

One way to open a two-frequency loop is to connect an open stub to the top, as shown in Figure 10A. If this stub is a quarter-wave long at the 1-wavelength loop frequency, it appears as a short circuit across the top two wires. Similarly, at the 2-wavelength frequency, it appears as an open circuit. A version of this has been described by Gullstad.[30]

The feed point resistance is low at the lower frequency and high at the upper. One matching method is to use a quarter-wave transformer for the upper frequency. Alternatively, a less-than-quarter-wave section can be left in the line for both frequencies, giving intermediate matching conditions. This immediately suggests connecting the two stubs together for the simple feed condition shown in Figure 10B.

With usual quad dimensions, an open-wire stub would be longer than the loop diagonal. A line with

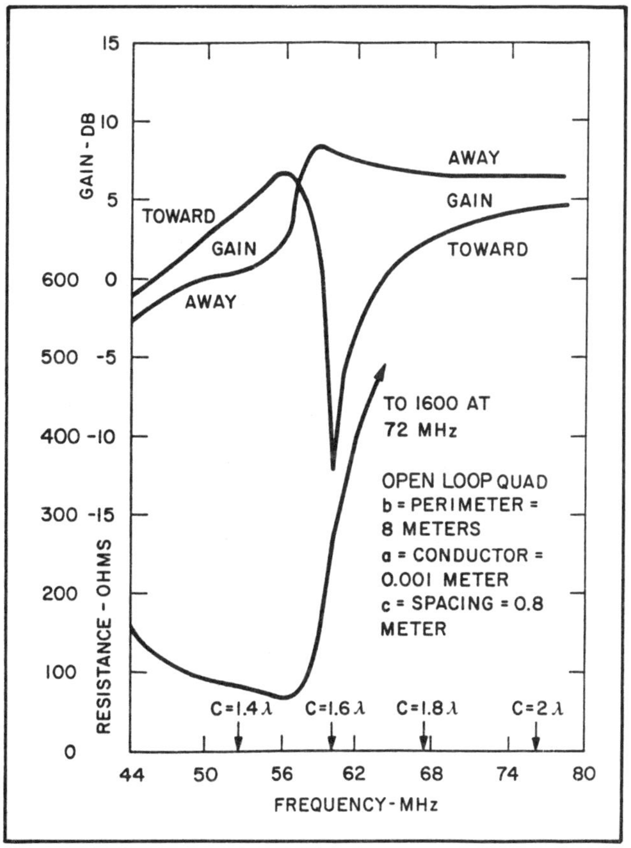

Figure 9. Performance of an open-loop quad. Gain and drive resistance as functions of frequency. There is essentially no parasitic action at the second overtone; perimeter equals two wavelengths. Instead, such action occurs for 3/2 wavelengths perimeter.

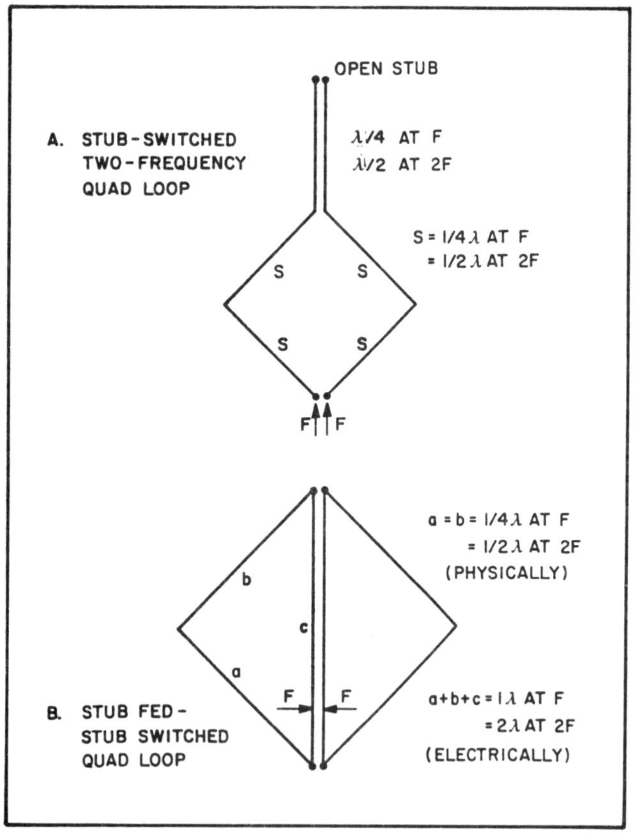

Figure 10. Stub-switched two-frequency loops. (A) Single stub closes loop on fundamental frequency and opens loop at opposite feed on second overtone. Drive resistance is low on fundamental; high on overtone. (B) Combining an open stub plus a matching stub to provide switching and impedance changes. Drive resistance and gain performance are functions of feed position along the stub. Stub propagation factor is ideally equal to 0.71.

a velocity factor of 70 percent would be essentially the same length. This mechanically convenient length could be obtained with two lengths of coax in parallel.

MININEC isn't intended for transmission-line calculations, nor will it handle conductors with insulation. Therefore, for an exploration of the general design, the model was established as two mirror-image triangle loops separated by a small distance. The feed was simulated by parallel pulses of opposite sign in the resulting simulated transmission line.

The result of this preliminary study is summarized in Figure 11. The broadside gain remains high over a very wide range of frequencies—above isotropic over a range of 3.5:1. The feed resistance is low over most of the range, and is essentially zero where each loop is less than one-half wavelength in perimeter. The resistance increase at 35 MHz is a resonance point, which also shows on the reactance curve in Figure 12. Reactance is quite variable. Other feed points along the transmission line section give greater variation in resistance and lower gain at most frequencies.

Because of the variation in reactance, one of these loops would make a poor parasitic element in a wideband beam. Two or more loops could be connected in the W8JK or ZL-Special configuration, or could be used in log-periodic fashion. There appears to be a good field for study of wideband loops in this part of the quad family.

Side-by-side Quads

All the multi-loop designs presented in this book so far have used coaxial loops, or loops strung along the axis at right angles to the loop plane. As with dipoles, loops may be placed in the same plane and separated horizontally or vertically. It's convenient to separate these designs into two families. One uses two or more "standard" arrays—usually to increase gain, but

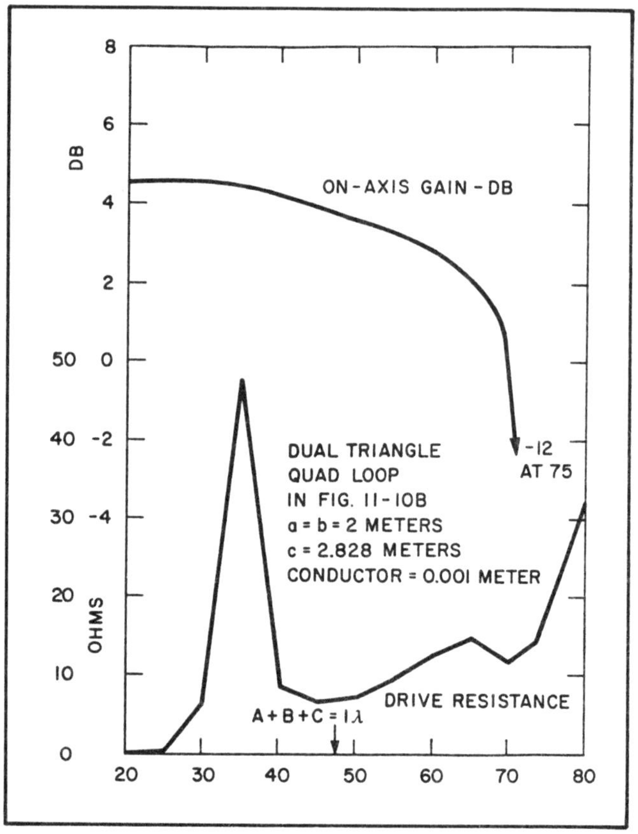

Figure 11. The stub-switched/fed loop of Figure 11 showing on-axis gain and drive resistance. The decrease at 70 MHz is due to lobe movement: the lobe angle and on-lobe gain were not studied.

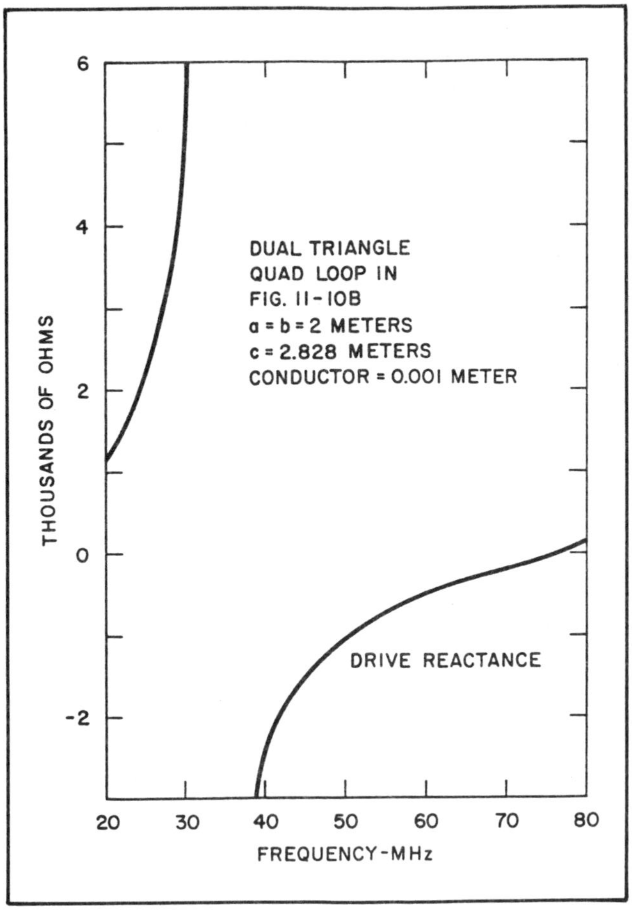

Figure 12. Reactance of the stub-switched/fed loop. The loop shows a parallel resonance at the one-wavelength frequency of the basic square loop and a series resonance at the second overtone. A 2:1 frequency range with good lobe angle stability can be covered with a gain variation of ±0.8 dB if a wide-range matching unit is used with open-wire line.

sometimes to give a particular pattern. This is called stacking. In the other family, elements are in the same plane for a particular purpose, often to simplify the feed system while attaining a particular performance.

Stacked Quad Loops

Because the effective area of a simple antenna usually exceeds its physical area, the gain of side-by-side antennas usually increases as spacing increases (up to a point). There's usually a simultaneous interaction in radiation resistance: a marked decrease at close spacing. Decreased radiation resistance means more current for a given power and thus higher gain. This component decreases as separation increases, often in a cyclic manner. The result is that there is usually a best separation for a given basic element gain.

Figure 13 shows the on-axis gain for a pair of quad loops as a function of separation, for two feeds. The stacking conditions for the two curves are as follows:

Side-driven curve:
Horizontally separated, vertically polarized.
Vertically separated, horizontally polarized.

Bottom-driven curve:
Horizontally separated, horizontally polarized.
Vertically separated, vertically polarized.

The curves apply to diamonds stacked along a line 45 degrees from the vertical.

At very close spacing, the gain may be lower than for a loop alone. Maximum increase is just over 3 dB for the side-driven condition, and about 2 dB for bottom driven. Note the suggestion of cyclic change in gain with element separation. This is the normal ef-

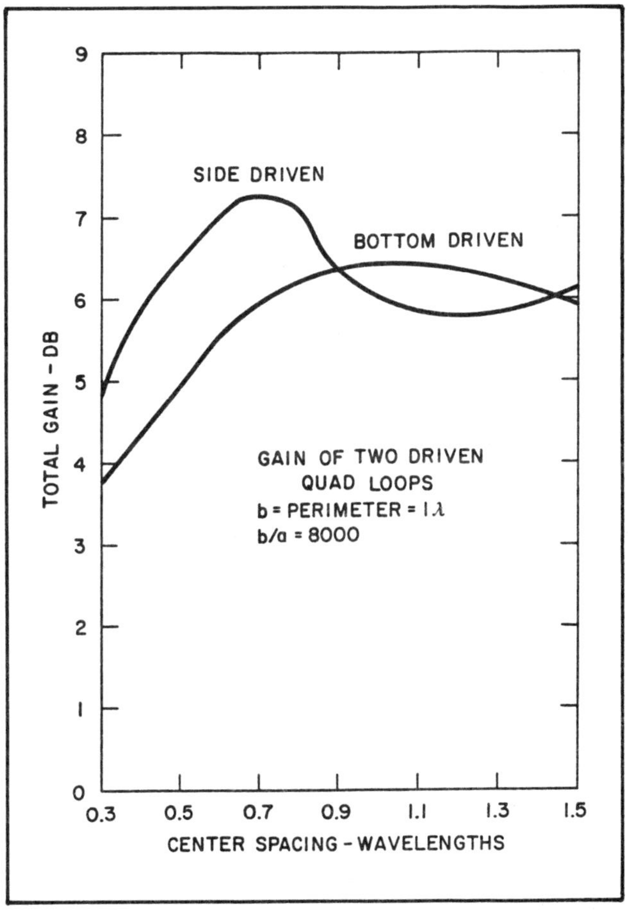

Figure 13. Stacking gain or lobe intensity of two spaced in-plane loops fed in phase. See text for application to horizontal and vertical stacking.

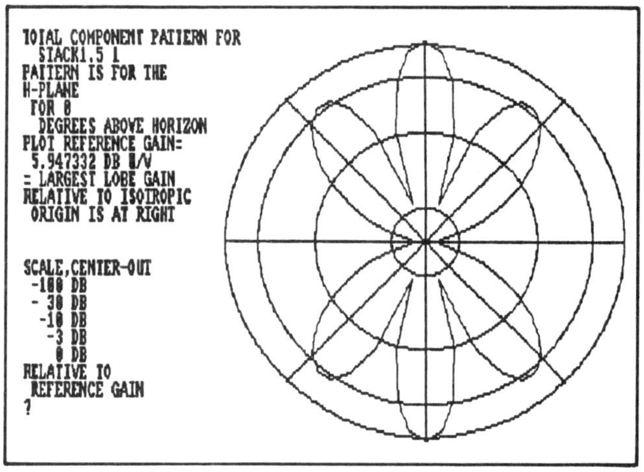

Figure 14. Horizontal plane pattern for two one-wavelength loops in the vertical plane, separated 1.5 wavelengths and fed in phase. The pattern is essentially the same as two collinear dipoles spaced 1.5 wavelengths on center.

fect of resistance interaction. Wider spacings weren't explored, because the main-lobe gain will converge to 3 dB at large separations.

There will also be pattern changes, with the main lobe narrowing and additional lobes appearing (see Reference 23 for the pattern factors). Figure 14 shows the pattern for the two drive conditions at 1.5-wavelengths separation.

This pattern is for two loops. For two-quad beams it becomes the multiplication factor for the beam pattern. Usually, the lobes at 40 degrees become smaller in beams, appearing as minor lobes. The gain pattern for stacked two or more element quads has not been studied. The results are easily estimated by multiplying the pattern of the individual array by the pattern factor for the spacing used. Usually spacings in the range of one to two wavelengths give reasonable gain with acceptable secondary lobe intensity.

All of the "tricks" of array design can be applied to quad arrays, in particular tapering the feed power. This may be found useful at UHF and above.

The gain curve of Figure 13 applies to loops only. Greater separations may be necessary if multi-element arrays are stacked, although there is a suggestion in the reactance variation (not shown here) that the quad won't require as much spacing increase as the Yagi. This is another area needing more study.

Dopple Quad

The basic dopple quad uses a pair of back-to-back diamonds connected at the drive point by a line of zero length (see Figure 15A). This gives an increase in loop separation, as required for improving gain. The name and history shows that it was developed in Germany. (In German, doppel means duplicate or double.) A history, description, and some test results are given in Reference 37. Figures 15B and C show a three-wavelength rectangular loop and symmetrical curtain, respectively.

The performance of the basic dopple loop is shown in Figure 16. At resonance, gain has increased by about 1.2 dB referred to a single diamond. Drive resistance is reasonable, about 150 ohms. Resonance occurs at about 41.8 MHz for the example. As for the regular quad loop, an increase in gain can be secured by operating above resonance.

The dopple loop can also be regarded as two opened loops connected together, working at twice the normal resonant frequency. The performance of this arrangement is shown in Figure 17. The gain has increased to over 7 dB. However, this increase is less than the relative increase in area. Drive resistance has increased markedly. In this mode the center- (or top/bottom) fed dopple loop is basically a voltage-fed antenna. Low impedance or current feed at the center of one side of the basic diamond would be possible.

In the original design, the dopple loop is used with Yagi elements as parasitic reflectors—reportedly 3,

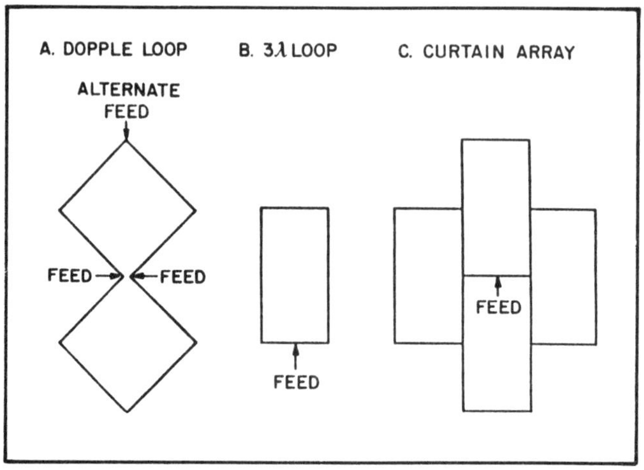

Figure 15. Additional stacking methods. (A) Dopple quad loop consisting of two diamonds facing each other. Use current feed if each loop is one wavelength in perimeter; voltage feed if two wavelengths in perimeter. (B) Three-wavelength rectangular loop. Horizontal arms are one-half wavelength long; vertical arms one wavelength. Typically bottom-center fed. (C) Symmetrical curtain array of three-wavelength rectangles. Each added section is voltage fed from the adjacent one.

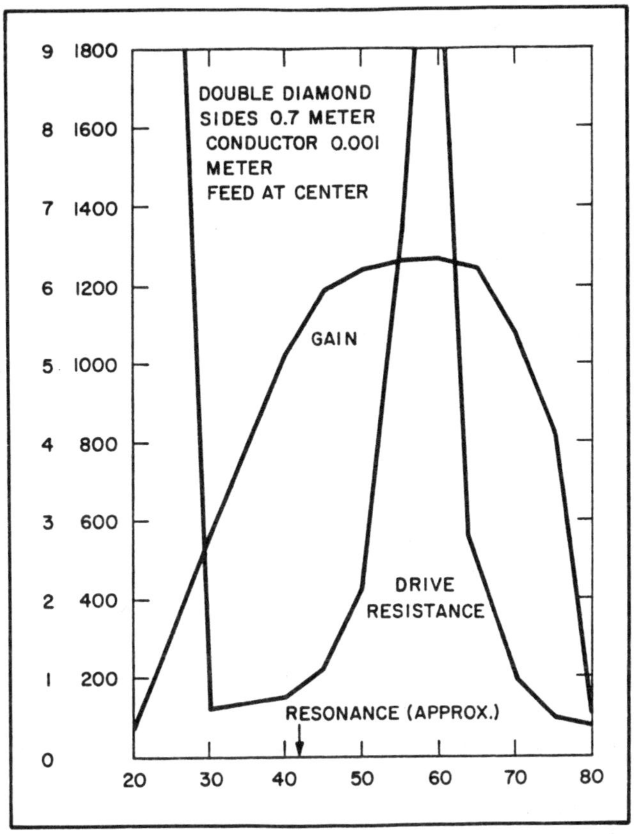

Figure 16. Current-fed dopple loop performance data. Gain and drive resistance are shown of a current-fed dopple loop with 0.7 meter sides.

5, or 7 being used. By symmetry, three reflectors are easily placed directly behind the high-current points of the basic design. More reflectors improve the gain, as in Yagi arrays. An even number would appear better for the opened-loop version.

For this study, the reflectors were approximated by mounting the antenna 3.3 feet (1 meter) above perfect earth. The gain and drive resistance of the opened-loop version is shown in Figure 18. Gain has increased to a respectable 10.4 dB. Main-lobe gain is more than 9 dB across the frequency range 60 to 120 MHz. However, the lobe shifts from the line perpendicular to the loops at low frequencies, and splits into two lobes at the higher end.

Drive resistance is high. The lower peak corresponds to the two-wavelength-per-loop resonance. Reactance below resonance is positive, (down to one wavelength), and is capacitive for higher frequencies. The second resistance peak marks the nominal three-wavelength resonance, but this isn't truly resonant; reactance remains negative. (This is the same behavior found earlier for "fat" conductor loops.) The opened-loop version can be stacked in two directions, becoming the Chireix-Mesny array mentioned previously.

Patterns for this antenna in two planes are shown in Figure 19A and B. The asymmetry and beginning minor lobe are associated with the lobe splitting mentioned earlier. The pattern is for a frequency of 90 MHz.

The combination of good gain, wide gain-bandwidth, and reasonably stable feed characteristics suggests that this antenna be studied further for UHF and VHF use.

Curtain Quad

The family of rectangular loops includes several useful properties. One of these is the three-wavelength loop, with the top and bottom sections one-half wavelength and the sides one wavelength, as shown in Figure 15B. Very closely, this radiates as two half waves in phase, stacked one wavelength. On the loop axis, radiation from the two sides cancels. There is a vertically polarized component in the loop plane, as for the conventional quad.

Design data for these loops is given in Figure 20. Drive resistance is relatively high, nearly ideal for a

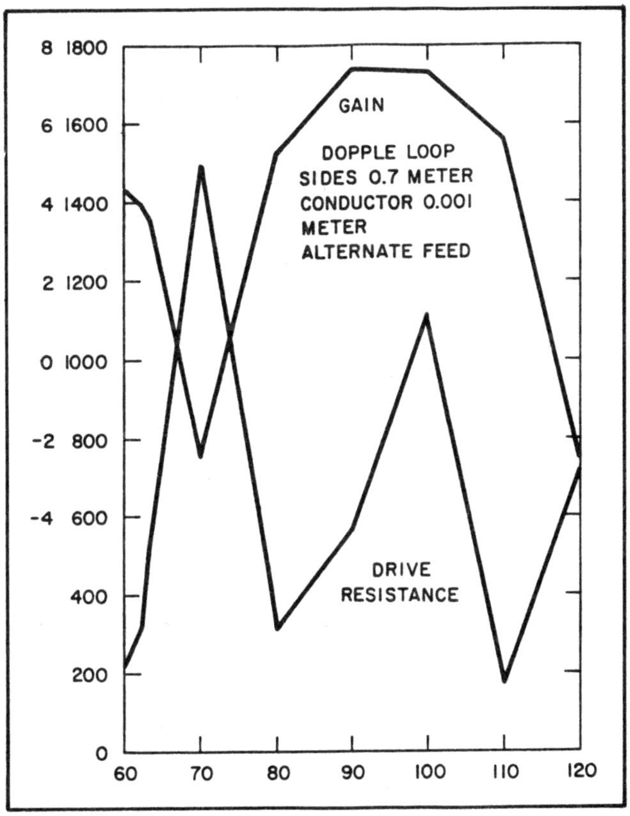

Figure 17. Voltage-fed dopple loop performance data. Same as Figure 16, but for voltage feed.

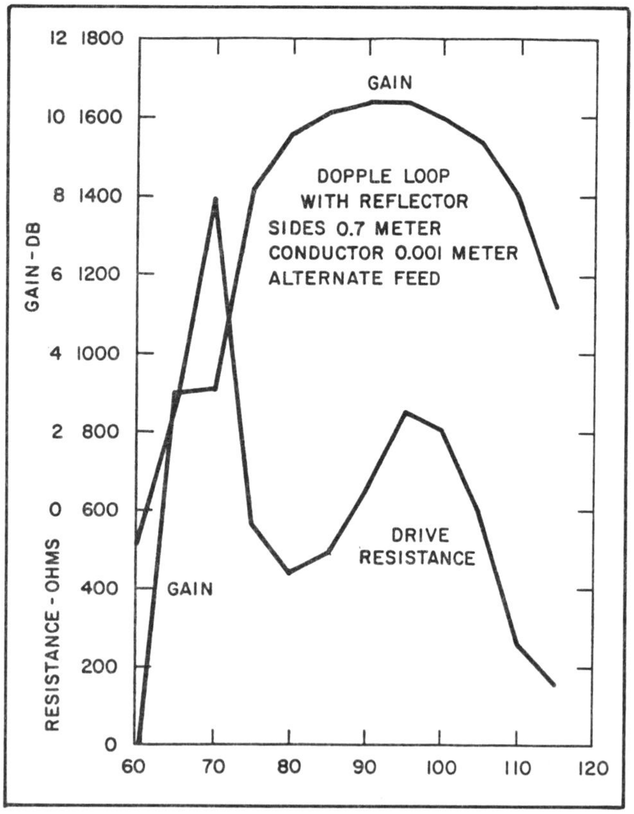

Figure 18. Performance of a voltage-fed dopple loop with reflector. Gain and drive resistance are shown for a voltage-fed dopple loop spaced 3.3 feet (1 meter) from an ideal reflector, approximating the performance of a loop with multiple Yagi reflector rods.

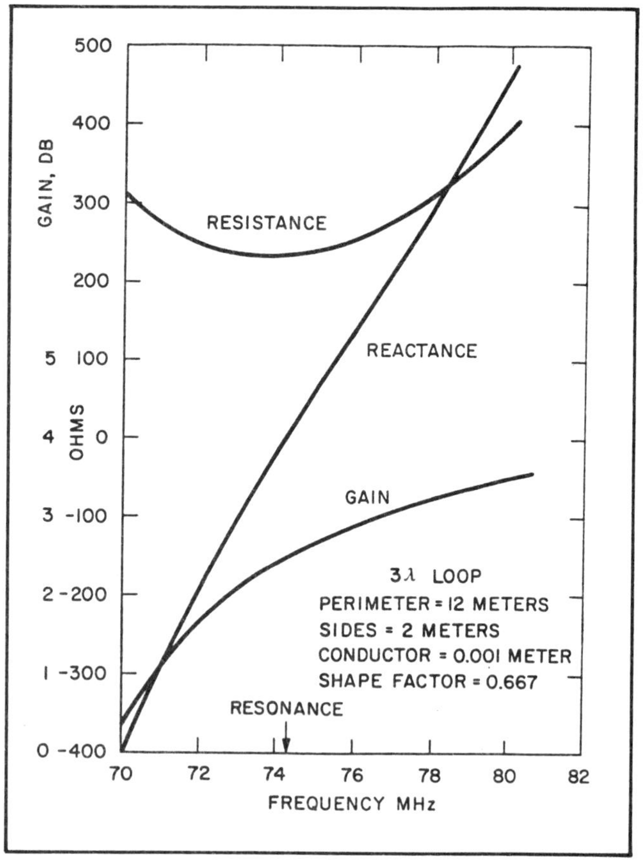

Figure 20. Design data for three-wavelength loops. Gain, drive resistance, and reactance are plotted for the loop of Figure 15B. Drive resistance is related to that of folded dipoles. Reactance will be a function of conductor size.

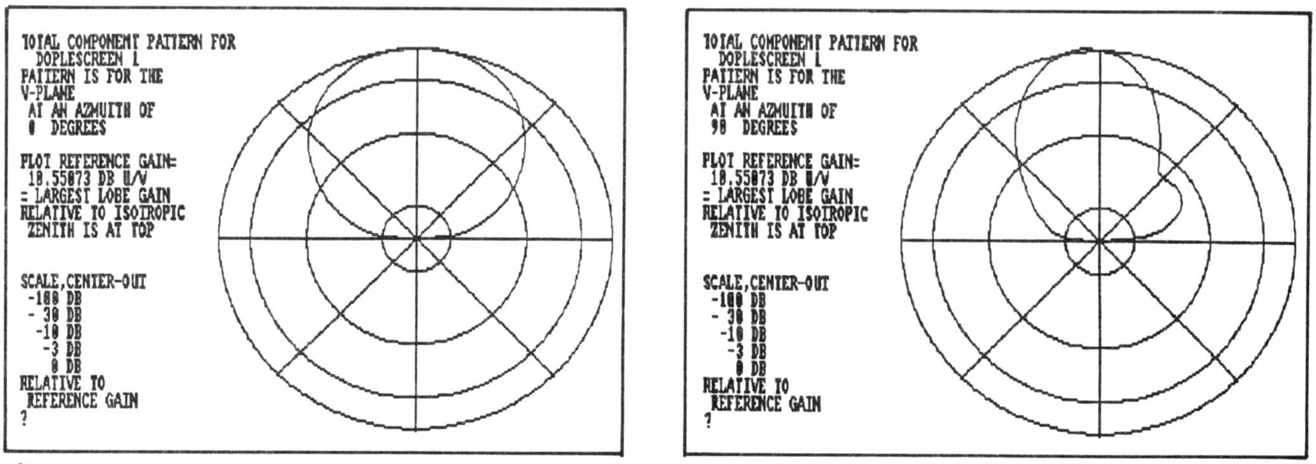

(A) (B)

Figure 19. Patterns for a dopple loop with reflector. (A) Pattern at right angles to the line of loops for the conditions of Figure 18, for the frequency of maximum gain, 90 MHz. (B) Same, but pattern is for the plane of the axis of loops. The lobe is narrowed because of the stacking effect. The asymmetry results from the frequency being above the two-wavelength frequency.

300-ohm line, or a 50-ohm line with a 4:1 balun. Resonance occurs below the three-wavelength frequency, at 98.9 percent for the perimeter/conductor ratio used. Main lobe gain is essentially that of two parallel dipoles in phase at resonance: it increases at higher frequencies, as usual in quad loops.

An interesting item of the resonant condition is that the top or bottom sections are voltage fed. This means that an additional U can be connected to the top or bottom to give vertical stacking. There's also a high-voltage point at the center of the vertical side, so a side-turned U can be connected to the two stacked sections. Other connections are possible but do not give proper phasing for main-lobe control. The symmetrical development is shown in Figure 15C. This curtain is described in Reference 38, along with concept, construction, and performance data (see also Reference 20).

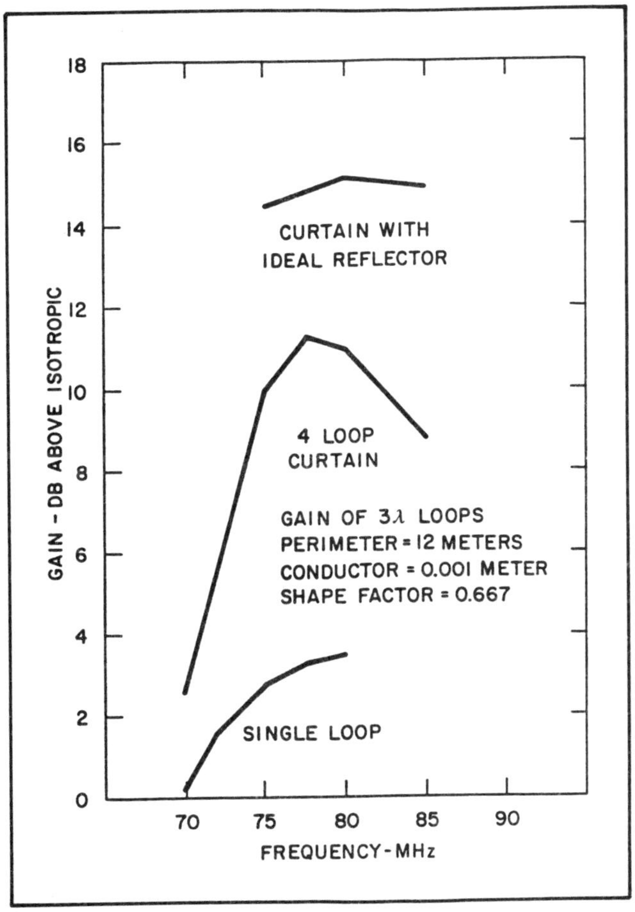

Figure 21. Gains of curtain arrays. Gain versus frequency curves are shown for the loop of Figure 15B, which is the array of Figure 15C and this array with an ideal reflector. A second curtain or a set of Yagi rods could be used as the reflector, with some change in gain (and drive impedance).

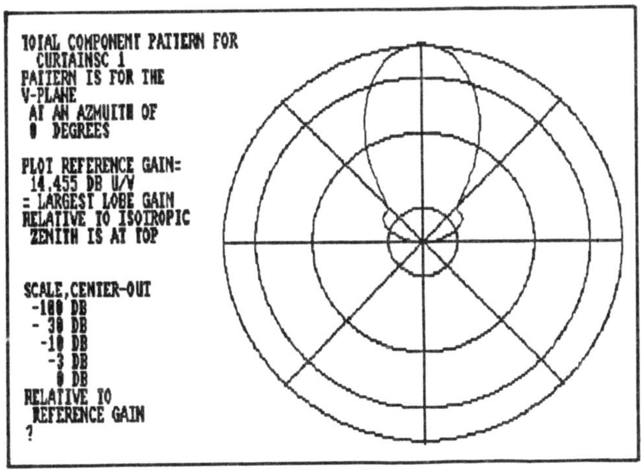

Figure 22. Pattern of curtain array with reflector. Pattern of the array of Figure 15C spaced 3.3 feet (1 meter) from an ideal reflector at 75 MHz. Pattern is for the plane parallel to the half-wave sections of the loop.

The result of a brief investigation of the curtain quad is summarized in Figure 21. A four-loop (or seven effective half-wave elements) array shows a main-lobe gain of 11.3 dB. This 7 dB higher than for single loop; the simplest array factor analysis gives 6-dB increase. Anderson[38] estimates gain as G = 10 log(n), where n is the number of effective half-wave elements, or 8.5 dB for the main lobe. With a reflector added, gain increases to 15.2 dB. This is greater than the 13.7 dB estimated by Anderson,[38] from the relationship G = 10 log(n) + 5.2.

The horizontal and vertical patterns for the curtain with a perfect reflector are shown in Figures 22 and 23. The secondary lobes found here are normal in cur-

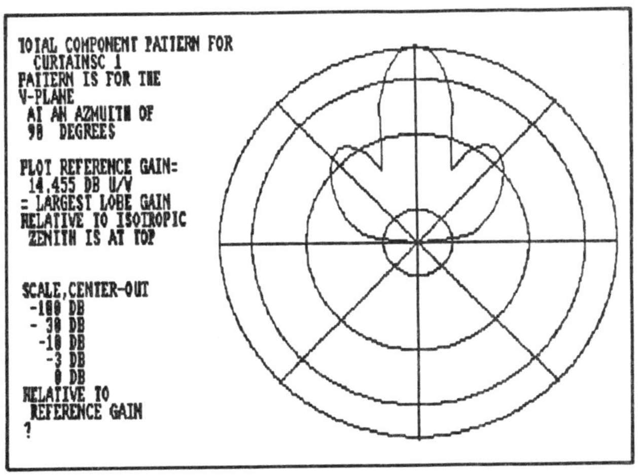

Figure 23. Pattern of curtain array with reflector. Same as Figure 22, but for the plane parallel to the one-wavelength sections. This component is polarized 90 degrees to that of Figure 22.

tain arrays. Control of their magnitude is often a major design goal, nearly as important as main-lobe gain.

A practical advantage of this array is that the feed system doesn't become more complex as the array is expanded. The array is also reasonably broadband—some 12 percent for this version, even though the conductor size is small. There are interesting possibilities for broadbanding and minor lobe control by varying side length and diameter.

This array has great possibilities for VHF and SHF designs. One can also visualize a pair of multi-element arrays as the antenna for meter-wave radio astronomy. High performance and mechanical simplicity indicate more use in the future.

Chapter 12

Optimizing a Quad Design

Tips on improving your design

There's a natural wish that any antenna perform at its best. This wish becomes more marked with large antennas, because the cost and labor involved in getting such an antenna into operation increases manyfold. Because paper designs are easy and inexpensive, it's worthwhile to go through the steps of using design optimization.

The material presented in earlier chapters has included both general design information and calculated dimensions for specific bands. Methods of scaling designs were given to allow transfer of the example-band designs to other bands.

Much of the data given previously, especially in Chapters 4, 5, and 6, is intended to present a reasonably easy-to-use optimization technique. As stated before, the techniques will give a design with optimum gain and satisfactory bandwidth and front-to-back ratio. However, tuning is recommended if the best F/B ratio is needed at a particular frequency. This may be required in any event if detuning objects are nearby.

The term optimization implies that a goal is stated. In antennas these include:

- gain
- front-to-back ratio
- bandwidth
- drive impedance

In amateur antennas gain is usually claimed as the goal, but drive impedance is the one most commonly discussed. Some practical designs are tuned to optimize F/B ratio, but this is rarely an overall design goal. The relative narrowness of amateur bands usually makes bandwidth of little importance.

Drive impedance is often allowed to be more important than is justified. Any antenna of reasonable bandwidth will have considerable impedance variation as frequency changes. While tuning the radiator is one way to introduce compensating reactances, it is not the only way, nor the best. Low-loss feed line and a match-box have been recommended as a better approach for normal amateur use. If true broad-band matched-impedance operation is necessary, matching at the antenna is best. When this is done, the exact value to be matched is relatively unimportant.

Matching is much simpler if the variation of drive resistance and reactance is regular, without sudden excursions in value. No discussion of matching techniques is included here, but some of the following examples will show easy and difficult match conditions. Other than this, drive impedance is allowed to vary at will

Note again that it is rarely possible to develop a design that is optimum for several design conditions. Practical antenna design is based on compromise among several choices.

Optimization by Computation

The steps in optimization can become tedious. Additionally, the graphical data used earlier can be improved by going to a numerical design. A full optimization procedure needs computer analysis.

To simplify these steps, I have made a number of modifications to the MININEC programs discussed in earlier chapters. These are in the form of patches, which make use of the MININEC code to automate much of the optimization process. The added routines are:

- Step the frequency over a range.
- Step the element tuning reactance over a range.
- Reset the element tuning reactance (loads) over a range.
- Optimize.

Computer Routines

With the MININEC3 built-in reset of environment, excitation, loading, and frequency, the routines are very flexible and simple to implement. If the computer system is reliable, no attention is needed—even for the 24 to 30 hours of run time that may be necessary to get optimum conditions for a multi-element beam. (The time is reduced if the starting point is first chosen by graphical approximation.)

The step and reset routines are simple loops. They set a variable and call the normal calculation loop. The optimize routine also includes a loop plus a test. This may be for improved gain or improved front-to-back ratio. The loop starts by increasing the reflector load by an input amount. If this increases the goal value, another increment is added. If the goal value decreases, the increment sign is changed. After the best condition for reflector tuning is found, the director loads change in the same way. Then another cycle may be started to refine values, for example, by changing the tuning of a particular element or changing step size.

While the process is automatic, it's worthwhile to check the progress occasionally. This is especially true if a design is to be optimized for front-to-back ratio. If the initial conditions are far from optimum, an intermediate medium F/B ratio condition may occur which meets a slope test. It seems best to optimize first for gain, then for front-to-back ratio.

Further Improvement?

An additional small increase in gain and front-to-back ratio is often possible to achieve by a second iteration of the optimization process. Also, a change in reflector tuning can be used to move the frequency of best F/B ratio to the center of the band for best average QRM immunity, or to a favorite operating frequency.

Some improvement in performance can be secured by such steps as increasing the number of directors while keeping the same boom length, or varying the spacing and size of the directors. It seems that the penalty in weight and complexity of these further variations is greater than is justified by the small improvement in performance. There is a virtue in antenna simplicity.

Small or Large Computers

Computing for large antennas requires a large memory. Good basics for work on multi-element beams include: two megabytes of memory, at least two disk drives, a compiler, and excellent line-voltage surge protection. A math coprocessor and an "uninterruptible" power supply are also worthwhile. However, the optimization techniques do not demand these "goodies." Small computers and less elaborate installations can obtain the same results by substituting the careful choice of initial conditions and by adding run time.

Some additional programs simplify input preparation and result evaluation. The "Geometry" program simply duplicates the MININEC input routine. It's normally used to create elements, rather than complete antennas. Another program, "Array," combines elements into antennas by manipulating the values of the MININEC input array. The element can be scaled in size, and/or moved. Wire size can be changed, as can the number of calculation segments. At present, there's no provision for element rotation; I use Geometry for this. Finally, the program "Plot" prepares a polar diagram for any horizontal or vertical plane, or for an arbitrary elevation angle. Of course, MININEC must be set to produce and save to disk the data to be plotted.

If you intend to do any extensive work, I recommend programs similar to these. An inspection of the MININEC source code will make the needed routines clear. These programs are available commercially.

The steps which follow show what happens during

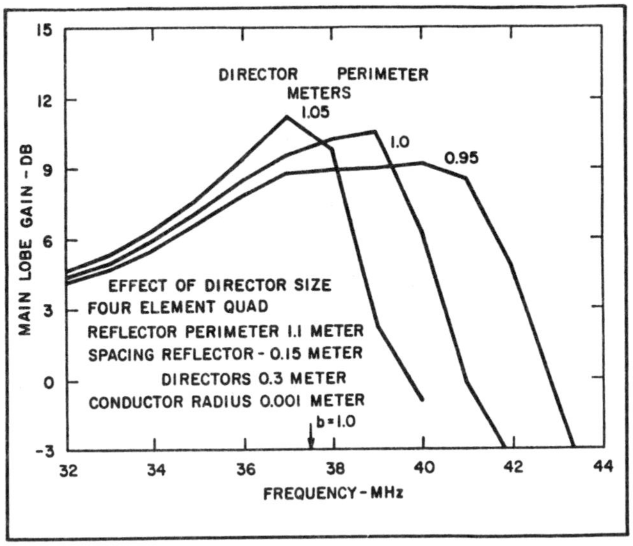

Figure 1. Gain-bandwidth curves for a four-element quad, director tuning varied, reflector tuning constant. Gain increases as bandwidth is reduced. Virtually all change in bandwidth is on the high-frequency side (the frequencies closest to the director resonant frequency).

the optimization process. A number of prototype designs are given which are optimized for a specific goal.

Interactions: Gain, Front-to-back Ratio, and Bandwidth

Because a beam is an assembly of close-coupled circuits, any change that occurs will affect the entire antenna. Figures 1 through 6 show some of the results of these interactions.

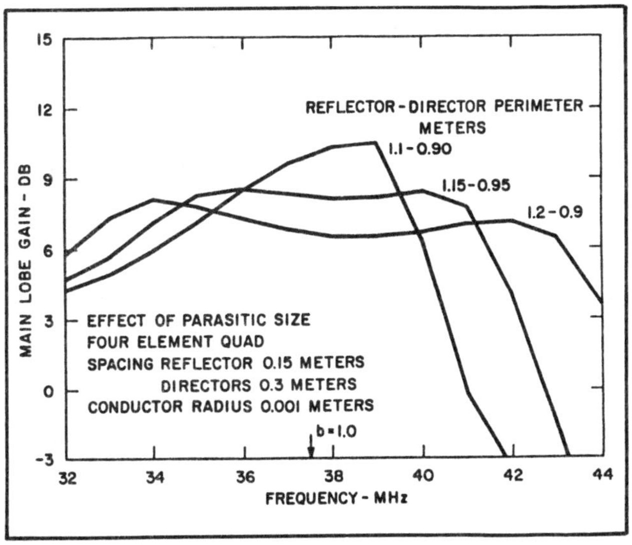

Figure 2. As for Figure 1, but with director and reflector tuning changed. Bandwidth change is greater, but most of the change is on the high-frequency side.

Figure 1 shows gain variation with frequency when the reflector/director size remains constant and the director size changes. Nearly all the effect occurs at the high-frequency side, where the director is closer to resonance. There is, of course, a reduction of gain as the bandwidth increases.

Figure 2 shows the gain variation with both reflector/antenna and director size changed by equal amounts, but in opposite directions. Now both the upper and lower-frequency sides show some effect of tuning, with the major effect occurring at the high side. Again, gain decreases as bandwidth increases. The gain curve begins to show two peaks as director and reflector tuning become further separated.

Figures 3 and 4 are the corresponding curves for front-to-back ratio. When comparing these curves, it is shown that:

- The position of best F/B ratio is determined primarily by reflector tuning.
- Bandwidth for a given F/B ratio increases as reflector/director tuning separates.
- Front/back ratio degrades markedly if reflector/director tuning is too close.
- There is a tuning combination that gives the best F/B ratio.

No attempt has been made to find guidelines for obtaining the very best F/B ratio.

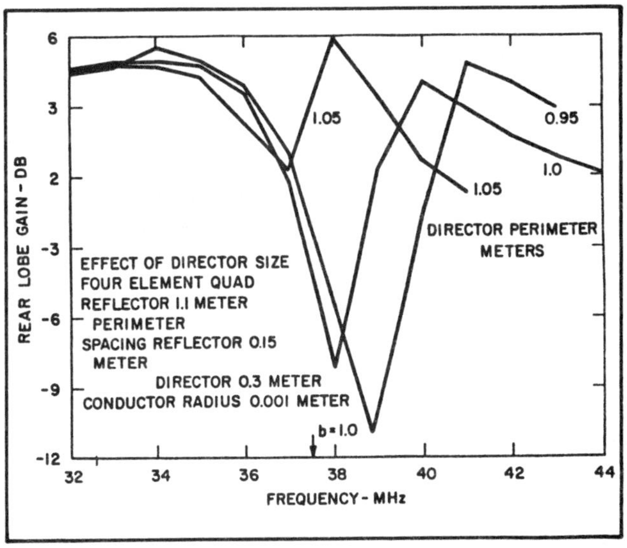

Figure 3. Back-lobe intensity corresponding to Figure 1. There is a tuning condition that gives minimum back lobe. Over the tuning range studied, this condition also gives greatest interference rejection beamwidth. The maximum-gain condition gives a large back lobe.

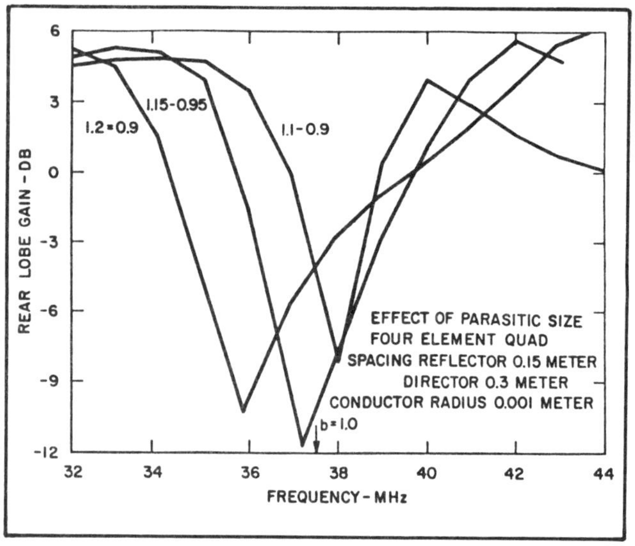

Figure 4. Back-lobe intensity corresponding to Figure 2. As in Figure 3, there is a condition giving minimum back lobe. The low back-lobe bandwidth can be increased with a relatively small penalty in back-lobe magnitude.

Figures 5 and 6 show the drive resistance and reactance variation with frequency for two values of director tuning. The reflector is the same size in both. While the 1.05-wavelength director gives appreciably more gain, it introduces marked changes in both resistance and reactance variation with frequency. With a narrowband signal, as for CW or SSB when using a matchbox, such marked changes are only a nuisance

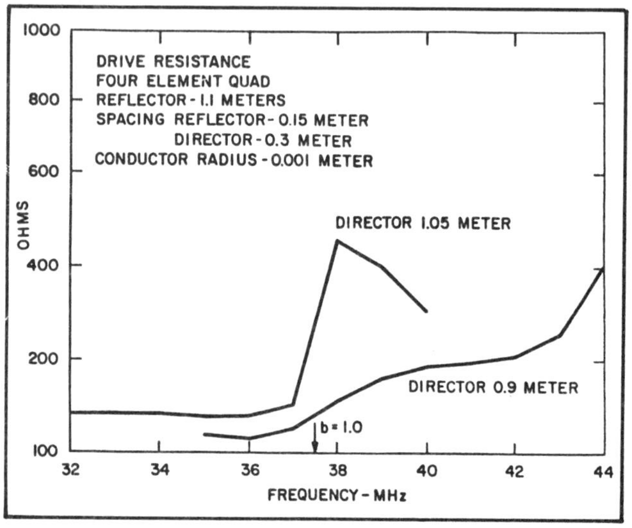

Figure 5. Drive-resistance variation for two tuning values of a four-element quad. The curve for the 1.05-wavelength director corresponds to maximum gain calculated for this antenna family.

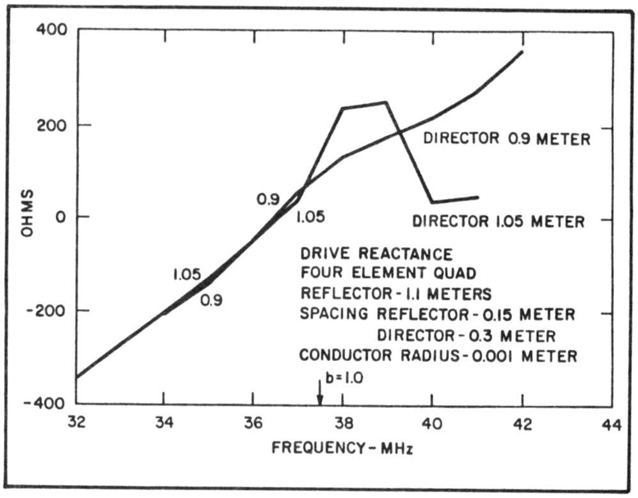

Figure 6. Drive reactance corresponding to Figure 5. The marked changes in tuning conditions complicate design of matching networks.

in tune up. But if a passive compensating network is to be used to secure wideband feed-line matching, the 1.05-wavelength director condition will require a more complex matching network.

Although more work is needed to refine values, it appears that a first-order estimate of the practical or working bandwidth of quads can be taken as the separation between director and parasitic self-resonant frequencies. This simple approximation fails in front-to-back ratio if the resonant frequencies are too close, and in gain variation if they are separated too much.

Further effects of antenna-element variations

Instead of changing element size, a parasitic element can be tuned by inserting a stub or lumped reactance at a current point, or a shunt capacitance across voltage points. Two of the effects of series tuning are shown in Figure 7, showing the gain as a function of frequency for two director tuning values in a four-element quad. With the 0.9-wavelength directors, the gain is about 9 dB and bandwidth about 15 percent of the center frequency. With director tuning of 300 ohms added, the gain increases to 11 dB and the bandwidth decreases to about 6 percent.

These stub-tuned values aren't exactly comparable to the size-varied values of the earlier figures. In part, this is due to the difference in director size, but a more important factor is that the radiator size was also

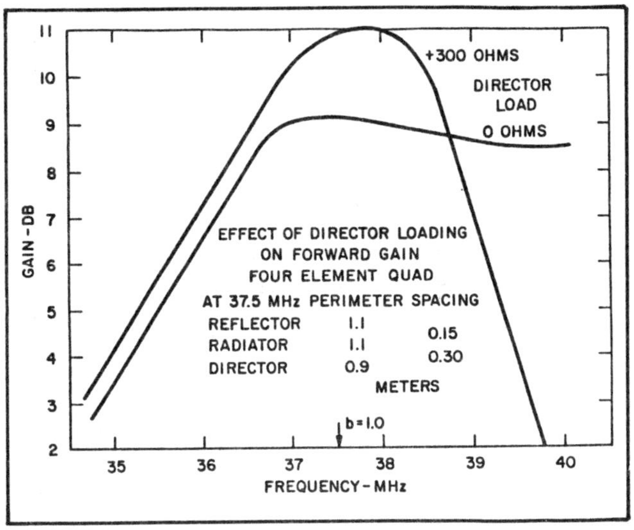

Figure 7. Effect of director tuning on gain. The 2-dB increase in gain is accompanied by a reduction of bandwidth (from about 8 to 4 MHz).

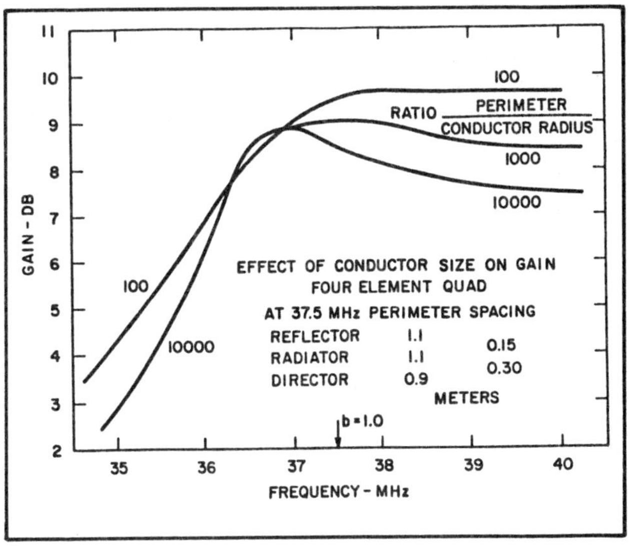

Figure 9. Apparent effect of conductor size on gain. The changes are due to the change in resonant frequency with conductor size.

changed in some cases. Approximately, the correspondence is:

25-ohm load: 0.95-wavelength director
200-ohm load: 1.00-wavelength director
375-ohm load: 1.05-wavelength director

Another effect of these two director tuning values is shown in Figure 8. There's very little change in the magnitude of the front-to-back ratio, but the frequen-

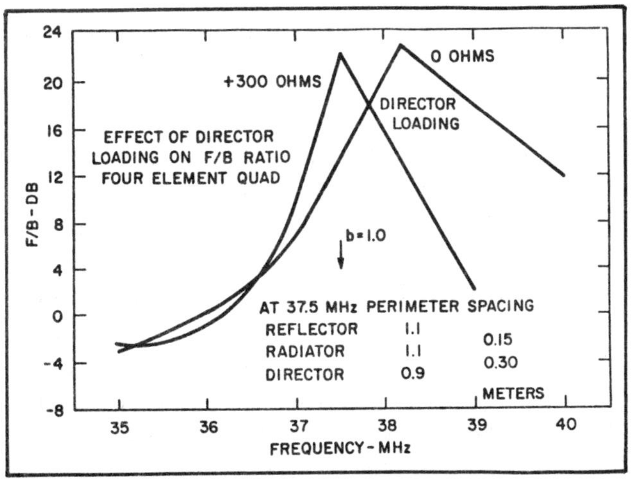

Figure 8. Effect of director tuning on front/back ratio. The bandwidth for a given front/back ratio decreases as the director is tuned for more gain. Overall front/back ratio remains nearly constant.

cy for the smallest back lobe has shifted. Also, the bandwidth for a given value of F/B ratio has changed by about the same ratio as the overall bandwidth.

The value of 300 ohms represents about the maximum practical tuning range. A small further increase in gain is possible, but front-to-back ratio becomes poor, and the impedance curve shows the changes found in Figures 5 and 6. To a close approximation, the effect of change in size or introduction of tuning reactance is identical over the usable tuning range. Figure 2 in Chapter 6 allows interchange between size and loading.

Figures 9 through 11 show the apparent effect of changing conductor size. All of the variations are due to the change in element reactance as conductor size changes, plus the way the reactance changes as the thickness factor omega (or loop perimeter divided by conductor radius) changes. Note that MININEC assumes perfect wire conductivity, so there will be a small additional change in gain due to actual conductor resistance. This is usually a fraction of a dB.

Figure 12 shows the performance effects of changing radiator location in an antenna of fixed-boom length. This example is for a short-boom quad which will be discussed later. There is very little effect on gain until the radiator-parasitic spacing is very close. However, there is a best spacing for good F/B ratio. While more study is needed, the indication is that antenna-reflector spacing of 0.12 to 0.20 wavelength

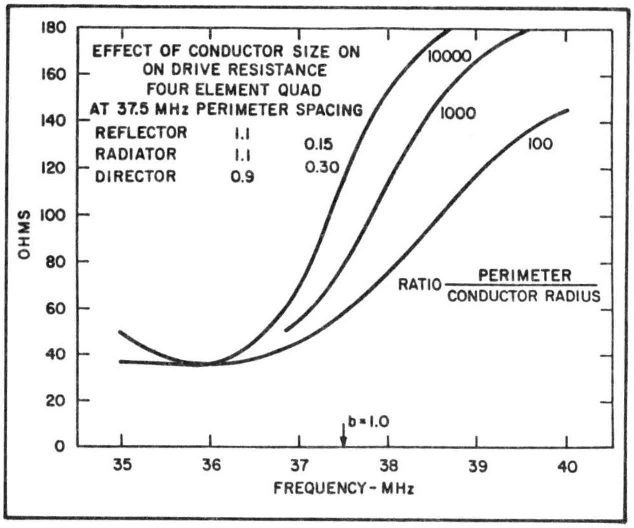

Figure 10. As for Figure 9, but for drive resistance.

plus adjustment of reflector will bring F/B ratio to or near the best possible measurement.

Short-boom Quads

Because space is often the main limitation on antenna design, I spent quite a bit of time on a three-element 14-MHz quad with a boom length of 21 feet. The major results are summarized in Figure 13 and Table 1. Some additional results appear in Table 2.

Design G in Table 1 gives high gain and good front-to-back ratio at the lower part of the band, but poorer performance at the high end. Design D is good near the band center. If front-to-back ratio across the band is needed, design E appears to be the best.

Still another possibility is shown in Figure 13. Gain is good across the band, with fair F/B ratio near the lower phone segment. This design assumes tuning stubs in the parasitic elements to allow adjustment during installation.

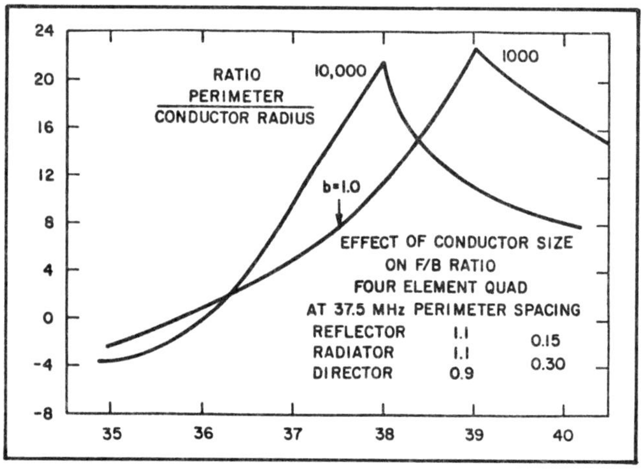

Figure 11. As for Figure 9, but for front/back ratio.

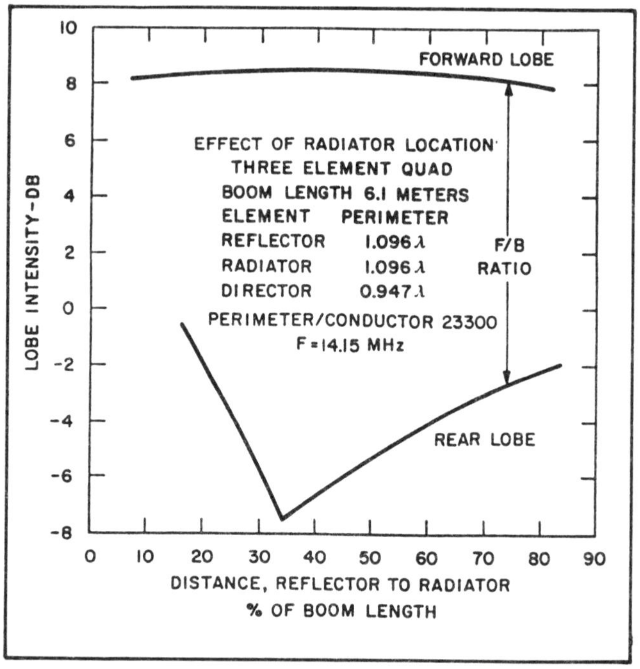

Figure 12. Effect of changing radiator position for a given spacing between reflector and director in a three-element quad. Maximum gain occurs with the radiator at the midpoint. Maximum front/back ratio occurs with the director spacing at twice that of the reflector. Exact values depend on overall boom length.

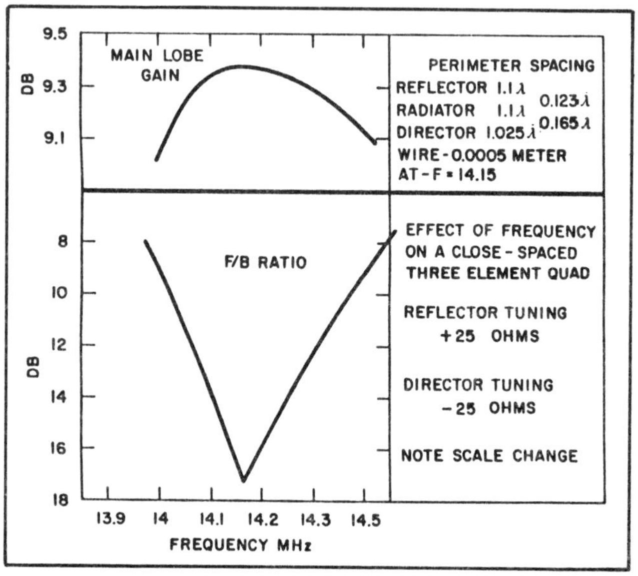

Figure 13. Performance of an optimized short-boom quad. See Tables 1 and 2, and the text.

length	frequency (MHz) gain and F/B ratio					design
Ref/Dir	14.00	14.10	14.20	14.30	14.40	
1.1/0.94	8.3/11.5	8.3/13.3	7.9/16.1	7.7/14.2	—	—
1.1/1.025	9.0/12.6	—	8.9/13.5	—	8.7/8.9	G
1.08/1.05	9.0/9.7	9.6/17.0	9.7/16.4	9.5/10.5	—	D
1.08/1.03	8.8/8.4	9.1/12.3	9.1/13.6	9.0/12.1	—	E
1.09/1.05	9.5/16.3	9.4/14.4	9.3/10.3	9.0/7.1	—	—

Boom length equals 2.88.
Design frequency is 14.15 MHz.
Lengths are in wavelengths.
Gain and front/back ratio are in dB.

Table 1. Performance of several short-boom three-element quads as a function of frequency. About 9 dB of gain and 17 dB front/back ratio are the limits of performance at this boom length.

It's apparent from these values that it's difficult to secure much change in gain. It's also difficult to keep the front-to-back ratio high. These examples show, again, that antenna design must be based on compromises. And if size is the primary goal, the range of choice in other factors becomes small.

Supergain in Quads

In principle, any desired gain can be secured from any size antenna. However, attempts to attain high gain are accompanied by a reduction in drive resistance and bandwidth.

Table 2 summarizes an investigation of the possibility of supergain in the same short-boom quad of Figure 13, as developed by the optimization routine. Starting at the top of Table 2, you will note the reflector is too large for best gain. Increasing the size of the director slowly increases the gain, until you reach the last table entry. The increase in gain is accompanied by a small change in F/B ratio and an appreciable re-

Tuning-stub reactance reflector (ohms)	director (ohms)	gain (dB)	F/B ratio	R (ohms)	jX (ohms)
0.00	0.00	8.37	16.47	84.10	104.60
25.00	0.00	8.03	16.83	106.80	98.50
−25.00	0.00	8.62	11.97	60.20	100.00
−25.00	25.00	8.65	12.14	59.60	98.50
−25.00	50.00	8.69	12.34	58.90	96.80
−25.00	75.00	8.74	12.58	57.80	94.90
−25.00	100.00	8.79	12.88	56.50	82.80
−25.00	125.00	8.87	13.27	54.60	90.40
−25.00	150.00	8.94	14.74	52.00	87.90
−25.00	175.00	9.04	14.36	48.50	85.20
−25.00	200.00	9.17	43.60	82.80	82.20
−25.00	225.00	9.32	16.46	37.30	81.20
−25.00	250.00	9.51	18.22	29.30	81.30
−25.00	273.00	9.73	19.80	20.40	84.30
−25.00	300.00	9.83	16.65	12.20	84.30
−25.00	325.00	9.18	9.62	8.80	91.00

dimensions	Loop circumference (B)	Element spacing (wavelengths) (C)
reflector	1.10	0.123
radiator	1.10	0.165
director	0.95	

Loop circumference to wire radius: 20,000.

Table 2. Performance of a short-boom quad calculated with a gain optimization program. The marked reduction of drive resistance near maximum gain is characteristic of an approach to super-gain conditions.

Tuning-stub reactance reflector (ohms)	director (ohms)	gain (dB)	F/B	R (ohms)	jX (ohms)
0.0	0.0	9.2	12.1	80.5	99.3
25.0	0.0	8.9	18.7	105.5	104.6
−25.0	0.0	9.2	9.0	60.9	85.7
0.0	25.0	9.2	12.2	80.2	98.8
0.0	50.0	9.3	12.3	79.9	98.2
0.0	75.0	9.4	12.4	79.6	97.5
0.0	100.0	9.4	12.6	79.1	96.7
0.0	125.0	9.5	12.8	78.6	95.7
0.0	150.0	9.6	13.0	77.9	84.4
0.0	175.0	9.7	13.4	77.0	92.9
0.0	200.0	9.9	13.7	75.7	91.1
0.0	225.0	10.0	14.2	73.9	88.6
0.0	250.0	10.2	15.0	71.1	85.7
0.0	275.0	10.4	16.2	66.6	82.2
0.0	300.0	10.7	18.4	59.3	78.5
0.0	325.0	11.1	22.8	47.8	77.0
0.0	350.0	11.5	21.9	32.2	81.3
0.0	375.0	11.4	11.4	118.1	95.8

dimensions	(B)	(wavelengths) (C)
reflector	1.1	0.15
Radiator	1.1	0.30
director	0.9	

Loop circumference to wire radius: 8,800

Table 3. Similar to Table 2, but for a four-element quad. The untuned condition is good for bandwidth, with tuning around 200 ohms representing a reasonable compromise design.

duction in drive resistance. Although not shown here, an increase in gain is also accompanied by a reduction in bandwidth, very nearly as shown in Figure 7.

While it appears that this tuning places this quad either close to or in the supergain region, I haven't explored the possibility further. This is due partly to the relatively low gain improvement found, but due mostly to loss of bandwidth and reduction of drive resistance. The best features of the quad are lost. If supergain is a goal, the curtain array or multiple-thread helix are better choices.

Long-boom Quads

Where high gain is the goal, boom length must be allowed to increase. Table 3 is the result of an optimization run on a near-optimum performance four-element quad. Maximum gain is found to be 11.5 dB; maximum front-to-back ratio is 22.8 dB. Either of the corresponding conditions give narrow bandwidth. For best bandwidth with good F/B ratio, the loaded reflector is slightly too small.

Figures 14A through 14C show the lobe behavior of a four-element quad as the operating frequency changes from a high front-to-back ratio point to higher frequencies. These are for the reflector and director units of Figures 1 and 3. The growth of the rear lobe is accompanied by reduction of the main-lobe gain. There are also changes in secondary-lobe structure. Changes toward lower frequencies are not as rapid for main-lobe gain, but are virtually identical for the rear lobe. This is a relatively narrowband antenna.

Tables 4 through 7 show the performance expected from six and eight elements on 1.35 and 1.95-wavelength booms, for both quads and quagis. These follow the pattern of Table 3 and are the results of optimization runs.

In these antenna sizes, gain increases of 2 to 3 dB are possible when maximum gain is the goal, as compared to the starting compromise design. Front-to-back ratio varies widely. Note that the single optimization-pass procedure does not attempt to control the frequency of best F/B ratio. Another optimization run could be used to get the best F/B ratio at a given point. It must be remembered that the bandwidth becomes progressively smaller as the director becomes larger.

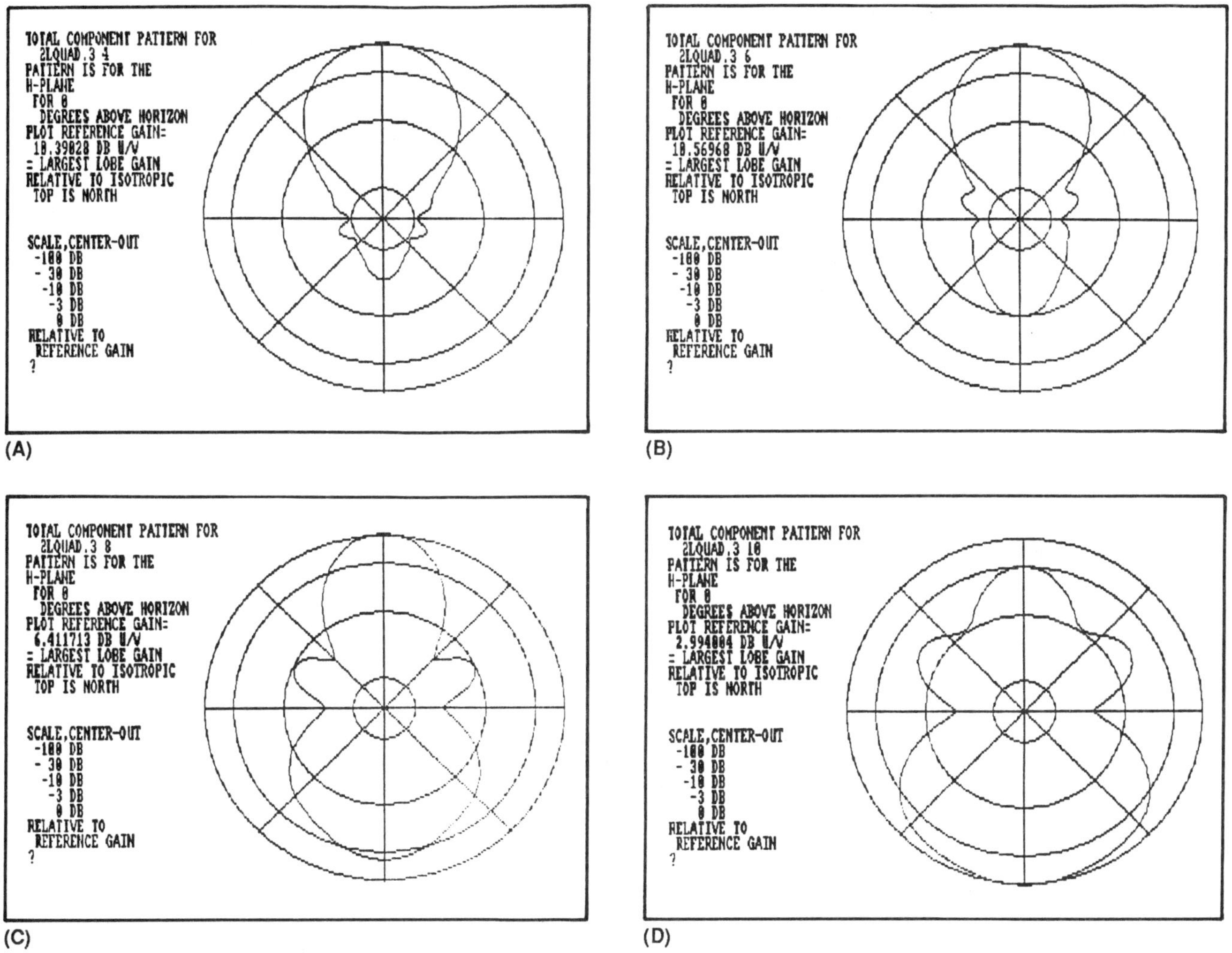

Figure 14. Variation of lobe structure with frequency, for a four-element quad with 1.0-wavelength directors spaced 0.3 wavelength and a 1.1-wavelength reflector spaced 0.15 wavelength. Curve (A) is at 38 MHz, (B) at 39 MHz, (C) at 40 MHz, and (D) at 41 MHz. This antenna is designed for high gain at the expense of bandwidth.

The pattern of these long-boom quads and quagis will show major lobe beamwidth decreasing as the gain increases. The back lobe may be a smaller version of the main lobe, with two side minima, or it may show a quasi-minimum directly opposite the main lobe. Under some conditions of considerable mistuning, the minor lobe will approach the gain of the main lobe and have greater bandwidth.

Typical horizontal-plane plots for conditions giving reasonably small back lobes are shown in Figure 15. Figures 15A through C are for director loads of 0, 150, and 300 ohms, respectively. Note the narrowing of the main lobe and the progressive changes in back-lobe structure. (The plots in Chapter 6 show patterns of some of the smaller beams.)

To use this table data, select the boom length which meets construction limits and the table value closest to performance goals. Scale all dimensions by the ratio of design-to-operating frequency, design being 37.5 MHz for all except Tables 1 and 2, where it is 14.15 MHz. If you aren't using tuning stubs, determine the change in element size for the given parasitic load from Figure 6 in Chapter 4 and Figure 2 in Chapter 6. The parasitic dimension should be adjusted by the amount of the change. Finally, if the scaled conductor size is unsuitable, determine the correction for the new size and apply it to all elements. If sufficient computer power is available, confirm all results by analysis, making further adjustments as needed.

If gain greater than that provided by the range of these tables is the goal, you may add additional directors. Figure 8 from Chapter 3 and Figure 1 from

Tuning-stub reactance reflector (ohms)	director (ohms)	gain (dB)	F/B ratio	R (ohms)	jX (ohms)
0.0	0.0	10.0	14.3	76.9	97.4
25.0	0.0	9.8	22.4	100.6	103.0
−25.0	0.0	10.0	9.6	58.4	84.0
0.0	25.0	10.2	14.6	76.3	96.9
0.0	50.0	10.3	15.0	75.5	96.3
0.0	75.0	10.4	15.4	74.6	95.7
0.0	100.0	10.5	15.9	73.5	94.9
0.0	125.0	10.7	16.5	72.1	94.2
0.0	150.0	10.9	17.3	70.5	93.4
0.0	175.0	11.1	18.3	68.5	92.6
0.0	200.0	11.3	19.8	65.9	91.8
0.0	225.0	11.5	22.8	62.5	91.4
0.0	250.0	11.8	24.4	58.4	91.6
0.0	275.0	12.2	24.7	54.4	93.0
0.0	300.0	12.5	20.8	48.2	96.7
0.0	325.0	12.7	15.2	44.9	103.8
0.0	350.0	12.8	11.3	51.6	113.6

		(wavelengths)
dimension	(B)	(C)
reflector	1.1	0.15
Radiator	1.1	0.15
director	0.9	0.20

Loop circumference to wire radius: 8,800.

Table 4. Similar to Table 2, but for a six element quad.

Tuning-stub reactance reflector (ohms)	director (ohms)	gain (dB)	F/B ratio (ohms)	R (ohms)	jX (ohms)
0.0	0.0	−23.2	−26.3	11.4	108.8
0.0	25.0	1.1	− 3.7	30.5	142.1
0.0	50.0	1.7	− 1.9	51.5	154.2
0.0	75.0	2.2	− 0.5	71.0	154.2
0.0	100.0	2.9	1.3	84.3	146.9
0.0	125.0	3.6	2.8	90.8	138.1
0.0	150.0	4.3	4.3	93.0	130.9
0.0	175.0	4.8	5.1	95.4	125.7
0.0	200.0	5.3	6.1	93.1	121.8
0.0	− 25.0	12.2	9.0	66.1	118.3
0.0	− 50.0	12.7	14.5	43.4	103.2
0.0	− 75.0	12.4	22.6	49.8	92.3
0.0	−100.0	11.9	26.9	57.9	88.9
0.0	−125.0	11.5	22.1	64.1	88.8
25.0	− 50.0	12.5	15.2	59.8	115.2

	Loop circumference	**Element spacing**	**Director length (wavelengths)**
dimension	(B)	(C)	(L)
reflector	1.1	0.15	
radiator	1.1	0.3	
director	—	—	0.46

Loop circumference to wire radius: 8,800.
Director length to rod radius: 182.

Table 5. Similar to Table 4, but for a six-element quagi. At this director spacing, the two are essentially identical in performance, but the tuning characteristics differ somewhat. The best tuning is capacitive, so the Yagi directors must be shortened.

Tuning-stub reactance reflector (ohms)	director (ohms)	gain (dB)	F/B ratio (ohms)	R (ohms)	jX (ohms)
0.0	0.0	10.7	15.7	74.9	98.9
25.0	0.0	10.4	26.6	98.4	105.2
−25.0	0.0	10.7	10.3	56.7	84.9
0.0	25.0	10.8	16.0	74.1	98.6
0.0	50.0	11.0	16.4	73.3	98.4
0.0	75.0	11.2	16.7	72.3	98.2
0.0	100.0	11.3	17.0	71.2	98.0
0.0	125.0	11.5	17.5	69.9	97.7
0.0	150.0	11.8	17.9	68.5	98.0
0.0	175.0	12.0	18.0	66.8	98.3
0.0	200.0	12.3	17.9	65.1	98.9
0.0	225.0	12.6	17.3	65.3	99.4
0.0	250.0	12.9	16.2	62.4	101.6
0.0	275.0	13.2	14.8	62.8	103.6
0.0	300.0	13.5	13.3	66.7	104.0
0.0	325.0	13.7	12.5	73.9	92.9
0.0	350.0	13.8	13.4	47.6	67.8
0.0	375.0	11.1	11.4	19.5	103.6

	Loop circumference	Element spacing (wavelengths)
dimension	(B)	(C)
reflector	1.1	0.15
radiator	1.1	0.3
director	0.9	

Table 6. Similar to Table 2, but for an eight-element quad. Tuning conditions are nearly identical for 2, 4, or 6 directors. Adding still more directors will further increase the gain with little effect on other parameters.

Chapter 6 give a good estimate of the required boom length. Drive resistance and reactance will be very close to the table values from six and eight-element beams.

A number of long-boom quads and quagis have been described in the literature. (You might like to read Reisert, W1JR, 1985 and 1986, and the product review section of *QST*, April 1987.) These designs appear to be based on the experimental work of G3JVL.

The use of dual reflectors in the 1986 design is interesting. However, a number of trials of dual-reflector designs using MININEC analysis has failed to show

Tuning-stub reactance reflector (ohms)	director (ohms)	gain dB	F/B ratio (ohms)	R (ohms)	jX (ohms)
0.0	0.0	1.4	− 2.2	10.5	106.7
0.0	25.0	0.2	− 2.4	17.3	121.7
0.0	−25.0	12.9	15.2	26.4	75.5
0.0	−50.0	13.5	11.6	78.3	95.0
0.0	−75.0	13.4	13.5	64.5	1.5.1

		(wavelengths)	
dimension	(B)	(C)	(L)
reflector	1.1	0.2	
radiator	1.1	0.3	
director	—		0.5

Loop circumference to wire radius: 8,800.
Director length to rod radius: 182.

Table 7.

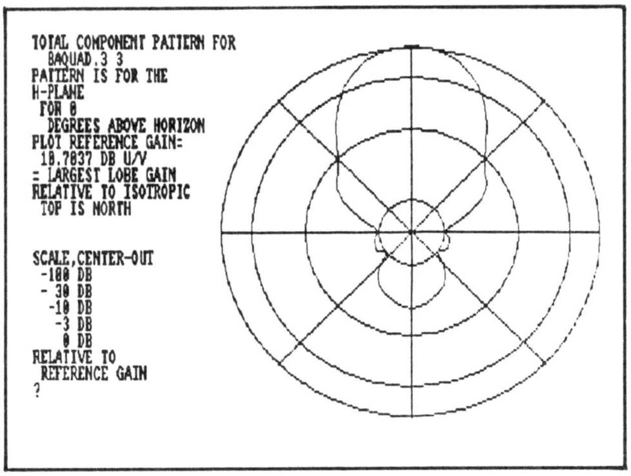

(A)

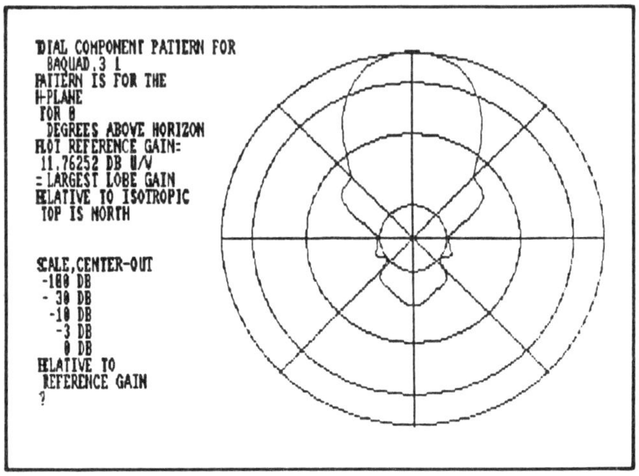

(B)

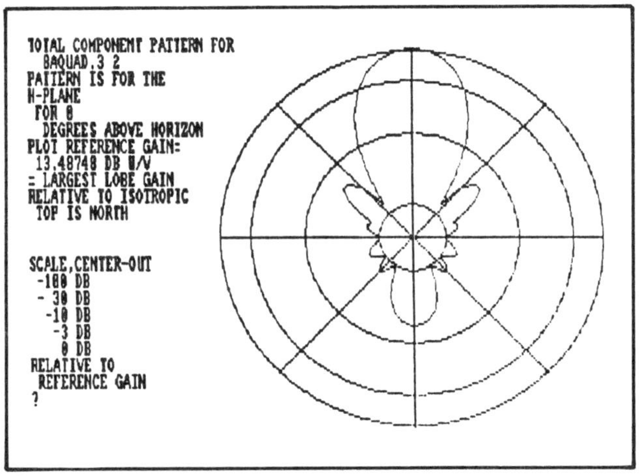

(C)

Figure 15. Variation of lobe structure of an eight-element quad with director tuning. Reflector 1.1 wavelengths at 0.5-wavelength spacing, directors 0.9 wavelength, spaced 0.3 wavelength. (A) Director loading zero, (B) directors loaded 150 ohms, (C) directors loaded 300 ohms. The values of (A) give good broadband performance, those at (C) high gain and narrow bandwidth.

more than a few dB improvement in front-to-back ratio. This is an area for further study. I hope that someone with access to a mainframe computer will analyze these extra-long boom designs and report their results.

Chapter 13

Summary

Goals of this study of quad antennas

In my study of quad antennas, I have tried to develop data and procedures to allow attainment of a quad design that meets stated performance goals. The achievement of these goals has been made possible by relatively recent developments in numerical analysis of wire antennas, in a form suitable for small personal computers. Other contributing factors have been developments in computers, bringing data storage and computing power cost to affordable ranges.

Much of the work reported here has been general, aimed at providing a base of understanding quad performance, and giving a jumping-off point for further work. At the same time, the need for easy-to-use data for typical bands and antenna sizes has been addressed.

Specific items covered include:

- A set of design values for quad antennas for common amateur bands, in common sizes.
- A scaling procedure to allow adjustment of physical construction factors without loss of performance.
- Another scaling procedure to allow adjustment of physical construction factors without loss of performance.
- Still other scaling procedures to allow any of the example designs at the example frequency to be transferred to other frequencies.

Also discussed are a number of variations in shape and arrangement, with an indication of the performance capabilities of these arrangements. A few of the variations have been studied in some detail and are suggested as possibilities for widespread use, as satisfying a general need. Others are found to fall into the family, "different, but not better."

In addition to specific and example designs, data is presented to allow calculation of performance by the theory of arrays. Hints on use of this data are examined, as are those on the use of the basic numerical technique. Other hints relate to design and construction.

To summarize in a different way, earlier parts of this series showed that good performance is obtained if a "standard" radiator-reflector combination is used, with directors being added to meet goals or to fit available space. Accordingly, the first design step is to choose the reflector size. Figure 3 in Chapter 5 gives a visual guide for this; Figures 8 and 19 in the same chapter provide data for sizing the reflector. Note that the frequency of the best front-to-back ratio is primarily set by the reflector dimension. A reflector perimeter around 1.1 wavelength is nearly optimum.

Chapters 3 and 5 recommend that the radiator be operated above resonance. Many of the suggested designs use 1.1 wavelength for the radiator circumference. A reflector-radiator spacing of 0.15 wavelength is good. However, both radiator size and spacing can be changed over a reasonable range with little change in performance.

In general, it must be stated that the study hasn't generated any surprises. The major difference from common practice is the repeated showing of the importance of conductor size on performance. Previous amateur practice neglected this, substituting a recommendation that element tuning is necessary to get the best performance. If recommended procedures are followed, this tuning process won't be needed for carefully constructed antennas in the clear, as shown by specific examples. However, there are cautions that site-specific factors may require compensation by tuning.

Some of the common lore of quads is confirmed. A two-element quad is found to be almost exactly the same in performance as a three-element Yagi beam. The Yagi has a slight edge in gain, and often in front-to-back ratio. The quad has a slight edge in low-radiation angle.

In larger antennas the difference is less marked. It appears to be more difficult to hold high front-to-back ratio in quads while the previously found interchangeability of quad and Yagi directors appears to be fully confirmed.

The common reputation of the quad as superior in performance at low heights appears to be fully confirmed. However, the claim that the quad is superior under poor conditions remains a question. Several factors have been suggested as relating to this claim and as guides to experimental study.

In a number of places throughout the study, suggestions were made for further investigation. No apologies are offered for the fact that these areas are not covered here. The information which is covered represents hundreds of hours of expanded personal-computer use, over five thousand pages of printed data results, and a considerable amount of investigation. But while this study is comprehensive, it is not exhaustive, and there is plenty more to be researched.

Appendix A

Availability of MININEC

The original version of MININEC appeared in Technical Document 516 of the Naval Ocean Systems Center, San Diego, California 92152. Copies are available from the NTIS, 5285 Port Royal Road, Springfield, Virginia 22161, as AD A121535.

The latest version of MININEC, (version 3), is in Technical Document 938 of NOSC. The best way to get a copy is to join the Applied Computational Electromagnetics Society. To do so, contact membership secretary R. Adler, Code 62AB, Naval Postgraduate School, Monterey, California 93943. Dues are $45 per year, $25 for students (as of 1991). This membership brings a newsletter, makes available a number of useful programs, and ensures that you will receive the latest update of MININEC3. (As of this writing, there have been 13 updates to the program.)

MININEC3 is written for the IBM-PC but is easily transferred to the Amiga, the Macintosh, or any computer which handles Microsoft Basic. It is a large program, above the capability of 64-K machines. In addition to MININEC itself, several preparatory and cleanup programs are available. Standard plotting routines can be used in many cases. If any extended work is to be done, compiling of MININEC programs is strongly recommended.

Commercial versions of the MININEC program are available and can be found by scanning the ads in amateur radio publications.

NOTES

Appendix B

References

1. Storer, "Impedance of Thin-Wire Loop Antennas," *Transactions of AIEE Communications*, November 1956.
2. Hallen, "Theoretical Investigations Into Transmitting and Receiving Qualities of Antennae," *Nova Acta Regiae So. Sci.*, Upsaliensis (Sweden), (4) 11 1938. (See Reference 20, Kraus, for summary.)
3. Storer, "Impedance of Thin-Wire Loop Antennas," *Cruft Laboratory Technical Report 212*, May 1955.
4. C.L. Chen and R.W.P. King, *The Small Bare Loop in an Infinite Conducting Medium*, 1963.
5. Phyllis A. Kennedy, "Loop Antenna Measurements," *IRE Transactions, Antenna Propagation*, October 1956.
6. R.W.P. King, C.W. Harrison, and D.G. Tingley, "The Current in Bare Circular Loop Antennas in a Dissipative Medium," *IEEE Transactions Antenna Propagation*, July 1965.
7. Wu, "Theory of Thin Circular Loop Antenna," *Journal of Mathematics and Physics, 3,* November, December 1962.
8. R.W.P. King, C.W. Harrison, and D.G. Tingley, "The Admittance of Bare Circular Loop Antennas in a Dissipative Medium," *Cruft Laboratory Technical Report no. 419,* August 1963.
9. R.W.P. King, C.W. Harrison, and D.G. Tingley, "The Admittance of Bare Circular Loops in a Dissipative Medium," *IEEE Transactions Antenna Propagation,* July 1964.
10. R.W.P. King and C.W. Harrison, "The Admittance of Bare Circular Loop Antennas in a Dissipative Medium," Sandia Corporation Monograph SCR-674, 1974.
11. Shin'ichi Ito, Naoki Inagaki, and Toshio Sekiguchi, "An Investigation of the Array of Circular-Loop Antennas," *IEEE Transactions Antenna Propagation,* July 1971.
12. James E. Lindsey, Jr., "A Parasitic End-fire Array of Circular Loop Elements," *IEEE Transactions Antenna Propagation*, September 1967.
13. J. Appel-Hansen, "The Loop Antenna with Director Arrays of Loops and Rods," *IEEE Transactions Antenna Propagation*, July 1972.
14. J.C. Logan and J.W. Rockway, "The New MININEC (Version 3): A Mini-numerical Electromagnetics Code," Naval Ocean System Center TD 938, September 1986.
15. A.J. Julian, J.C. Logan, and J.W. Rockway, "MININEC: a Mini-numerical Electromagnetics Code," Naval Oceans System Center TD 516, September 1982.
16. F.E. Terman, *Radio Engineers' Handbook*, McGraw-Hill, New York, 1943. (Any edition will do.)
17. A. Shoamanesh and L. Shafai, "Properties of Coaxial Yagi Loop Arrays," *IEEE Transactions Antenna Propagation*, July 1978.
18. A. Shoamanesh and L. Shafai, "Design Data for Coaxial Yagi Array of Circular Loops," *IEEE Transactions Antenna Propagation,* September 1979.

19. James L. Lawson, *Yagi Antenna Design, American Radio Relay League,* Newington, Connecticut, 1986.
20. John D. Kraus, *Antennas,* McGraw-Hill, New York, 1988.
21. Ronald King, "Theory of the Corner-Driven Square Loop Antenna," *IEEE Antennas and Propagation AP-4*, July 1956.
22. William I. Orr, W6SAI, *All About Cubical Quad Antennas*, Radio Publications, Wilton, Connecticut, 1971.
23. *The ARRL Antenna Book, 13th edition*, American Radio Relay League, Newington, Connecticut.
24. *NAB Engineering Handbook*, A. Prose Walker, editor, McGraw-Hill, New York, 1960.
25. W.J. Byron, W7DHD, "Designing a No-Tune Amplifier With Your Personal Computer," *Ham Radio*, September 1987, page 8.
26. Smith and Scott, Jr., *IEEE Transaction AP-35*, August 1978.
27. Sheila Prasad, "Theory of Corner-Driven Coupled Square Loop Antennas," *ONR Report NR-371-016*, Harvard University, Cambridge, February 1959.
28. Jerry Sevick, "Transmission Line Transformers," *American Radio Relay League*, Newington, Connecticut, 1987.
29. Fanckboner, "Using Baluns in Transmatches with High-Impedance Lines," *QST*, April 1981.
30.Gullstad, "The Phantom Stub," *QST*, December 1979.
31. James, "A Seven-Band Boomless Quad," *CQ*, April 1982.
32. Pichitino, "A New Principle in Two-band Rotary Beam Design," *QST*, October 1948.
33. R.P. Haviland, W4MB, "Design Data for Pipe Masts, *Ham Radio*, September 1974.
34. Keith Henney, Editor, *Radio Engineering Handbook*, McGraw-Hill, New York, New York, 1959.
35. Dick Bird, G4ZU, "The G4ZU 'Bird Cage' Aerial," *CQ*, April 1960.
36. William I. Orr, W6SAI, "Antennas," *CQ Magazine*, January 1974.
37. Keith Kunde, "Try a 'Dopplequad' Beam Antenna for 2 Meters," *QST*, February 1985.
38. J. Ross Anderson, "Meet the Curtain-Quad Antenna," *QST*, November 1984. (See also Reference 20, Kraus.)

Bibliography

1. Belrose, "The Effect of Supporting Structure on Simple Wire Antennas," *QST*, December 1982.
2. Dorbuck, "Matching-Network Design," *QST*, March 1979.
3. Johns, "Dual-Frequency Antenna Traps," *QST*, November 1983.

Index